Simulated and Virtual Realities

Elements of Perception

Simulated and Virtual Realities

Elements of Perception

Edited by

Karen Carr and Rupert England

Sowerby Research Centre, British Aerospace Operations Ltd, Bristol,
and
Interface Technology Research Ltd, Bristol

Taylor & Francis
Publishers since 1798

UK   Taylor & Francis Ltd, 4 John Street, London WC1N 2ET

USA   Taylor & Francis Inc., 1900 Frost Road, Suite 101, Bristol, PA 19007

British Library Cataloguing in Publication data

A catalogue record for this book is available from the British Library

ISBN 0-7484-0128-8 (cased)
   0-7484-0129-6 (paper)

Library of Congress Cataloging in Publication Data are available

Cover design by Hybert Design and Type

Typeset by Euroset, Alresford, Hampshire SO24 9PG

Printed in Great Britain by Burgess Science Press, Basingstoke, on paper which has a specified pH value on final paper manufacture of not less than 7·5 and is therefore 'acid free'.

Contents

Contributors   ix

1 Introduction   **1**
Karen Carr
1.1 Virtual reality: tool or concept?   1
1.2 Synthesizing perception   3
1.3 Themes   5
1.4 The human basis for virtual reality   7

2 Virtual environments and environmental instruments   **11**
Stephen R. Ellis
2.1 Communication and environments   11
2.2 Virtualization   15
2.3 Origins of virtual environments   22
2.4 Virtual environments: performance and trade-offs   39

3 Visual realism and virtual reality: a psychological perspective   **53**
Chris Christou and Andrew Parker
3.1 Introduction: what is visual realism?   53
3.2 Historical overview of visual psychology   55
3.3 Perceiving three dimensions   58
3.4 Natural images   59
3.5 Artificial images and requirements for optimal presentation   62
3.6 Computer graphics   67
3.7 Global illumination and virtual reality   73
3.8 Conclusion   80

4 Vision and displays   **85**
Graham K. Edgar and Peter J. Bex
4.1 The 'real' and the 'virtual'   85
4.2 Temporal aliasing   86
4.3 The accommodation response and virtual reality displays   93
4.4 Conclusions   98

5 Head-coupled virtual environment with display lag   **103**
Richard H.Y. So and Michael J. Griffin
   5.1 Introduction   103
   5.2 Effects of lags on head-coupled virtual environments   105
   5.3 Lag compensation   107
   5.4 Summary   110

6 Perceptual cues and object recognition   **113**
John M. Findlay and Fiona N. Newell
   6.1 Introduction   113
   6.2 Approaches to recognition   116
   6.3 Experimental studies of visual recognition   120
   6.4 Conclusion   127

7 Sensory-motor systems in virtual manipulation   **131**
Rupert England
   7.1 Introduction   131
   7.2 Perceptuo-motor behaviour   133
   7.3 Sensory neurophysiology   137
   7.4 The nature of perceptual integration   154
   7.5 The nature of response   158
   7.6 The nature of manipulation—classification   163
   7.7 Manipulating virtual objects   164
   7.8 Conclusions and recommendations   167

8 Auditory virtual environments   **179**
Mark Williams
   8.1 Introduction   179
   8.2 Human auditory system   180
   8.3 Virtual speech   182
   8.4 Creating an auditory virtual environment   184
   8.5 Auditory icons and virtual environments   185

9 Designing in virtual reality: perception–action coupling and affordances   **189**
Gerda J. F. Smets, Pieter Jan Stappers, Kees J. Overbeeke and
Charles van der Mast
   9.1 Introduction   189
   9.2 Definitions   190
   9.3 CAD in VR   193
   9.4 Conclusions   203

10 Social dimensions of virtual reality   **209**
Deborah Foster and John F. Meech
   10.1 Introduction   209
   10.2 A technological disclaimer   210
   10.3 Simulation and reality   211
   10.4 Reality control and hyperrealism   211
   10.5 How many realities are there?   212
   10.6 Who controls the virtual reality?   214

10.7   Is virtual reality a 'neutral' technology?   214
10.8   How might technology (and virtual reality) influence society?   215
10.9   Images and ideology   216
10.10 The user's relationship with virtual reality technology   217
10.11 Who decides the content of virtual reality?   219
10.12 Ethics and application   219
10.13 Directions for future research   220
10.14 Concluding comments   221

11 Summary and conclusions   **225**
Karen Carr

First author index   **229**

Subject index   **235**

Contributors

PETER J. BEX
School of Psychology, University of Wales, College of Cardiff, PO Box 901, Cardiff CF1 3YG, UK

KAREN CARR
Sowerby Research Centre, British Aerospace (Operations) Ltd, FPC 267, PO Box 5, Bristol BS12 7QW, UK and *Interface Technology Research Ltd, CRII, Du Pont Centre, Bristol BS16 1QD, UK*

CHRIS CHRISTOU
Utrecht Biophysics Research Institute, Faculteit der Natuur- en Sterrenkunde, Princeton-plein 5, Postbus 80000, 3508 TA Utrecht, The Netherlands

GRAHAM K. EDGAR
Sowerby Research Centre, British Aerospace (Operations) Ltd, FPC 267, PO Box 5, Bristol BS12 7QW, UK

STEPHEN R. ELLIS
NASA Ames Research Center, Moffett Field, California 94035, USA, and *University of California, Berkeley, California 94720, USA*

RUPERT ENGLAND
Sowerby Research Centre, British Aerospace (Operations) Ltd, FPC 267, PO Box 5, Bristol BS12 7QW, UK, and *Interface Technology Research Ltd, CRII, Du Pont Centre, Bristol BS16 1QD, UK*

JOHN M. FINDLAY
Department of Psychology, University of Durham, South Road, Durham DH1 3LE, UK

DEBORAH FOSTER
Faculty of Economics and Social Sciences, University of the West of England (Bristol), Bristol BS15 1QW, UK

MICHAEL J. GRIFFIN
Human Factors Research Unit, Institute of Sound and Vibration Research, University of Southampton, Southampton SO17 1BJ, UK

CHARLES VAN DER MAST
Delft University of Technology, Jaffalaan 9, NL-2628BX Delft, The Netherlands

JOHN F. MEECH
Department of Computing, University of the West of England (Bristol), Bristol BS15 1QW, UK, and *Interface Technology Research Ltd, CRII, Du Pont Centre, Bristol BS16 1QD, UK*

FIONA N. NEWELL
Department of Psychology, University of Durham, South Road, Durham DH1 3LE, UK

KEES J. OVERBEEKE
Delft University of Technology, Jaffalaan 9, NL-2628BX Delft, The Netherlands

ANDREW PARKER
University Laboratory of Physiology, Parks Road, Oxford OX1 3PT, UK

GERDA J. F. SMETS
Delft University of Technology, Jaffalaan 9, NL-2628BX Delft, The Netherlands

RICHARD H. Y. SO
Human Factors Research Unit, Institute of Sound and Vibration Research, University of Southampton, Southampton SO17 1BJ, UK

PIETER JAN STAPPERS
Delft University of Technology, Jaffalaan 9, NL-2628BX Delft, The Netherlands

MARK WILLIAMS
Sowerby Research Centre, British Aerospace (Operations) Ltd, FPC 267, PO Box 5, Bristol BS12 7QW, UK

1

Introduction

Karen Carr

1.1 Virtual reality: tool or concept?

One of the most important aspects of human thought has been the quest to distinguish between external events, such as changes in our environment, and internal events, such as perceptions. This is the distinction between the objective and the subjective, the distinction which would allow us to establish the certainty of knowledge by removing the mediation of human ways of thinking and perceiving. Philosophers have argued about whether it is ever possible to have objective, certain knowledge, and if so, how it is obtained.[1] Empirical scientists believe that we can gain certain knowledge from methodical observations; that we can, through our perceptions, know reality. This is particularly interesting in the context of this book, because virtual reality, considered by many to be an important development in human culture, is the achievement of the opposite: fooling people into accepting as real what is only perceived.

It is also a curious paradox that virtual reality is hailed as an important advance in helping us to visualize and control more complex information, such as abstract computational data, when it does so by moving us 'backwards' into our primitive, subjective viewpoint, manipulating our perceptions so that we use an egocentric way of thinking. Virtual reality reduces the need for abstract, extero-centric thinking by presenting processed information in an apparent three-dimensional space, and allowing us to interact with it as if we were part of that space. In this way our evolutionarily derived processes for understanding the real world can be used for understanding synthesized information. For example, the physical skills of our bodies can be used to help our understanding of spatial relationships, so that spatial problem solving which was previously achieved through internal abstract thought can now be physically acted out. Scientific visualization and teleoperation are prime examples of tasks which such 'physical thought' can benefit.

To some extent, it is this concept for presenting information which has driven the development of virtual reality: the concept of making a synthetic perceptual experience match a real perceptual experience, thus making the abstract appear concrete and intuitive. But there are also technological advances which are driving our use of virtual reality, such as head-mounted displays and high-performance graphic engines which are seized upon as new tools for which many uses can be found. Thus a difference of focus between concept-driven and tool-driven developments is possible, which echoes the dichotomy of concept-based and tool-based scientific revolutions described by Dyson (1993). This distinction can usefully be applied to virtual reality, because the ways in which we use concepts and

1

tools are different. A concept provides potential solutions (which need to be implemented and tested), while a tool increases our capability to implement and test. If tools are applied in an ad hoc manner, they may turn out to be useful, but they are more likely to be so if they are used methodically to test a concept. Should virtual reality then be considered as a technologically defined tool which we can try out on various applications; or should it rather be considered as a concept for the presentation and use of artificial information? If virtual reality is to be considered a tool, we should expect its use to be determined by technological capability; if it is to be considered a concept, cultural and theoretical factors will drive the development of the technology and its use. This difference is illustrated by the examples of an electric motor and of 'alternative energy': the first is used as a tool for many different applications, while the second is a concept which different technologies could try to fulfil. Clearly 'virtual reality' has been considered as both a tool and a concept; perhaps we should differentiate more carefully between the technology that gives us head-slaved images and the concept of stimulating egocentric perceptual processes, if we are to optimize and control the development of virtual reality.

In his discussion of the origins of virtual environments, Stephen Ellis (Chapter 2) demonstrates that there was a good deal of conceptual impetus to the early development of the technology. It is to be hoped that further development will be driven by the concept of what we want virtual reality to let us do. The work described by Gerda Smets and colleagues (Chapter 9) is an important and exemplary attempt to implement virtual reality to fit very specific human requirements: an interface for computer-aided design (CAD). Smets and her colleagues have a clear concept of human perception and behaviour and use this to guide the design of both the task and the tools in the CAD interface. They use what technology is available to support this approach, but do not let the capabilities of the technology obscure the requirements of the interface.

We should be aware that blurring the subjective/objective divide by making information more subjective, while reducing the need (and perhaps ultimately the ability) for the abstract solution of problems, might also unwittingly encourage metaphysical models of thought and understanding. Thus, through the use of virtual reality more people may come to believe that knowledge and understanding can come entirely from within themselves. Opinions will differ as to the desirability of this. Currently, there are some people who believe in 'paranormal' events, such as telepathy and clairvoyance. Other people do not, and look for internal explanations of such experiences. Similarly, experiences in virtual reality may be more readily distinguished from 'external' reality by some people than others. Some may even use virtual reality as a 'gateway' to paranormal experiences. Decisions about such matters will be made by different people who have different ways of thinking. What is disturbing, however, is the unknown effect which virtual reality might have on children who are still developing their ways of thinking (although our attitudes may develop throughout our lives). Deborah Foster and John Meech, in Chapter 10 of this book, suggest that children's psychological development may be detrimentally affected by the media, and extrapolate to include virtual reality as one of such media. Ellis concludes Chapter 2 with a discussion of the new forms of human expression, desirable or otherwise, which virtual reality technology might allow. It seems that, whether virtual reality can affect cognitive development for the good or for the bad, it will in any case be able to make cognitive development different from that taking place at present, and may shape our models of the world and hence the direction that humankind takes in the future.

Thus, as tool or as concept, virtual reality should be taken very seriously, as it has the potential, in some sense, to reshape our minds.

1.2 Synthesizing perception

The contention behind this book is that taking virtual reality seriously means understanding the process by which technology can fool our perceptions. The book was written as a result of discussions and themes raised at the 'Simulated and Virtual Realities' conference organized by the Applied Vision Association at Bristol University in March 1993.[2] The title of this conference was chosen to reflect the fact that simulation tries to imitate reality, whereas virtual reality does not necessarily do so, though both are creating a synthetic environment. Thus simulation is a simulated reality, which is one type of virtual reality. Whether imitating the real world or not, however, the perceptual system which will perceive this artificially created environment both evolved and learned to perceive in the real world. Thus there are fundamental aspects to perception which must be understood if we are to control the experience of the perceiver of virtual reality. Perceptual processes include the stimulation of the sensory receptors and the processing and interpretation of that stimulation. If we could identify how the perceptual processes work and what aspects of the environment provide the critical stimulation, we could use this knowledge to design the synthetic perceptual environments which give us virtual reality. We would then, in effect, be using a 'perceptual language' to communicate with perceptual experiences (Carr and England, 1993).

The best way we know to understand the perceptual processes is to use a scientific method, and indeed the very specific discipline of psychophysics evolved precisely to deal with subjective, intangible experiences in a systematic way. Chris Christou and Andrew Parker describe the psychophysical approach in their chapter of this book (Chapter 3, section 3.2.5), and show how the degree of realism in the computation of computer graphics affects the perception of the intended image (section 3.7). For example, the interpretation of an object's shape is partly governed by rules based on how the object is illuminated. In the terms used above, Christou and Parker show how to identify and use the critical elements of the perceptual language.

Attaching devices to stimulate our senses is not a simple matter, as our senses are finely tuned and behave differently in different environmental conditions and under different psychological conditions. For example, perceptual processes at all levels, from basic sensation to the perception of value can be affected by contexts. Contexts can provide extra information which can help or hinder the perception of an object. Many visual search studies have shown that the composition of the background has a large effect on people's ability to locate a target (see, for example, Gale *et al.*, 1995). A frame of reference can influence the judgement of true vertical (Witkin, 1949), or the identification of a rectangle as a diamond or a square (Mach, 1897). The position of the moon in the sky affects our perception of its size (see Hershenson, 1989).

Visual recognition can depend upon physical and semantic contexts which provide rules for interpretation, as John Findlay and Fiona Newell show in Chapter 6 (6.3.3). Auditory localization can also be affected by visual context (see Williams, 8.3.2 and England, 7.4.2). There are also numerous examples of semantic context affecting perceptual judgements. Petzold (1992) describes a range of studies showing the effects of context on the perception of both qualitative and quantitative factors. 'Context' in this case includes such factors as the range of samples from which a judgement must be made, or priming people with ideas before a judgement task. These semantic contexts can shift judgements in various ways, for example in an opposite direction to the context (contrast effect) or towards the context (assimilation). Thus how beautiful one might judge a landscape to be will depend to some extent on what was perceived or thought about previously.

Some attempts have been made to quantify the effects of context, but the most successful are generally limited to lower levels of perceptual processing. For example, Petzold (1992) discusses different mathematical models quantifying the Ebbinghaus illusion (which shows an effect of surround context on size perception). It is possible to quantify how the perception of a colour changes according to the colour it is surrounded by (see Walraven, 1992).

An important aspect of virtual reality, which also provides part of the context, is that the viewpoint is subjective and presented as the perceiver's own. This has many advantages, some of which are discussed by Ellis (2.2) and Smets *et al.* (9.3.2.1). Different viewpoints give different perceptions and contexts, and variation of viewpoint is a factor much used in art, photography, and films as a form of communication to give different perceptual effects and experiences (see Foster and Meech, 10.4). Viewpoint in this way provides a context for perceptual interpretation. It may be important to consider the implications of making these different viewpoints appear to be the perceiver's own, particularly if the virtual environment being presented does not reflect reality. Foster and Meech's discussion of simulation and the perception of reality (10.3 and 10.5) indicates what some of the implications might be.

Part of our experience of the real world is that not only can we see, but we can often also be seen. This knowledge can affect the way we behave, our self-image, and consequently the way we perceive ourselves to interact with our environment. This may be an important reason for providing the users of virtual environments with virtual bodies and registering their movements. Being able to see one's own body may contribute to the visual perception of scale, by providing both a reference and a defined viewpoint. One of the papers presented at the 'Simulated and Virtual Realities' conference described a pioneering attempt to formulate some measure of 'presence' in a virtual environment, and the way in which it is achieved (Slater and Usoh, 1993).

The chapters which follow cover a wide range of perceptual issues, including sensation, different levels of perceptual processing, the integration of different senses, and cognition. Perception is often described as having different stages, usually hierarchical (especially 'bottom-up' processing with the capability for 'top-down' processing, such as intention, to influence low-level stages). For examples of models of perception, see Marr (1982), Gregory (1980), and the review of models of visual recognition in Chapter 6. There is, however, as yet no clear distinction between some stages, and in some cases it is still unknown at which stage a perceptual process takes place. It is not established, for example, which information available to our senses has to be processed to what level before some selective mechanism can ensure that only useful information is processed further (see 7.4.3). An important aim in perceptual research is to link the findings from physiological studies with the results of psychophysical studies, as is demonstrated by Findlay and Newell (6.1.1) and England (7.3). Unfortunately, most physiological studies relate to relatively early stages of perception, so few robust physiological correlates have been found for higher-level processes, such as the effects of semantic context. Research has been carried out to try to associate electrical activity and magnetic fields in the brain with high-level processes (e.g. Chapman *et al.*, 1988). The future progress of these lines of research may be of interest to those hoping to create virtual reality by direct stimulation of the brain.

Knowing how to model the real world and how we perceive it is not sufficient to allow us to provide a virtual reality with robust and believable perceptual experiences. The simulation of the perceptual stimuli has to be accomplished with technology, and that is not an easy task, as Graham Edgar and Peter Bex discuss in Chapter 4. They show how the technology itself has a direct effect on the perceptual experience, and this prevents simulation of a perceptually realistic environment. This may not always be a problem, depending upon what the desired effect is; it will, however, always be a problem if the effect of the technology is to cause misperceptions, discomfort, or even health risks. Edgar and Bex discuss in some detail how research into two display technology characteristics can help us understand how the effects occur and how we might overcome the problems they cause. In the short following chapter (Chapter 5), Richard So and Michael Griffin also address a display technology problem which they are very familiar with: the problem of lag in the update rate of a head-mounted display. They summarize some of their extensive work in the area, and examine some possible solutions. Christou and Parker (3.5) compare the capabilities of visual displays with the characteristics of human vision, showing that there is a 'window of visibility' within which visual displays must perform if the low-level visual processes are to be correctly activated. It is to be hoped, of course, that technology will improve over time to minimize or eliminate some of the problems described in these chapters. The development of the technology will proceed more effectively, however, if it is known how the technology interacts with perceptual mechanisms and which problems are most important. Often the understanding of the perceptual mechanism can lead to a simpler and less expensive solution than that sought by the drive for increasingly higher-performance technology. For example, in order to eliminate undesirable temporal aliasing, display technology could be developed to have higher update rates; alternatively an appropriate filtering (blurring) algorithm could provide the desired perceptual experience at less cost (Edgar and Bex, Chapter 4).

1.3 Themes

There are several themes which recur throughout the chapters of this book, and they are presented in different ways. Some of these themes are introduced here so that the reader will be alert to these different treatments.

Definitions

Whether you wish to use the term 'virtual reality', 'VR', 'cyberspace', 'synthetic environments' or 'virtual environments' (and there are plenty of strongly held views on this), the one factor common to these definitions and conceptions is that they are all concerned with the stimulation of human perceptual experience to create an impression of something which is not really there. Purists at one extreme will define virtual reality in terms of an experience, which can encompass dreams, hallucinations, *trompes-l'oeil*, and even books and films which absorb the attention. Purists at the other extreme will insist on using definitions which derive from technology, such as a computer-generated environment, or a head-mounted display. Within this book several different approaches to virtual reality are evident. Stephen Ellis (Chapter 2) wishes to avoid the term virtual reality altogether and very usefully presents a structured definition of three types of 'virtualization'. Deborah Foster and John Meech (Chapter 10) analyse virtual reality as a media technology with social implications. Gerda Smets *et al.* (Chapter 9) suggest that virtual reality is a simulation which can be configured in any way. Rupert England (Chapter 7) differentiates between virtual reality as a general term which might include images obtained indirectly from reality (e.g. from cameras) and cyberspace which is a subset of virtual reality, and entirely computer-generated.

As mentioned earlier (section 1.1), there are good reasons to differentiate between technology and concept. There are also logical reasons for distinguishing process from result. If virtual environments provide the process, then perhaps we can call the resulting experience virtual reality. The current inconsistencies in the use of the term 'virtual reality' would probably not arise with the term 'simulated reality', as to simulate is clearly a process.

Simulation

Simulation is a term which can be interpreted as broadly as can virtual reality. In fact, in one sense, virtual reality is one kind of simulation (Smets, 9.1; Ellis, 2.3.2), and in another sense, a simulation is one kind of virtual reality (Foster and Meech, 10.3). In the former, simulations do not necessarily require a human observer, and a computational model of a process is a self-contained simulation. In the latter, a simulation is a pretence which depends upon interpretation by a person who is familiar with the rules of representation. Christou and Parker illustrate this distinction effectively when they differentiate between computational computer graphics and pictorial art as simulations (3.6.1).

Realism

There are at least two ways in which realism can be considered with respect to virtual reality. On the one hand, virtual reality can try to create a perceptual experience which would be believable if it were experienced in the real world, and in this case realism in virtual reality is a simulation of possible real worlds. On the other hand, even if virtual reality is creating an experience which would not be possible in the real world, it can still only be perceived with the same perceptual mechanisms we use in the real world; the more accurately these mechanisms are stimulated, the greater the perceptual realism. In this case, realism is the accurate construction of patterns of information important for perception, as shown by Christou and Parker (3.7). That this is sometimes difficult to achieve, is shown by Edgar and Bex (Chapter 4). In addition, those who would like to create realism of either kind by using their own judgement and artistic skills would gain a useful insight into artistic 'realism' by studying Gombrich's (1977) discussion of representational style and illusion in art.

In contrast, there are examples of intentional 'unrealism', such as the distortion of representations in order to convey information more effectively. Ellis refers to the distortion employed in cartography to exaggerate features (2.3.2) and there are examples of artistic distortion used in order to represent a realistic 'appearance' rather than an accurate perspective projection (see Gombrich, 1977, p. 262 and p. 217). Thus realism may not always be the best approach, and informed deviation from realistic patterns of information may sometimes allow virtual reality to communicate more effectively than reality itself.

It is important to distinguish between the perception of realism and the perception of reality. Perceiving something as reality requires belief in its existence. For example, if two people perceive a ghostly shape at night, and one believes in ghosts while the other does not, the former will see the form as real while the latter will think it is a trick of light. Foster and Meech (10.5) provide an interesting discussion about how many realities there can be. A 'sense of reality' (Christou and Parker, 3.1) does not necessarily imply belief in a reality, but rather that the perception is sufficient to convey reality, even though the perceiver's own knowledge prevents belief in it.

Ecological factors in perception

Although 'ecological perception' is widely used to describe the particular school of thought instigated by Gibson (1979) (also known as 'direct perception': see 3.2.4 and 9.3.2), it would be wrong to restrict the term to this use. However we believe perception is achieved, there can be no doubt that ecological factors had some influence on how perception has evolved. Physiological evidence indicates some commonality across animals in the different physiological processes developed to deal with particular aspects of the environment (see 7.3.1), and the fact that perception developed to allow action in the environment has been emphasized (Christou and Parker, 3.1 and 3.3; Williams, 8.2.3; Smets *et al.*, 9.3.1.3). Findlay and Newell (6.1.2) and England (7.3.2) both describe the current theory that there may be two distinct visual systems, one for action and one for recognition. Indeed, the concept of distinct single systems for each perceptual sense is proving to be difficult to justify (England, 7.2.2, 7.3.5.9, 7.4.3).

With respect to the theory of ecological perception proposed by Gibson (1979), this is one of several models of perception, and one which has received both support (Smets *et al.*, 9.3.2.3) and criticism (Christou and Parker, 3.2.4). Other models are briefly reviewed by Christou and Parker (3.2) and Findlay and Newell (6.2).

Biology and technology

It has already been argued here that an understanding of how the perceptual mechanisms have evolved will help us to use technology to create virtual reality. But technology also develops by a process of evolution (see Ellis, 2.3.5), in which periods of gradual development are interspersed with innovation. Sometimes innovation in technology is biologically inspired (this is the main aim of the scientific discipline of bionics). Several authors in this book mention the possibility of developing a visual display with variable resolution which matches the variable resolution of the retina (Ellis, 2.4.3; Christou and Parker, 3.5.4; Findlay and Newell, 6.3.2). As England advocates (7.1.2), it is the task of the human factors engineer to try to identify the requirements of a virtual reality system in such a way that technology can evolve along the most efficient route: the route of adapting technology to humans. Some of the chapters reveal that the development of new technology and the understanding of human senses go hand in hand. For example, Findlay and Newell show how biologically inspired computational neural networks can be used for understanding vision (6.2.6) and Smets *et al.* describe how, inspired by the perceptual concept of 'affordances', they plan to use such networks to create the evolution of virtual tools for design (9.3.3). Findlay and Newell also look forward to using virtual reality technology itself to help study human perception (6.4).

1.4 *The human basis for virtual reality*

This book is intended to provide students and researchers involved in the development of virtual reality, tool or concept, with an overview of human characteristics which affect the way in which virtual reality can be achieved. Discussions of psychology, philosophy, physiology, psychophysics, human factors and sociology offer a broad-ranging statement of current knowledge. By describing the human basis, we can allow developers to understand and use the foundations upon which virtual reality can be built. This is not intended as a reference book, as there are insufficient data available to provide a comprehensive guide. It is intended rather as a baseline from which researchers who have little or no background in perceptual psychology and physiology can generate their own ideas of how to drive the development of technology, and hopefully incorporate psychological experimentation into their own research programmes.

As already mentioned above, an Applied Vision Association conference provided the impetus for this book, and thus the original theme was vision and visual virtual realities. Four of the chapters, therefore, discuss only aspects of vision. As England's review of neurophysiology (Chapter 7) shows, however, vision should not be considered in isolation from the other senses. In fact, the senses and all their modalities are highly integrated. Also, if we wish to allow a natural interaction with virtual reality, we must provide for visuo-motor behaviour which requires both vision and taction. England therefore reviews tactual perception in some detail, as this sense is often neglected or little understood as a contributor to virtual reality. Mark Williams (Chapter 8) has made the review of the senses more complete by providing a brief description of auditory perception and the role of audition in virtual reality.

Acknowledgements

I would like to thank Steve Ellis for a brief but useful discussion about the use of the term 'virtual reality' (and for good-naturedly allowing his chapter to be included in a book bearing that term in its title). I also gratefully acknowledge the constructive criticism of this chapter from my father, Jeff Carr, and the various helpful comments from my colleagues at British Aerospace.

Notes

1. See, for example: David Hume (*A Treatise of Human Nature*, 1734), Immanuel Kant (*Critique of Pure Reason*, 1781), Karl Popper (*The Logic of Scientific Discovery*, London: Hutchinson, 1972 (3rd, revised, Edn); *Objective Knowledge: An Evolutionary Approach*, Oxford: Oxford University Press, 1975 (corrected Edn.)). Chris Christou and Andrew Parker (Chapter 3, section 3.2) provide an overview of the debate between nativist and empiricist theories of perception, which is essentially a debate about whether perceptual knowledge is provided genetically, or whether it is learned from perceptual experience.

2. The abstracts from this conference were published in *Ophthalmic and Physiological Optics* in October 1993 (Vol. 13 (4), pp. 434–440). The papers presented were as follows:

   Colour constancy under terrestrial and alien daylights, *D. Foster and K. Linnell*. Physical and psychophysical evaluation of a radiosity-based method for generating images, *A. J. Parker, C. G. Christou, B. G. Cumming and G. Jones*. Temporal aliasing: investigating multiple imaging, *P. J. Bex, G. K. Edgar and A. T. Smith*. Curvature discrimination with low resolution computer generated imagery, *M. J. Cook*. Effects of display lags on head-coupled virtual reality systems, *R. H. Y. So and M. Griffin*. Visual accommodation with virtual images, *G. K. Edgar, J. C. D. Pope and I. Craig*. Real problems with virtual worlds, *M. Mon-Williams, J. P. Wann, S. Rushton and R. Ackerley*. An investigation of visual and tactual feedback in a simple visuo-motor task, *R. England*. Visual object recognition: what information, is used?, *F. Newell and J. Findlay*. Stereo visual information and mental rotation, *K. Carr*. A simulation study of the effects of night-vision goggles on depth perception, *P. Bagnall, S. Selcon and P. Wright*. Do we have a symbolic pipeline architecture for the encoding and representation of 3D shape?, *A. Johnston and P. J. Passmore*. Individual differences and limits in the perception of spatial representations, *J. Springer, H. Falter and M. Rötting*. Nature and origins of virtual environments (there's nothing new to cyberspace!), *S. Ellis*. Virtual environments for architectural walkthrough, *M. Slater and M. Usoh*. Virtual reality for designers, *G. Smets, C. J. Overbeeke and P. J. Stappers*. An economical radiosity method capable of representing curved surfaces and their shadows, *A. Zisserman, G. Jones, C. G. Christou, B.G. Cumming and A.J. Parker*. Applying virtual environments to trainers and simulators, *G. J. Jense and F. Kuijper*. Sociological effects of virtual reality, *J. F. Meech and M. Baker*.

References

Carr, K. and England, R., 1993, The role of realism in virtual reality, in *Proceedings of Virtual Reality '93, The Third Annual Conference on Virtual Reality*, pp. 24–33, London: Meckler.

Chapman, R. M., Modena, I., Narici, L., Pizzella, V., Romani, G. L., Salustri, C., McCrary, J. W. and Garnsey, S., 1988, Electric and magnetic brain activity related to cognitive performance, in *Electric and Magnetic Activity of the Central Nervous System: Research and Clinical Applications in Aerospace Medicine, AGARD Proceedings No. 432*, pp. 10-1–8, Washington, DC: NASA Scientific and Technical Information Branch.

Dyson, F., 1993, George Green and physics, *Physics World*, **6** (8), 33–8.

Gale, A., Astley, S., Carr, K. and Moorhead, I. (Eds), 1995, *Visual Search 3*, Proceedings of the Third International Conference on Visual Search, Nottingham, London: Taylor & Francis, in press.

Gibson, J. J., 1979, *The Ecological Approach to Visual Perception*, Boston, MA: Houghton Mifflin.

Gombrich, E. H., 1977, *Art and Illusion: A Study in the Psychology of Pictorial Representation*, 5th Edn, Oxford: Phaidon Press.

Gregory, R. L., 1980, Perceptions as hypotheses, *Philosophical Transactions of the Royal Society, B*, **290**, 181–97.

Hershenson, M. (Ed.), 1989, *The Moon Illusion*, Hillsdale, NJ: Lawrence Erlbaum.

Mach, E., 1959, *The Analysis of Sensations* (translated from the German edition, 1897), New York: Dover.

Marr, D., 1982, *Vision*, San Francisco, CA: W. H. Freeman.

Petzold, P., 1992, Context effects in judgments of attributes: an information integration approach, in Geissler, H.-G., Link, S. W. and Townsend, J. T. (Eds), *Cognition, Information Processing and Psychophysics: Basic Issues*, pp. 175–205, Hillsdale, NJ: Lawrence Erlbaum.

Walraven, J., 1992, Colour basics for the display designer, in Widdell, H. and Post, D. L. (Eds), *Colour on Electronic Displays, Defense Research Series*, Vol. 3, pp. 3–38, New York: Plenum Press.

Witkin, H. A., 1949, The nature and importance of individual differences in perception, *Journal of Personality*, **18**, 145–60.

2

Virtual environments and environmental instruments[1]

Stephen R. Ellis

2.1 Communication and environments

2.1.1 Virtual environments are media

Virtual environments created through computer graphics are communications media (Licklider *et al.*, 1978). They have both physical and abstract components like other media (see Chapter 10 for a discussion of the sociological aspects of media). Paper, for example, is a communication medium but the paper is itself only one possible physical embodiment of the abstraction of a two-dimensional surface onto which marks may be made. Consequently, there are alternative instantiations of the same abstraction. As an alternative to paper, for example, the Apple Newton series of intelligent information appliances resemble handwriting-recognizing magic slates on which users write commands and data with a stylus (see Apple Computer Co., 1992). The corresponding abstraction for head-coupled, virtual image, stereoscopic displays that synthesize a coordinated sensory experience is an environment. Recent advances and cost reductions in the underlying technology used to create virtual environments have made possible the development of new interactive systems that can subjectively displace their users to real or imaginary remote locations.

Different expressions have been used to describe these synthetic experiences. Terms like 'virtual world' or 'virtual environment' seem preferable since they are linguistically conservative, less subject to journalistic hyperbole and easily related to well-established usage as in the term 'virtual image' of geometric optics. These so-called 'virtual reality' media several years ago caught the international public imagination as a qualitatively new human–machine interface (Pollack, 1989; D'Arcy, 1990; Stewart, 1991; Brehde, 1991), but they, in fact, arise from continuous development in several technical and non-technical areas during the past 25 years (Ellis, 1990; 1993; Brooks, 1988; Kalawsky, 1993). Because of this history, it is important to ask why displays of this sort have only recently captured public attention.

The reason for the recent attention stems mainly from a change in the perception of the accessibility of the technology. Though its roots, as discussed below, can be traced to the beginnings of flight simulation and telerobotics displays, recent drops in the cost of interactive three-dimensional graphics systems and miniature video displays have made

11

it realistic to consider a wide variety of new applications for virtual environment displays. Furthermore, many video demonstrations in the mid-1980s gave the impression that indeed this interactive technology was ready to go. In fact, at that time, considerable development was needed before it could be practicable and these design needs still persist for many applications. Nevertheless, virtual environments can become Ivan Sutherland's 'ultimate computer display'; but in order to ensure that they provide effective communications channels between their human users and their underlying environmental simulations, they must be designed.

2.1.2 Optimal design

A well designed human–machine interface affords the user an efficient and effortless flow of information between the device and its human operator. When users are given sufficient control over the pattern of this interaction, they themselves can evolve efficient interaction strategies that match the coding of their communications to the machine to the characteristics of their communication channel (Zipf, 1949; Mandelbrot, 1982; Ellis and Hitchcock, 1986; Grudin and Norman, 1993). Successful interface design should strive to reduce this adaptation period by analysis of the users' task and their performance limitations and strengths. This analysis requires understanding of the operative design metaphor for the interface in question, i.e. the abstract or formal description of the interface in question.

The dominant interaction metaphor for the human–computer interface changed in the 1980s. Modern graphical interfaces, like those first developed at Xerox PARC (Smith *et al.*, 1982) and used for the Apple Macintosh, have transformed the 'conversational' interaction from one in which users 'talked' to their computers to one in which they 'acted out' their commands within a 'desk-top' display. This so-called desk-top metaphor provides the users with an illusion of an environment in which they enact system or application program commands by manipulating graphical symbols on a computer screen. Smets *et al.* in Chapter 9 of this book discuss one approach to optimizing interface design for a virtual environment.

2.1.3 Extensions of the desk-top metaphor

Virtual environment displays represent a three-dimensional generalization of the two-dimensional desk-top metaphor.[2] The central innovation in the concept, first stated and elaborated by Ivan Sutherland (1965; 1970) and Myron Krueger (1977; 1983) with respect to interactive graphics interfaces, was that the pictorial interface generated by the computer could became a palpable, concrete illusion of a synthetic but apparently physical environment. In Sutherland's terms, this image would be the 'ultimate computer display'. These synthetic environments may be experienced either from egocentric or exocentric viewpoints. That is to say, the users may appear to actually be immersed in the environment or see themselves represented as a 'You are here' symbol (Levine, 1984) which they can control through an apparent window into an adjacent environment.

The objects in this synthetic space, as well as the space itself, may be programmed to have arbitrary properties. But, the successful extension of the desk-top metaphor to a full 'environment' requires an understanding of the necessary limits to programmer creativity in order to ensure that the environment is comprehensible and usable. These limits derive from human experience in real environments and illustrate a major connection between work in telerobotics and virtual environments. For reasons of simulation fidelity, previous telerobotic and aircraft simulations, which have many of the aspects of virtual environments, also have had to take explicitly into account real-world kinematic and dynamic constraints in ways now usefully studied by the designers of totally synthetic environments (Hashimoto *et al.*, 1986; Bussolari *et al.*, 1988; Kim *et al.*, 1988; Tachi *et al.*, 1989; Bejczy *et al.*, 1990; Sheridan, 1992; Cardullo, 1993).

2.1.4 Environments

Successful synthesis of an environment requires some analysis of the parts that make up the environment. The theatre of human activity may be used as a reference for defining an *environment* and may be thought of as having three parts: a *content*, a *geometry*, and a *dynamics* (Figure 2.1) (Ellis, 1991).

Content

The *objects* and *actors* in the environment are its content. These objects may be described by *vectors* which identify their position, orientation, velocity, and acceleration in the environmental space, as well as other distinguishing characteristics such as their colour, texture, and energy. This vector is thus a description of the *properties* of the objects. The subset of all the terms of the characteristic vector which is common to every actor and object of the content may be called the *position vector*. Though the *actors* in an environment may for some interactions be considered objects, they are distinct from objects in that in addition to characteristics they have *capacities* to initiate interactions with other objects. The basis of these initiated interactions is the storage of energy or information within the actors, and their ability to control the release of this stored information or energy after a period of time. The *self* is a distinct actor in the environment which provides *a point of view* establishing the frame of reference from which the environment may be constructed. All parts of the environment that are exterior to the self may be considered the field of action. As an example, the balls on a billiard table may be considered the content of the billiard table environment and the cue ball controlled by the pool player may be considered

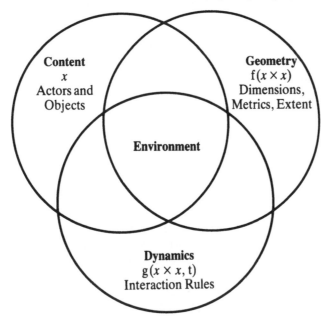

Figure 2.1 Decomposition of an environment into its abstract functional components.

the self. The additional energy and information that makes the cue ball an actor is imparted to it by the cue controlled by the pool player and his knowledge of game rules.

Geometry

The geometry is a description of the environmental field of action. It has *dimensionality,* *metrics*, and *extent*. The dimensionality refers to the number of independent descriptive terms needed to specify the position vector for every element of the environment. The metrics are systems of rules that may be applied to the position vector to establish both an ordering of the contents and the concept of geodesic lines or the loci of minimal distance paths between points in the environmental space. The extent of the environment refers to the range of possible values for the elements of the position vector. The environmental space or field of action may be defined as the cartesian product of all the elements of the position vector over their possible ranges. An environmental trajectory is a time-history of an object through the environmental space. Since kinematic constraints may preclude an object from traversing the space along some paths, these constraints are also part of the environment's geometric description.

Dynamics

The dynamics of an environment are the *rules of interaction* among its contents describing their behaviour as they exchange energy or information. Typical examples of specific dynamical rules may be found in the differential equations of newtonian dynamics describing the responses of billiard balls to impacts of the cue ball. For other environments, these rules also may take the form of grammatical rules or even of look-up tables for pattern-match-triggered action rules. For example, a syntactically correct command typed at a computer terminal can cause execution of a program with specific parameters. In this case the meaning and information of the command plays the role of the energy, and the resulting rate of change in the logical state of the affected device, plays the role of acceleration.

This analogy suggests the possibility of developing a *semantic* or *informational mechanics* in which some measure of motion through the state space of an information processing device may be related to the meaning or information content of the incoming messages. In such a mechanics, the proportionality constant relating the change in motion to the message content might be considered the *semantic* or *informational* mass of the program. A principal difficulty in developing a useful definition of 'mass' from this analogy is that information processing devices typically can react in radically different ways to slight variations in the surface structure of the content of the input. Thus it is difficult to find a technique to analyse the input to establish equivalence classes analogous to alternate distributions of substance with equivalent centres of mass. The centre-of-gravity rule for calculating the centre of mass is an example of how various apparently variant mass distributions may be reduced to a smaller number of equivalent objects in a way simplifying consistent theoretical analysis as might be required for a physical simulation on a computer.

The usefulness of analysing environments into these abstract components, content, geometry, and dynamics, primarily arises when designers search for ways to enhance operator interaction with their simulations. For example, this analysis has organized the search for graphical enhancements for pictorial displays of aircraft and spacecraft traffic (McGreevy and Ellis, 1986; Ellis *et al.*, 1987; Grunwald and Ellis, 1988, 1991, 1993). But it can also help organize theoretical thinking about what it means to be in an environment through reflection concerning the experience of physical reality.

2.1.5 Sense of physical reality

Our sense of physical reality is a construction derived from the symbolic, geometric, and dynamic information directly presented to our senses. But it is noteworthy that many of the aspects of physical reality are only presented in an incomplete, noisy form. For example, though our eyes provide us only with a fleeting series of snapshots of only parts of objects present in our visual world, through *a priori* 'knowledge' brought to perceptual analysis of our sensory input, we accurately interpret these objects to continue to exist in their entirety[3] (Gregory, 1968, 1980, 1981; Hochberg, 1986). Similarly, our goal-seeking behaviour appears to filter noise by benefiting from internal dynamical models of the objects we may track or control (Kalman, 1960; Kleinman *et al.*, 1970). Accurate perception consequently involves considerable *a priori* knowledge about the possible structure of the world (see also the discussion of top-down and bottom-up processing in section 6.4.3, and the different theories of perception described in section 7.4). This knowledge is under constant recalibration based on error feedback. The role of error feedback has been classically mathematically modelled during tracking behaviour (McRuer and Weir, 1969; Jex *et al.*, 1966; Hess, 1987) and notably demonstrated in the behavioural plasticity of visual-motor coordination (Welch, 1978; Held *et al.*, 1966; Held, and Durlach, 1991) and in vestibular and ocular reflexes (Jones *et al.*, 1984; Zangemeister and Hansen, 1985; Zangemeister, 1991).

Thus, a large part of our sense of physical reality is a consequence of internal processing rather than being something that is developed only from the immediate sensory information we receive. Our sensory and cognitive interpretive systems are predisposed to process incoming information in ways that normally result in a correct interpretation of the external environment, and in some cases they may be said to actually 'resonate' with specific patterns of input that are uniquely informative about our environment (Gibson, 1950; Heeger, 1989; Koenderink and van Doorn, 1977; Regan and Beverley, 1979). Other internalized processes which affect the perception of reality are considered in section 10.5, in the contexts of epistemology and sociology.

These same constructive processes are triggered by the displays used to present virtual environments. Since the incoming sensory information is mediated by the display technology, however, these constructive processes will be triggered only to the extent the displays provide high perceptual fidelity. Accordingly, virtual environments can come in different stages of completeness, which may be usefully distinguished by their extent of what may be called 'virtualization'.

2.2 *Virtualization*

2.2.1 Definition of virtualization

Virtualization may be defined as 'the process by which a viewer interprets patterned sensory impressions to represent objects in an environment other than that from which the impressions physically originate'. A classical example would be that of a virtual image as defined in geometrical optics. A viewer of such an image sees the rays emanating from it as if they originated from a point that could be computed by the basic lens law rather than from their actual location (Figure 2.2).

Virtualization most clearly applies to the two sense modalities associated with remote stimuli, vision and audition. In audition as in vision, stimuli can be synthesized so as to appear to be originating from sources other than their physical origin (Wightman and Kistler, 1989a, 1989b; see also section 8.4.2). But carefully designed haptic stimuli that provide illusory senses of contact, shape and position clearly also show that virtualization can be applied to other sensory dimensions (Lackner, 1988; see also section 7.7). In fact, one could consider the normal functioning of the human sensory systems as the special case in which the interpretation of patterned sensory impressions results in the perception of real objects in the surrounding physical environment, which are in fact the physical energy sources. In this respect perception of reality resolves to the case in which, through a process of cartesian systematic doubt, it is impossible for an observer to refute the hypothesis that the apparent source of the sensory stimulus is indeed the physical source.

Virtualization, however, extends beyond the objects to the spaces in which they themselves may move. Consequently, a more detailed discussion of what it means to 'virtualize' an environment is required. This discussion will only use visual examples but analogous remarks could be made concerning other sense modalities associated with spatial perception.

2.2.2 Levels of virtualization

Three levels of virtualization may be distinguished: virtual space, virtual image, and virtual environments. These levels represent identifiable points on a design continuum of virtualization as synthesized sensory stimuli more and more closely acquire the sensory and motor characteristics of a real environment. As more and more sources of sensory

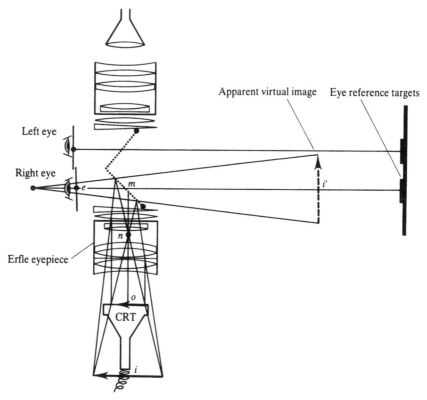

Figure 2.2   A virtual image i *created by a simple lens (or in this case, an Erfle eyepiece) placed at* n *and viewed from* e *through a half-silvered mirror at* m *appears to be straight ahead of the viewer at* i'. *The visual direction and accommodation required to see the virtual image clearly are quite different from what would be needed to see the real object at* o. *An optical arrangement similar to this would be needed to superimpose synthetic computer imagery on a view of a real scene as in a heads-up display.*

information are available, the process of virtualization can be more and more complete until at the extreme the resulting impression is indistinguishable from that originating in physical reality.

Virtual space

The first form, construction of a virtual space, refers to the process by which a viewer perceives a three-dimensional layout of objects in space when viewing a flat surface presenting the pictorial cues to space: perspective, shading, occlusion, and texture gradients (Figure 2.3). This process, which is akin to map interpretation or picture viewing, is the most abstract of the three. Viewers must literally learn to interpret pictorial images (Gregory and Wallace, 1974; Senden, 1932; Jones and Hagen, 1980). It is also not an automatic interpretive process because many of the physiological reflexes associated with the experience of a real three-dimensional environment are either missing or inappropriate for the patterns seen on a flat picture. The basis of the reconstruction of virtual space must be the optic array: the patterned collection of relative lines of sight to significant features in the image, that is, contours, vertices, lines, and textured regions. Since scaling does not affect the relative position of the features of the optic array, perceived size or scale is not intrinsically defined in a virtual space.

Virtual image

The second form of virtualization is the perception of a virtual image. In conformance with the use of this term in geometric optics, it is the perception of an object in depth

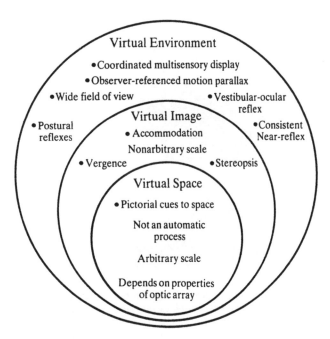

Figure 2.3   Levels of virtualization. As displays provide richer and more varied sources of sensory information, they allow users to virtualize more and more complete theatres of activity. In this view the virtual space is the most restrictive and the virtual environment is the most inclusive having the largest variety of information sources (indicated by bullet points).

in which cues from accommodation, vergence, and (optionally) stereoscopic disparity are present, though not necessarily consistent (Bishop, 1987). Since virtual images can incorporate stereopsis and vergence cues, the actual perceptual scaling of the constructed space is not arbitrary but, somewhat surprisingly, nor is it simply related to viewing geometry (Foley, 1980, 1985; Collewijn and Erkelens, 1990; Erkelens and Collewijn, 1985a, 1985b) (Figure 2.4).

Virtual environment

The final form is the virtualization of an environment. In this case, the key added sources of information are observer-slaved motion parallax, depth-of-focus variation, and wide field-of-view without visible restriction of the field of view. If properly implemented, these additional features can be consistently synthesized to provide stimulation of major space-related psychological responses and physiological reflexes such as accommodative vergence and vergence accommodation of the 'near response' (Hung *et al.*, 1984; Deering, 1992; and section 4.3 of this book), the optokinetic reflex, the vestibular-ocular reflex (Feldon and Burda, 1987), and postural reflexes (White *et al.*, 1980). These features when embellished by synthesized sound sources (Wenzel *et al.*, 1988; Wenzel, 1991; Wightman and Kistler, 1989a, 1989b; see also section 8.4.1) can substantially contribute to an illusion of telepresence (Bejczy, 1980), that is, actually being present in the synthetic environment.

Figure 2.4 See-through, head-mounted, virtual image, stereoscopic displays allow users to interact with virtual objects synthesized by computer graphics superimposed in their field of vision; however, the perceived depth of the stereo overlay must be adjusted for perceptual biases and distortions (Ellis and Bucher, 1994). The above electronic haploscope, redesigned in collaboration with Ramon Alarcon, is currently being used to study these biases (photograph courtesy of NASA). Similar see-through displays intended for medical applications have been studied at the University of North Carolina (Rolland, 1994) and related displays for mechanical assembly have been developed by Boeing Computer Services (Janin et al., 1993).

Measurements of the degree to which a virtual environment display convinces its users that they are present in the synthetic world can be made by measuring the degree to which these environmental responses can be triggered in it (Figure 2.5) (Nemire and Ellis, 1991; see also section 7.8.1). This approach provides an alternative to the use of subjective scales of 'presence' to evaluate the simulation fidelity of a virtual environment display. Subjective evaluation scales such as the Cooper–Harper rating (Cooper and Harper, 1969) have been used to determine simulation fidelity of aircraft. Related subjective scales have also been used for workload measurement, but these techniques should be used judicially since different scaling techniques can provide inconsistent results which do not generalize well across different raters (e.g. Hart and Staveland, 1988). Though they have utility for design, such subjective rating scales are unlikely to provide measurements for development of simple explanatory concepts and stable equivalence classes because of individual variability across the raters.

The fact that actors in virtual environments interact with objects and the environment by hand, head, and eye movements, tightly restricts the subjective scaling of the space so that all system gains must be carefully set. Mismatch in the gains or position measurement offsets will degrade performance by introducing unnatural visual-motor and visual-vestibular correlations. In the absence of significant time lags, humans can adapt to these unnatural correlations. Time lags do interfere with complete visual-motor adaptation, however (Held

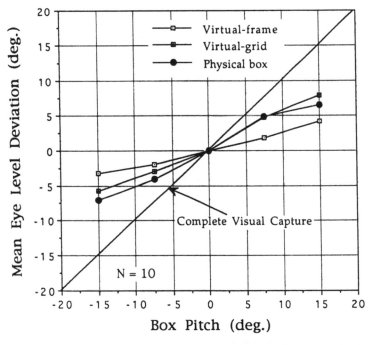

Figure 2.5 *Observers who view into a visual frame of reference, such as a large room or box that is pitched with respect to gravity, will have their sense of the horizon biased towards the direction of the pitch of the visual frame (Martin and Fox, 1989). An effect of this type is shown for the mean of 10 subjects by the trace labelled 'Physical box'. When a comparable group of subjects experienced the same pitch in a matched virtual environment using a stereo head-mounted display, the biasing effect as measured by the slope of this displayed function was about half that of the physical environment (see the trace labelled 'Virtual-frame'). Adding additional grid texture to the virtual surfaces (see the trace labelled 'Virtual-grid') increased the amount of visual-frame-induced bias, i.e. the so-called 'visual capture' (Nemire and Ellis, 1991; see also section 7.4.2).*

and Durlach, 1991; Jones *et al.*, 1984) and when present in the imaging system can cause motion sickness (Crampton, 1990). (See also Chapter 5 of this book.)

2.2.3 Environmental viewpoints and controlled elements

Virtual spaces, images or environments may be experienced from two kinds of viewpoints: egocentric viewpoints, in which the sensory environment is constructed from the viewpoint actually assumed by users, and exocentric viewpoints in which the environment is viewed from a position other than that where users are represented to be. In this case, they can literally see a representation of themselves (McGreevy and Ellis, 1986; Barfield and Kim, 1991). This distinction in frames of reference results in a fundamental difference in movements users must make to track a visually referenced target. Egocentric viewpoints classically require compensatory tracking, and exocentric viewpoints require pursuit tracking. This distinction also corresponds to the difference between inside-out and outside-in frames of reference in the aircraft simulation literature. The substantial literature on human tracking performance in these alternative reference frames, and the general literature on human manual performance, may be useful in the design of synthetic environments (Poulton, 1974; Wickens, 1986).

2.2.4 Breakdown by technological functions

The illusion of immersion in a virtual environment is created through the operation of three technologies which provide a functional breakdown as an alternative to the preceding abstract analysis. (1) Sensors, such as head position or hand shape sensors, to measure operators' body movements, (2) effectors, such as a stereoscopic displays or headphones, to stimulate the operators' senses and (3) special-purpose hardware and software to interlink the sensors and effectors to produce sensory experiences resembling those encountered by inhabitants immersed in a physical environment (Figure 2.6). In a virtual environment this linkage is accomplished by a simulation computer. In a head-mounted teleoperator display the linkage is accomplished by the robot manipulators, vehicles, control systems, sensors and cameras at a remote worksite.

Though the environment experienced with a teleoperator display is real and that experienced via the virtual environment simulation is imaginary, digital image processing allows the merging of both real and synthetic data making intermediate environments of real and synthetic objects also possible. Truly remarkable displays will be possible fusing sensor data and geographic databases.

The successful interaction of a human operator with virtual environments presented by head and body referenced sensory displays depends upon the fidelity with which the sensory information is presented to the user. The situation is directly parallel to that faced by the designer of a vehicle simulator. In fact, since virtual environments extend flight simulation technology to cheaper, accessible forms, developers can learn much from the flight simulation literature (Cardullo, 1993).

Virtual environments are simulators that are generally worn rather than entered. They are personal simulators. They are intended to provide their users with a direct sense of presence in a world or space other than the physical one in which they actually are (see section 10.3 of this book for a discussion of 'simulation'). Their users are not in a cockpit that is within a synthetic environment, they are in the environment themselves. Though this illusion of remote presence is not new, as mentioned earlier, the diffusion and rapid drop in the cost of the basic technology has raised the question of whether such displays

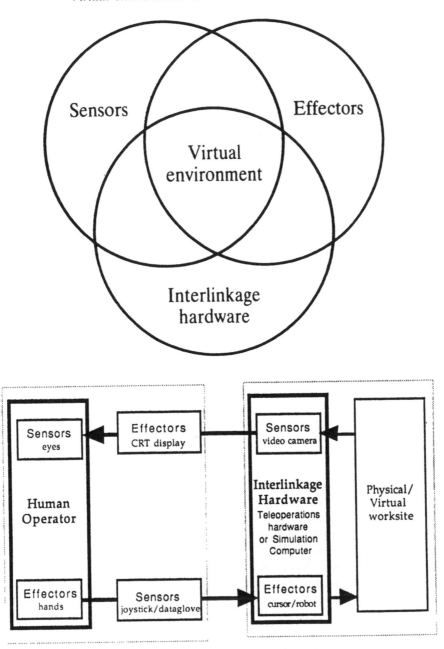

Figure 2.6 Technological breakdowns of virtual environments.

can become practically useful for a wide variety of new applications ranging from video games to laparoscopic surgical simulation.

Unfortunately, at the composition time of this chapter, users of most display systems staying in virtual environments for more than a few minutes have been legally blind (i.e. have 20/200 vision), stuck with an awkwardly narrow field-of-view (~ 30°), suffering from motion sickness, heading for a stereoscopic headache, and have been suffering from an unmistakable pain in the neck from the helmet's weight. Newer displays promise to reduce some of these problems, but performance targets necessary for a variety of possible applications are only currently being determined (see Chapters 3 to 8 of this book).

2.2.5 Spatial and environmental instruments

Like the computer graphics pictures drawn on a display surface, the enveloping synthetic environment created by a head-mounted display may be designed to convey specific information. Thus, just as a spatial display generated by computer graphics may be transformed into a spatial instrument by selection and coupling of its display parameters to specific communicated variables, so too may a synthetic environment be transformed into an environmental instrument by design of its content, geometry, and dynamics (Ellis and Grunwald, 1989a,b). Transformations of virtual environments into useful environmental instruments, however, are more constrained than those used to make spatial instruments because the user must actually inhabit the environmental instrument. Accordingly, the transformations and coupling of actions to effects within an environmental instrument must not diverge too far from those transformations and couplings actually experienced in the physical world, especially if the instrument is to be used without disorientation, poor motor coordination, and motion sickness. Thus, spatial instruments may be developed from a greater variety of distortions in the viewing geometry and scene content than environmental instruments. Environmental instruments, however, may be well-designed if their creators have appropriate theoretical and practical understanding of the constraints. Thus, the advent of virtual environment displays provides a veritable cornucopia of opportunity for research in human perception, motor-control, and interface technology.

2.3   Origins of virtual environments

2.3.1   Early visionaries

The obvious, intuitive appeal that virtual environment technology has is probably rooted in the human fascination with vicarious experiences in imagined environments. In this respect, virtual environments may be thought of as originating with the earliest human cave art (Fagan, 1985), though Lewis Carroll's *Through the Looking-Glass* (1883) certainly is a more modern example of this fascination.

Fascination with alternative, synthetic realities has been continued in more contemporary literature. Aldous Huxley's 'feelies' in *Brave New World* (1932) were also a kind of virtual environment, a cinema with sensory experience extended beyond sight and sound. A similar fascination must account for the popularity of microcomputer role-playing adventure games such as *Wizardry* (Greenberg and Woodhead, 1980). Motion pictures, and especially stereoscopic movies, of course, also provide examples of noninteractive spaces (Lipton, 1982). Theatre provides an example of a corresponding performance environment which is more interactive and has been discussed as a source of useful metaphors for human interface design (Laural, 1991).

The contemporary interest in imagined environments has been particularly stimulated by the advent of sophisticated, relatively inexpensive, interactive techniques allowing the inhabitants of these environments to move about and manually interact with computer graphics objects in three-dimensional spaces. This kind of environment was envisioned in the science fiction plots (Daley, 1982) of the movie *TRON* (1981) and in William Gibson's *Neuromancer* (1984) yet the first actual synthesis of such a system using a head-mounted stereo display was made possible much earlier in the middle 1960s by Ivan Sutherland who developed special-purpose fast graphics hardware specifically for the purpose of experiencing computer-synthesized environments through head-mounted graphics displays (Sutherland, 1965, 1970). The relationship between culture and technology is given a sociological analysis in Chapter 10, section 10.7.

Another early synthesis of a synthetic, interactive environment was implemented by Myron Krueger using back-projection and video processing techniques (Krueger, 1977, 1983, 1985) in the 1970s. Unlike the device developed for Sutherland, Krueger's environment was projected onto a wall-sized screen. In Krueger's *VIDEOPLACE*, the users' images appeared in a two-dimensional graphic video world created by a computer. The *VIDEOPLACE* computer analysed video images to determine when an object was touched by an inhabitant, and it could then generate a graphic or auditory response. One advantage of this kind of environment is that the remote video-based position measurement does not necessarily encumber the user with position sensors. A more recent and sophisticated version of this mode of experience of virtual environments is the implementation from the University of Illinois called, with apologies to Plato, the 'Cave' (Cruz-Neira *et al.*, 1992).

2.3.2 Vehicle simulation and three-dimensional cartography

Probably the most important source of virtual environment technology comes from previous work in fields associated with the development of realistic vehicle simulators, primarily for aircraft (Rolfe and Staples, 1986; CAE Electronics, 1991; McKinnon and Kruk, 1991; Cardullo, 1993) but also automobiles (Stritzke, 1991) and ships (Veldhuyzen and Stassen, 1977; Schuffel, 1987). The inherent difficulties in controlling the actual vehicles often require that operators be highly trained. Since acquiring this training on the vehicles themselves could be dangerous or expensive, simulation systems synthesize the content, geometry, and dynamics of the control environment for training and for testing of new technology and procedures.

These systems usually cost millions of dollars and have recently involved helmet-mounted displays to recreate part of the environment (Lypaczewski *et al.*, 1986; Barrette *et al.*, 1990; Furness, 1986, 1987; Kaiser Electronics, 1990). Declining costs have now brought the cost of a virtual environment display down to that of an expensive workstation and made possible 'personal simulators' for everyday use (Foley, 1987; Fisher *et al.*, 1986; Kramer, 1992; Bassett, 1992).

The simulator's interactive visual displays are made by computer graphics hardware and algorithms. Development of special-purpose hardware, such as matrix multiplication devices, was an essential step that enabled generation of real-time, that is, greater than 20 Hz, interactive three-dimensional graphics (Sutherland, 1965, 1970; Myers and Sutherland, 1968). More recent examples are the 'geometry engine' (Clark, 1982, 1980) and the 'reality engine' in Silicon Graphics *IRIS* workstations. These 'graphics engines' now can project literally millions of shaded or textured polygons, or other graphics primitives, per second (Silicon Graphics, 1993). Though this number may seem large, rendering of naturalistic objects and surfaces can require rendering hundreds of thousands of polygons. Efficient software techniques are also important for improved three-dimensional graphics performance. 'Oct-tree' data structures, for example, have been shown to dramatically improve processing speed for inherently volumetric structures (Jenkins and Tanimoto, 1980; Meagher, 1984). Additionally, special variable resolution rendering techniques for head-mounted systems also can be implemented to match the variable resolution of the human visual system and thus not waste computer resources rendering polygons that the user would be unable to see (Netrovali and Haskell, 1988; Cowdry, 1986; Hitchner and McGreevy, 1993). (See also section 6.4.2.)

Since vehicle simulation may involve moving-base simulators, programming the appropriate correlation between visual and vestibular simulation is crucial for a complete simulation of an environment. Moreover, failure to match these two stimuli correctly can lead to motion sickness (AGARD, 1988). Paradoxically, however, since the effective travel of most moving-base simulators is limited, designers must learn to introduce subthreshold visual-vestibular mismatches to produce illusions of greater freedom of movement. These allowable mismatches are built into so-called 'washout' models (Bussolari *et al.*, 1988; Curry *et al.*, 1976) and are key elements for creating illusions of extended movement. For example, a slowly implemented pitch-up of a simulator can be used as a dynamic distortion to create an illusion of forward acceleration. Understanding the tolerable dynamic limits of visual-vestibular miscorrelation will be an important design consideration for wide-field-of-view head-mounted displays.

The use of informative distortion is also well established in cartography (Monmonier, 1991) and is used to help create convincing three-dimensional environments for simulated vehicles. Cartographic distortion is also obvious in global maps which must warp a spherical surface into a plane (Cotter, 1966; Robinson *et al.*, 1984) and three-dimensional maps, which often use significant vertical scale exaggeration (6–20×) to clearly present topographic features. Explicit informative geometric distortion is sometimes incorporated into maps and cartograms presenting geographically indexed statistical data (Tobler, 1963, 1976; Tufte, 1983, 1990; Bertin, 1967/1983), but the extent to which such informative distortion may be incorporated into simulated environments is constrained by the user's movement-related physiological reflexes. If the viewer is constrained to actually be in the environment, deviations from a natural environmental space can cause disorientation and motion sickness (Crampton, 1990; Oman, 1991). For this reason, virtual space or virtual image formats are more suitable when successful communication of the spatial information may be achieved only through spatial distortions (Figure 2.7). Even in these formats, however, the content of the environment may have to be enhanced by aids such as graticules to help the user discount unwanted aspects of the geometric distortion (McGreevy and Ellis, 1986; Ellis *et al.*, 1987; Ellis and Hacisalihzade, 1990).

In some environmental simulations the environment itself is the object of interest. Truly remarkable animations have been synthesized from image sequences taken by NASA spacecraft which mapped various planetary surfaces. When electronically combined with surface altitude data, the surface photography can be used to synthesize flights over the surface through positions never reached by the spacecraft's camera (Hussey, 1990). Recent developments have made possible the use of these synthetic visualizations of planetary and Earth surfaces for interactive exploration and they promise to provide planetary scientists with the new capability of 'virtual planetary exploration' (NASA, 1990; Hitchner, 1992; McGreevy, 1993)

2.3.3 Physical and logical simulation

Visualization of planetary surfaces suggests the possibility that not only the substance of the surface may be modelled but also its dynamic characteristics. Dynamic simulations for virtual environments may be developed from ordinary high-level programming languages like Pascal or C, but this usually requires considerable time for development. Interesting alternatives for this kind of simulation have been provided by simulation and modelling languages such as SLAM II, with a graphical display interface, and TESS (Pritsker, 1986). These very high languages provide tools for defining and implementing

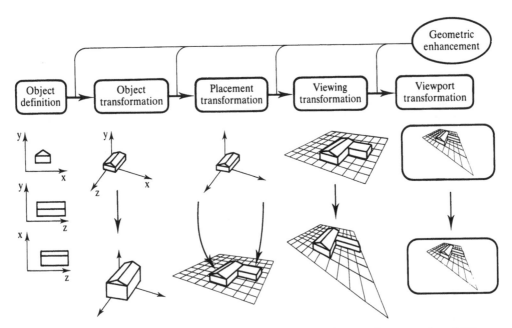

Figure 2.7   The process of representing a graphic object in virtual space allows a number of different opportunities to introduce informative geometric distortions or enhancements. These either may be a modification of the transforming matrix during the process of object definition or may be modifications of an element of a model. These modifications may take place (1) in an object relative coordinate system used to define the object's shape, or (2) in an affine or even curvilinear object shape transformation, or (3) during the placement transformation that positions the transformed object in world coordinates, or (4) in the viewing transformation or (5) in the final viewport transformation. The perceptual consequences of informative distortions are different depending on where they are introduced. For example, object transformations will not impair perceived positional stability of objects displayed in a head-mounted format, whereas changes of the viewing transformation, such as magnification, will.

continuous or discrete dynamic models. They can facilitate construction of precise systems models (Cellier, 1991).

Another alternative made possible by graphical interfaces to computers is a simulation development environment in which the simulation is created through manipulation of icons representing its separate elements, such as integrators, delays, or filters, so as to connect them into a functioning 'virtual machine'. A microcomputer program called *Pinball Construction Set* published in 1982 by Bill Budge is a widely distributed early example of this kind of simulation system. It allowed the user to create custom-simulated pinball machines on the computer screen simply by moving icons from a toolkit into an 'active region' of the display where they would become animated. A more educational, and detailed example of this kind of simulator was written as educational software by Warren Robinett. This program, called *Rocky's Boots* (Robinett, 1982), allowed users to connect icons representing logic circuit elements, that is, and-gates and or-gates, into functioning logic circuits that were animated at a slow enough rate to reveal their detailed functioning. More complete versions of this type of simulation have now been incorporated into graphical interfaces to simulation and modelling languages and are available through widely distributed object-oriented interfaces such as the interface builder distributed with NeXt® computers.

The dynamical properties of virtual spaces and environments may also be linked to physical simulations. Prominent, noninteractive examples of this technique are James Blinn's physical animations in the video physics courses, *The Mechanical Universe* and *Beyond the Mechanical Universe* (Blinn, 1987, 1991) These physically correct animations are particularly useful in providing students with subjective insights into dynamic three-dimensional phenomena such as magnetic fields. Similar educational animated visualizations have been used for courses on visual perception (Kaiser *et al.*, 1990) and computer-aided design (Open University and BBC, 1991). Physical simulation is more instructive, however, if it is interactive, and if interactive virtual spaces have been constructed which allow users to interact with nontrivial physical simulations by manipulating synthetic objects whose behaviour is governed by realistic dynamics (Witkin *et al.*, 1987, 1990). Particularly interesting are interactive simulations of anthropomorphic figures moving according to realistic limb kinematics and following higher level behavioural laws (Zeltzer and Johnson, 1991).

Some unusual natural environments are difficult to work in because their inherent dynamics are unfamiliar and may be nonlinear. The immediate environment around an orbiting spacecraft is an example. When expressed in a spacecraft-relative frame of reference known as 'local-vertical-local–horizontal', the consequences of manoeuvring thrusts becomes markedly counter-intuitive and nonlinear (NASA, 1985). Consequently, a visualization tool designed to allow manual planning of manoeuvres in this environment has taken account of these difficulties (Grunwald and Ellis, 1988, 1991, 1993; Ellis and Grunwald, 1989b). This display system most directly assists planning by providing visual feedback of the consequences of the proposed plans. Its significant features enabling interactive optimization of orbital manoeuvres include an 'inverse dynamics' algorithm that removes control nonlinearities. Through a 'geometric spreadsheet', the display creates a synthetic environment that provides the user control of thruster burns which allows independent solutions to otherwise coupled problems of orbital manoeuvring (Figures 2.8 and 2.9). Although this display is designed for a particular space application, it illustrates a technique that can be applied generally to interactive optimization of constrained nonlinear functions.

2.3.4 Scientific and medical visualization

Visualizing physical phenomena may be accomplished not only by constructing simulations of the phenomena but also by animating graphs and plots of the physical parameters themselves (Blinn, 1987, 1991). For example, multiple time functions of force and torque at the joints of a manipulator or limb while it is being used for a test movement may be displayed (see, for example, Pedotti *et al.*, 1978), or a simulation of the test environment in question itself may be interactively animated (Figure 2.10).

One application, for which a virtual space display already has been demonstrated some time ago in a commercial product has been in visualization of volumetric medical data (Meagher, 1984). These images are typically constructed from a series of two-dimensional slices of CAT, PET, or MRI images in order to allow doctors to visualize normal or abnormal anatomical structures in three dimensions. Because the different tissue types may be identified digitally, the doctors may perform an 'electronic dissection' and selectively remove particular tissues. In this way remarkable skeletal images may be created which currently aid orthopaedic and cranio-facial surgeons to plan operations (Figure 2.11). These volumetric databases also are useful for shaping custom-machined prosthetic bone implants and for directing precision robotic boring devices for precise fit between implants and

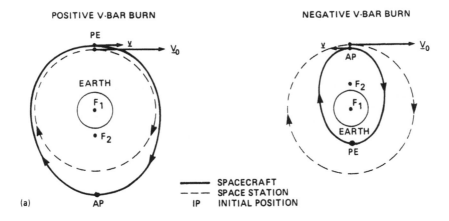

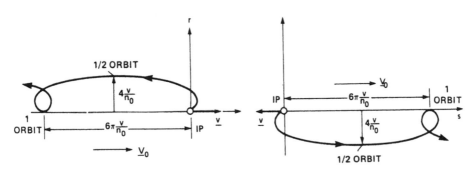

Figure 2.8 Unusual environments sometimes have unusual dynamics. The orbital motion of a satellite in a low earth orbit (upper panels) changes when thrust $\underline{v}$ is made either in the direction of orbital motion, $\underline{V}_O$, (left) or opposed to orbital motion (right) and indicated by the change of the original orbit (dashed lines) to the new orbit (solid line). When the new trajectory is viewed in a frame of reference relative to the initial thrust point on the original orbit (Earth is down, orbital velocity is to the right, see lower panels), the consequences of the burn appear unusual. Forward thrusts (left) cause nonuniform, backward, trochoidal movement. Backward thrusts (right) cause the reverse.

surrounding bone (Taylor *et al.*, 1990). Though these static databases have not yet been presented to doctors as full virtual environments, existing technology is adequate to develop improved virtual space techniques for interacting with them and may be able to enhance the usability of the existing displays for teleoperated surgery (Green *et al.*, 1992; UCSD Medical School, 1994; Satava and Ellis, 1994). Related scene-generation technology can already render detailed images of this sort based on architectural drawings and can allow prospective clients to visualize walkthroughs of buildings or furnished rooms that have not yet been constructed (Greenberg, 1991; Airey *et al.*, 1990; Nomura *et al.*, 1992).

2.3.5 Teleoperation and telerobotics and manipulative simulation

The second major technical influence on the development of virtual environment technology is research on teleoperation and telerobotic simulation (Goertz, 1964; Vertut and Coiffet, 1986; Sheridan, 1992). Indeed, virtual environments existed before the name itself, as telerobotic and teleoperations simulations. The display technology, however, in these cases

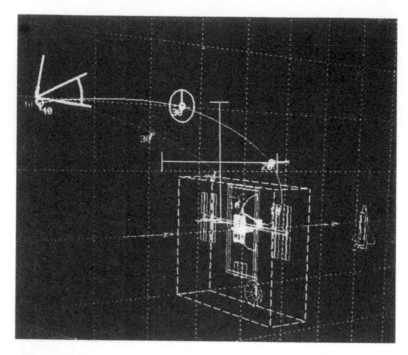

Figure 2.9   A proximity operations planning display presents a virtual space that enables operators to plan orbital manoeuvre despite counter-intuitive, nonlinear dynamics and operational constraints, such as plume impingement restrictions. The operator may use the display to visualize his proposed trajectories. Violations of the constraints appear as graphics objects, i.e. circles and arcs, which inform him of the nature and extent of each violation. This display provides a working example of how informed design of a planning environment's symbols, geometry, and dynamics can extend human planning capacity into new realms. (Photograph courtesy of NASA.)

was usually panel-mounted rather than head-mounted. Two notable exceptions were the head-controlled/head-referenced display developed for control of remote viewing systems by Raymond Goertz at Argonne National Laboratory (Goertz *et al.*, 1965) and a head-mounted system developed by Charles Comeau and James Bryan of Philco (Figure 2.12) (Comeau and Bryan, 1961). The development of these systems anticipated many of the applications and design issues that confront the engineering of effective virtual environment systems. Their discussions of the field-of-view/image resolution trade-off is strikingly contemporary. A key difficulty, then and now, was the lack of a convenient and precise head tracker. The current popular, electromagnetic, six-degree-of-freedom position tracker developed by Polhemus Navigation (Raab *et al.*, 1979; also see Ascension Technology Corp., 1990; Polhemus Navigation Systems, 1990; Barnes, 1992) consequently was an important technological advance. Interestingly, this was anticipated by similar work at Philco (Comeau and Bryan, 1961) which was limited, however, to electromagnetic sensing of orientation. In other techniques for tracking the head position, accelerometers, optical tracking hardware (CAE Electronics, 1991; Wang *et al.*, 1990), or acoustic systems (Barnes, 1992) may be used. These more modern sensors are much more convenient than those used by the pioneering work of Goertz and Sutherland, who used mechanical position sensors, but the important, dynamic characteristics of these sensors have only recently begun to be fully described (Adelstein, Johnston and Ellis, 1992).

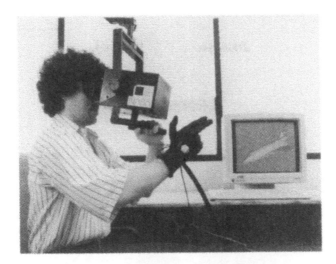

Figure 2.10 Virtual environment technology may assist visualization of the results of aerodynamic simulations. Here a DataGlove is used to control the position of a 'virtual' source of smoke in a wind-tunnel simulation so the operator can visualize the local pattern of air flow. In this application the operator uses a viewing device incorporating TV monitors (McDowall et al., 1990) to present a stereo view of the smoke trail around the test model also shown in the desk-top display on the table (Levit and Bryson, 1991). (Photograph courtesy of NASA.)

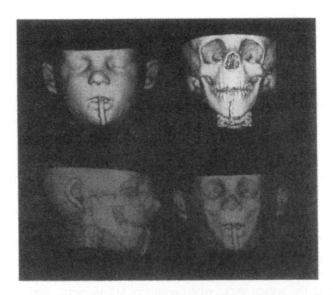

Figure 2.11 Successive CAT scan X-ray images may be digitized and used to synthesize a volumetric data set which then may be electronically processed to identify specific tissue. Here bone is isolated from the rest of the data set and presents a striking image that even non-radiologists may be tempted to interpret. Forthcoming hardware will give physicians access to this type of volumetric imagery for the cost of a car. Different tissues in volumetric data sets from CAT scan X-ray slices may be given arbitrary visual properties by digital processing in order to aid visualization. In this image, tissue surrounding the bone is made partially transparent so as to make the skin surface as well as the underlying bone of the skull clearly visible. This processing is an example of enhancement of the content of a synthetic environment. (Photograph courtesy of Octree Corporation, Cupertino, CA.)

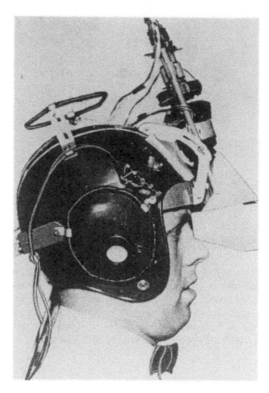

Figure 2.12 Visual virtual environment display systems have three basic parts: a head-referenced visual display, head and/or body position sensors, and a technique for controlling the visual display based on head and/or body movement One of the earliest systems of this sort, shown above, was developed by Philco engineers (Comeau and Bryan, 1961) using a head-mounted, biocular, virtual image viewing system, a Helmholtz coil electromagnetic head-orientation sensor, and a remote TV camera slaved to head orientation to provide the visual image. Today this would be called a telepresence viewing system. The first system to replace the video signal with a totally synthetic image produced through computer graphics, was demonstrated by Ivan Sutherland for very simple geometric forms (Sutherland, 1965).

A second key component of a teleoperation workstation, or of a virtual environment, is a sensor for coupling hand position to the position of the end-effector at a remote worksite. The earlier mechanical linkages used for this coupling have been replaced by joysticks or by more complex sensors that can determine hand shape, as well as position. Modern joysticks are capable of measuring simultaneously all three rotational and three translational components of motion. Some of the joysticks are isotonic (BASYS, 1990; CAE Electronics, 1991; McKinnon and Kruk, 1991) and allow significant travel or rotation along the sensed axes, whereas others are isometric and sense the applied forces and torques without displacement (Spatial Systems, 1990). Though the isometric sticks with no moving parts benefit from simpler construction, the user's kinematic coupling in his hand make it difficult for him to use them to apply signals in one axis without cross-coupled signals in other axes. Consequently, these joysticks use switches for shutting down unwanted axes during use. Careful design of the breakout forces and detents for the different axes on the isotonic sticks allow a user to minimize cross-coupling in control signals while separately controlling the different axes (CAE Electronics, 1991; McKinnon and Kruk 1991).

Although the mechanical bandwidth might have been only of the order of 2–5 Hz, the early mechanical linkages used for telemanipulation provided force-feedback conveniently and passively. In modern electronically coupled systems force-feedback or 'feel' must be actively provided, usually by electric motors. Although systems providing six degrees of freedom with force-feedback on all axes are mechanically complicated, they have been constructed and used for a variety of manipulative tasks (Bejczy and Salisbury, 1980; Hannaford, 1989; Jacobson *et al.*, 1986; Jacobus *et al.*, 1992; Jacobus, 1992). Interestingly, force-feedback appears to be helpful in the molecular docking work at the University of North Carolina, in which chemists manipulate molecular models of drugs in a computer graphics physical simulation in order to find optimal orientations for binding sites on other molecules (Figure 2.13) (Ouh-young *et al.*, 1989).

High-fidelity force-feedback requires electromechanical bandwidths over 30 Hz. Most manipulators do not have this high a mechanical response. A force-reflecting joystick with these characteristics, however, has been designed and built (Figure 2.14) (Adelstein and Rosen, 1991, 1992). Because of the required dynamic characteristics for high fidelity, it is not compact and is carefully designed to protect its operators from the strong, high-frequency forces it is capable of producing (see Fisher *et al.* (1990) for some descriptions of typical manual interface specifications; also see Brooks and Bejczy (1986) for a review of control sticks).

Manipulative interfaces may provide varying degrees of manual dexterity. Relatively crude interfaces for rate-controlled manipulators may allow experienced operators to

Figure 2.13 A researcher at the University of North Carolina uses a multi-degree-of-freedom manipulator to manoeuvre a computer graphics model of a drug molecule to find binding sites on a larger molecule. A dynamic simulation of the binding forces is computed in real time so the user can feel these forces through the force-reflecting manipulator and use this feel to identify the position and orientation of a binding site. (Photograph courtesy of University of North Carolina, Department of Computer Science.)

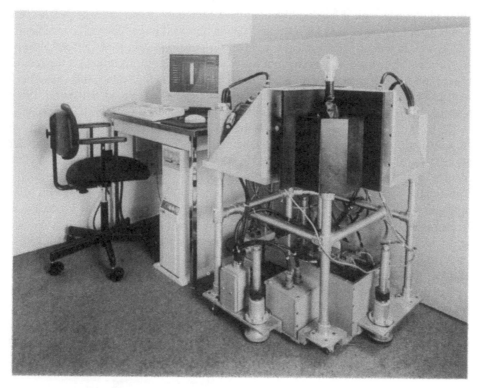

Figure 2.14   A high-fidelity, force-reflecting two-axis joystick designed to study human tremor. (Photograph courtesy of B. Dov Adelstein.)

accomplish fine manipulation tasks (Figure 2.15). Access to this level of proficiency, however, can be aided by coordinated displays of high visual resolution, by use of position control derived from inverse kinematic analysis of the manipulator, by more intuitive control of the interface, and by more anthropomorphic linkages on the manipulator.

An early example of a dexterous, anthropomorphic robotic end-effector is the hand by Tomovic and Boni (Tomovic and Boni, 1962). A more recent example is the Utah/MIT hand (Jacobson *et al.*, 1984). Such hand-like end-effectors with large numbers of degrees of freedom may be manually controlled directly by hand-shape sensors, for example, the Exos exoskeleton hand (Exos, 1990) (Figure 2.16).

Significantly, the users of the Exos hand often turn off a number of the joints, raising the possibility that there may be a limit to the number of degrees of freedom usefully incorporated into a dexterous master controller (Marcus, 1991). Less bulky hand-shape measurement devices have also been developed using fibre optic or other sensors (Zimmerman *et al.*, 1987; W Industries, 1991) (Figure 2.17). Use of these alternatives, however, involves significant trade-offs of resolution, accuracy, force-reflection and calibration stability as compared with the more bulky sensors (as shown in Figure 2.16). A more recent hand-shape measurement device had been developed that combines high static and dynamic positional fidelity with intuitive operation and convenient donning and doffing (Kramer, 1992). Chapter 7 reviews the perceptuo-motor issues involved in manipulative simulation (section 7.7 reviews virtual manipulation).

As suggested by the informal comments of Exos hand-master users who shut down apparently unneeded degrees of freedom on their hand-shape sensor, the number of degrees

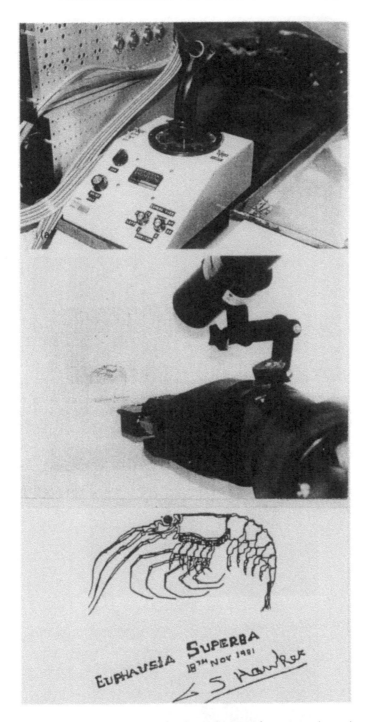

Figure 2.15 Experienced operators of industrial manipulator arms (centre) can develop great dexterity (see drawing at bottom) even with ordinary two-degree-of-freedom, joystick interfaces (top) for the control of robot arms with adequate mechanical bandwidth. Switches on the control box shift control to the various joints on the arm. The source of the dexterity illustrated here is the high dynamic fidelity of the control, a fidelity that needs to be reproduced if supposedly more natural haptic virtual environment interfaces are to be useful. (Photographs courtesy of Deep Ocean Engineering, San Leandro, CA.)

Figure 2.16   An exoskeleton hand-shape measurement system in a dexterous hand-master using accurate Hall-effect flexion sensors which is suitable to drive a dexterous end-effector. (Photograph courtesy of Exos, Inc., Burlington, MA.)

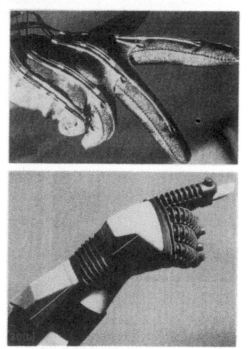

Figure 2.17   Less bulky hand-shape measuring instruments using flexible sensors (upper panel: courtesy of VPL, Redwood City, CA; lower panel courtesy of W Industries, Leicester, UK).

of freedom that need to be monitored by sensors used for virtual environment displays can become the subject of formal investigations. For example, the head-position sensors used on the Fakespace Boom constrain natural head roll in the coronal plane. Furthermore, anecdotal observations of architectural walkthroughs with closed head-mounted displays have indicated that it was better to disable head-roll tracking because, combined with sensor lag, it seemed to make the use of the display unpleasant (J. Nomura, Personal communication, 1993). Accordingly, one might reasonably ask what benefits in position and orientation display head-roll tracking provides. This question has been investigated for manipulative interaction with targets placed within arm's length of a user (Figure 2.18). The results show that if large head rotations with respect to the torso are not required ($> \sim 50°$), head-roll tracking provides only minor improvements in the users' ability to rotate objects into alignment with targets presented in a virtual environment (Adelstein and Ellis, 1993). Thus, roll-tracking and roll-compensation on some telepresence camera platforms may be unnecessary if the user's interface to the control of the remote device does not require large head–torso rotations.

2.3.6 Photography, cinematography, video technology

Since photography, cinema, and television are formats for presenting synthetic environments, it is not surprising that technology associated with special effects for these media have been applied to virtual environments. The LEEP optics, which are commonly used in many 'virtual reality' stereo-viewers, were originally developed for a stereoscopic camera system using matched camera and viewing optics to cancel the aberrations of the wide angle lens. The LEEP system field of view is approximately $110° \times 55°$, but it depends on how the measurement is taken (Howlett, 1991). Though this viewer does not allow adjustment for interpupillary distance, its large entrance pupil (30 mm radius) removes the need for such adjustment. The stereoscopic image pairs used with these optics, however, are presented 62 mm apart, closer together than the average interpupillary distance. This

Figure 2.18 Apparatus used to study the benefits of incorporating head-roll tracking into a head-mounted telepresence display. The left panel shows a stereo video camera mounted on a 3-degree-of-freedom platform that is slaved in orientation to the head orientation of an operator wearing a head-mounted video display at a remote site. The operator sees the video images from the camera and uses them to reproduce the orientation and position of rectangular test objects distributed on matching cylindrical work surfaces.

choice is a useful design feature which reduces the likelihood that average users need to diverge their eyes to achieve binocular fusion.

An early development of a more complete environmental illusion through cinematic virtual space was Morton Heilig's 'Sensorama'. It provided a stereo, wide-field-of-view, egocentric display with coordinated binaural sound, wind, and odour effects (Heilig, 1955). A more recent, interactive virtual space display was implemented by the MIT Architecture Machine Group in the form of a video-disk-based, interactive map of Aspen, Colorado (Lippman, 1980). The interactive map provided a video display of what the user would have seen had he actually been there moving through the town. Similar interactive uses of video-disk technology have been explored at the MIT Media Lab (Brand, 1987). One feature that probably distinguishes the multimedia work mentioned here from the more scientific and engineering studies reported previously, is that the media artists, as users of the enabling technologies, have more interest in synthesizing highly integrated environments including sight, sound, touch, and smell. A significant part of their goal is the integrated experience of a 'synthetic place'. On the other hand, the simulator designers are only interested in capturing the total experience insofar as this experience helps specific training and testing. Realism is itself not their goal, but effective communication and training are.

2.3.7 Role of engineering and physiological models

Since the integration of the equipment necessary to synthesize a virtual environment represents such a technical challenge in itself, there is a tendency for groups working in this area to focus their attention only on collecting and integrating the individual technologies for conceptual demonstrations in highly controlled settings. The videotaped character of many of these demonstrations of early implementation often has suggested system performance far beyond actually available technology. The visual resolution of the cheaper, wide field displays using LCD technology has often been, for example, implicitly exaggerated by presentation techniques using overlays of users wearing displays and images taken directly from large-format graphics monitors. In fact, the users of many of these displays are, as noted above in section 2.2.4, for practical purposes legally blind.

Accomplishment of specific tasks in real environments, however, places distinct real performance requirements on the simulation of which visual resolution is just an example. These requirements may be determined empirically for each task, but a more general approach is to use human performance models to help specify them. There are good general collections that can provide this background design data (e.g. Borah *et al.*, 1978; Boff *et al.*, 1986; Elkind *et al.*, 1989) and there are specific examples of how scientific and engineering knowledge and computer-graphics based visualization can be used to help designers conform to human performance constraints (Monheit and Badler, 1990; Phillips *et al.*, 1990; Larimer *et al.*, 1991). Useful sources on human sensory and motor capacities relevant to virtual environments are also available (Brooks and Bejczy, 1986; Howard, 1982; Blauert, 1983; Goodale, 1990; Durlach *et al.*, 1991; Ellis *et al.*, 1993, and within this book). (See also section 7.1 for further reference to ergonomic databases and a description of human factors in the design process.)

Because widely available current technology limits the graphics and simulation update rate in virtual environments to less than 20 Hz, understanding the control characteristics of human movement, visual tracking, and vestibular responses is important for determining the practical limits to useful work in these environments. Theories of grasp (see section 7.6), manual tracking (Jex *et al.*, 1966) spatial hearing (Blauert, 1983; Wenzel, 1991;

see section 8.2.3), vestibular response (see section 7.3.4), and visual-vestibular correlation (Oman, 1991; Oman *et al.*, 1986) all can help to determine performance guidelines.

Predictive knowledge of system performance is not only useful for matching interfaces to human capabilities, but it is also useful in developing effective displays for situations in which human operators must cope with significant time lags, for example those >250 ms, or other control difficulties. In these circumstances, accurate dynamic or kinematic models of the controlled element allow the designer to give the user control over a predictor which he may move to a desired location and which will be followed by the actual element (Hashimoto *et al.*, 1986; Bejczy *et al.*, 1990) (Figure 2.19).

Another source of guidelines is the performance and design of existing high-fidelity systems themselves. Of the virtual environment display systems, probably the one with the best visual display is the CAE Fiber Optic Helmet Mounted Display or the 'FOHMD' (Lypaczewski *et al.*, 1986; Barrette *et al.*, 1990) which is used in military aircraft simulators (Figure 2.20). It presents two 83·5° monocular fields of view with an adjustable binocular overlap, typically of about 38° in early versions, giving a full horizontal field-of-view of up to 162°. Similarly, the Wright-Patterson Air Force Base Visually Coupled Airborne Systems Simulator or 'VCASS' display, also presents a very wide field of view, and has been used to study the consequences of field-of-view restriction on several visual tasks (Wells and Venturino, 1990). Their results support other reports that indicate that visual performance is influenced by increased field-of-view, but that this influence wanes as fields-of-view greater than 60° are used (Hatada *et al.*, 1980).

Figure 2.19 A graphic model of a manipulator arm is electronically superimposed on a video signal from a remote worksite to assist users who must contend with time delay in their control actions. (Photograph courtesy of JPL, Pasadena, CA.)

Figure 2.20   Though very expensive, the CAE Fiber Optic Helmet Mounted Display, FOHMD (upper panel), is one of the highest-performance virtual environment systems used as a head-mounted aircraft simulator display. It can present an overall visual field $162° \times 83 \cdot 5°$ with 5-arcmin resolution with a high resolution inset of $24° \times 18°$ of $1 \cdot 5$ arcmin resolution. It has a bright display: 30 foot-lambert, and a fast, optical head-tracker: 60-Hz sampling, with accelerometer augmentation. The Kaiser WideEye® display (lower panel) is a roughly comparable, monochrome device designed for actual flight in aircraft as a head-mounted heads-up display. It has a much narrower field of view (monocular: 40°, or binocular with 50% overlap $40° \times 60°$; visual resolution is approximately 3 arcmin). (Photographs courtesy of CAE Electronics, Montreal, Canada; Kaiser Electronics, San José, CA.)

A significant feature of the FOHMD is that the 60-Hz sampling of head position had to be augmented by signals from helmet-mounted accelerometers to perceptually stabilize the graphics imagery during head movement. Without the accelerometer signals, perceptual stability of the enveloping environment requires head-position sampling over 100 Hz, as illustrated by well-calibrated teleoperations viewing systems developed in Japan (Tachi *et al.*, 1989, 1984). In general, it is difficult to calibrate the head-mounted, virtual image displays used in these integrated systems (see Chapter 5). One solution is to use a see-through system and to compare the positions of real objects and superimposed computer-generated objects (Hirose *et al.*, 1990, 1992; Ellis and Bucher, 1994; Janin *et al.*, 1993; Rolland, 1994).

Technical descriptions with performance data for fully integrated systems have not been generally available or accurately detailed (Fisher *et al.*, 1986; Stone, 1991a, 1991b), but this situation should change as reports are published in a number of journals, i.e. *IEEE Computer Graphics and Applications*; *Computer Systems in Engineering*; *Presence: the Journal of Teleoperations and Virtual Environments*; *Pixel: the Magazine of Scientific Visualization*; *Ergonomics*; *Human Factors*. Compendiums of the human factors design issues are available (e.g. Ellis *et al.*, 1993), and there are books collecting manufacturers' material which ostensibly describes the performance of the component technology (e.g. Kalawsky, 1993). But due to the absence of standards and the novelty of the equipment, developers are likely to find these descriptions still incomplete and sometimes misleading. Consequently, users of the technology must often measure the basic performance measurements of components themselves (e.g. Adelstein *et al.*, 1992).

2.4 Virtual environments: performance and trade-offs

2.4.1 Performance advances and customizability of the technology

With the state of off-the-shelf technology, it is unlikely that a fully implemented virtual environment display will today uniquely enable useful work at a price accessible to the average researcher. Those systems that have solved some of the major technological problems, namely, adequate head-tracking bandwidth and viewing resolution comparable to existing CRT technology, do so through special-purpose hardware that is very expensive. The inherent cost of some enabling technologies, however, is not high and development continues, promising improved performance and flexibility (e.g. optical head-tracking (Wang *et al.*, 1990) and high-quality detailed volumetric display display hardware for medium-cost workstations stations (Octree Corporation, 1991)). hardware for medium-cost workstations (Octree Corporation, 1991)). Medium-cost complete systems costing in the order of US$200 000 have currently proved commercially useful for visualizing and selling architectural products such as custom kitchens (Nomura *et al.*, 1992). No matter how sophisticated or cheap the display technology becomes, however, there will always be some costs associated with its use. With respect to practical applications, the key question is to identify those tasks that are so enabled by use of a virtual environment display, that users will choose this display format over alternatives.

Clearly, there are many possible applications, beside flight simulation and cockpit display design, which may soon become practical. In addition to the obvious video game applications, the technology could help CAD/CAM designers visualize their final product (as shown in Chapter 9), telerobotic programmers debug the planning algorithms, teleoperation controllers deal with time lags, laparoscopic surgeons plan and execute procedures, customers design and select custom kitchens. The list of possibilities could go on and on. The difficulties with such lists is that for each application a very particular set of static and dynamic task demands needs to be satisfied and its satisfaction involves the definition of a livable, but specially tuned environment.

The apparent general applicability of virtual environments to practically anything is, in reality, a significant difficulty. Technology derives its power from its specificity and its customizability (Basalla, 1988). The first LINK trainer which taught how an aircraft responds to control stick and pedal inputs was a commercial failure, but the later model designed for the more specific purpose of relating control inputs to flight instrument readings was a success. An aircraft simulator is valuable not because it can simulate an aircraft but because it can simulate a Boeing 747SP!

This requirement for specificity is really quite an order as the developers of flight simulators know full well, and it is where a good portion of development resources need to be committed. In fact, even in flight simulation we are only beginning to be able to create head-mounted simulators with adequate fidelity to avoid nausea, allow protracted use, and enable useful training. The motion sickness problem is only one of many human factors issues yet to be resolved with this technology as both the serious applications as well as entertainment applications are fielded.

This wide array of possibilities can be a significant design issue. In fact, virtual environments are probably almost uniquely useful for telerobotic-like tasks that require several coordinated foci of control within a 3-D database; a typical such task is encountered in laparoscopic surgery. But for other applications the egocentric frame of reference provided by head-mounted displays may not provide significant added performance over much cheaper, publicly viewable, panel-mounted alternatives. Research needs to be done to identify what level of virtualization is most appropriate to which applications and to understand how to match specific aspects of the display to needs of the human users. Some suggestions of appropriate matches for a variety of sensory and motor communication channels exist (Figure 2.21) but these have not been systematically tested or validated and in reality only represent informal engineering opinion. Some of the difficulties of matching human visual needs in virtual environment, not dicussed elsewhere in this book, are considered below. In general, the visual specifications for aircraft heads-up displays (HUDs) are a good first place to look for visual specifications for virtual environments (Weintraub and Ensing, 1992).

2.4.2 Stereoscopic visual strain

Designers of helmet-mounted displays for military applications have known that field use of stereoscopic displays is difficult because careful alignment is required to avoid problems with visual fatigue (O. J. Edwards, Personal communication, 1991; J. Melzer, Personal communication, 1991). Accordingly, stereo eye-strain is a likely difficulty for long-term use of stereo virtual environments, especially because most stereo displays of near objects present significant violations of normal relationships between vergence and accommodation (as specified in section 4.3). But new devices for measuring acuity, accommodation, and eye position (Takeda *et al.*, 1986) may help improve designs. Development of a self-compensating display that adjusts to the refractive state and position of the user's eyes is one possibility but currently well beyond current fieldable technology. As with eye-strain, the neck-strain caused by the helmet's mass is likely to be relieved by technical advances such as miniaturization. But, as Krueger has consistently emphasized, there will always be a cost associated with required use of head gear and the simple solution to this problem may be to avoid protracted use, as is possible with boom-mounted displays which may be quickly donned and doffed.

2.4.3 Resolution/field-of-view trade-off

Another cost associated with head-mounted displays is that, though they may generally have larger fields of view than the panel-mounted alternative, they will typically have correspondingly lower spatial resolution. Eye-movement recording technology has been used to avoid this trade-off by tracking the viewer's current area of fixation so that a high-resolution graphics insert can be displayed there. This technique can relieve the graphics processor of the need to display high-resolution images in the regions of the peripheral visual field that cannot resolve it (Cowdry, 1986). Reliable and robust eye-tracking technology is still, however, costly, but fortunately may be unnecessary if a high-resolution insert of approximately 30° diameter may be inserted. Since in the course of daily life most eye movements may be less than 15° (Bahill *et al.*, 1975), a head-mounted display system which controls the viewing direction of the simulation need not employ eye tracking if the performance environment does not typically require large amplitude eye movements.

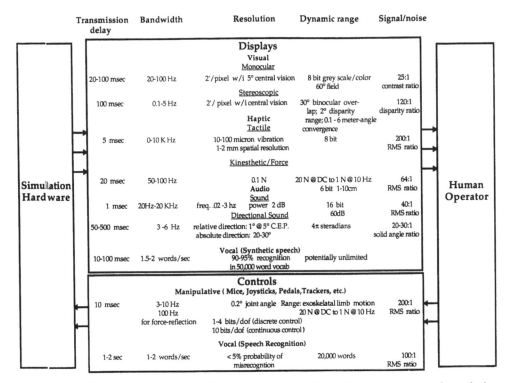

	Transmission delay	Bandwidth	Resolution	Dynamic range	Signal/noise
Displays					
Visual					
Monocular	20-100 msec	20-100 Hz	2'/pixel w/i 5° central vision	8 bit grey scale/color 60° field	25:1 contrast ratio
Stereoscopic	100 msec	0.1-5 Hz	2'/pixel w/i central vision	30° binocular over-lap; 2° disparity range; 0.1 - 6 meter-angle convergence	120:1 disparity ratio
Haptic					
Tactile	5 msec	0-10 K Hz	10-100 micron vibration 1-2 mm spatial resolution	8 bit	200:1 RMS ratio
Kinesthetic/Force	20 msec	50-100 Hz	0.1 N	20 N @ DC to 1 N @ 10 Hz 6 bit 1-10cm	64:1 RMS ratio
Audio					
Sound	1 msec	20Hz-20 KHz	freq. .02 -3 hz power 2 dB	16 bit 60dB	40:1 RMS ratio
Directional Sound	50-500 msec	3 -6 Hz	relative direction: 1° @ 5° C.E.P. absolute direction: 20-30°	4π steradians	20-30:1 solid angle ratio
Vocal (Synthetic speech)	10-100 msec	1.5-2 words/sec	90-95% recognition in 50,000 word vocab	potentially unlimited	
Controls					
Manipulative (Mice, Joysticks, Pedals,Trackers, etc.)	10 msec	3-10 Hz 100 Hz for force-reflection	0.2° joint angle 1-4 bits/dof (discrete control) 10 bits/dof (continuous control)	Range: exoskelatal limb motion 20 N @ DC to 1 N @ 10 Hz	200:1 RMS ratio
Vocal (Speech Recognition)	1-2 sec	1-2 words/sec	< 5% probability of misrecogntion	20,000 words	100:1 RMS ratio

Simulation Hardware — Human Operator

Figure 2.21 This block diagram describes some characteristics of the communication channels that must be opened between a user of a virtual environment and the underlying simulation program. These are simulation-specific and will vary for different applications, but the values in the table are based on human perceptual and motor performance characteristics that have allowed useful performance in at least one specific application. They enumerate a vocabulary of performance targets that simulation hardware may need to achieve for the development of useful virtual environments (Ellis, 1994).

2.4.4 Unique capabilities

In view of these and certainly other costs of virtual environment displays, what unique capabilities do they enable? Since these systems amount to a communications medium, they are intrinsically applicable to practically anything, as already shown. One unique feature of the medium, however, is that it can enable multiple, simultaneous, coordinated, real-time foci of control in an environment. Tasks that involve the manipulation of objects in complex visual environments and also require frequent, concurrent changes in viewing position, for example laparoscopic surgery (SAGES, 1991), are tasks that are naturally suited for virtual environment displays. Other tasks that may be mapped into this format are also uniquely suitable, for example, maintenance training for complex machinery such as jet engines which may be manually 'disassembled' using virtual environment displays. In selecting a task for which virtual environment displays may provide useful interfaces it is important to remember that effective communication is the goal, and that consequently one need not aspire to create a fully implemented virtual environment: a virtual space or a virtual image might even be superior. For non-entertainment applications, the illusion of an alternative reality is not necessarily the goal of the interface design. The case of the Mattel PowerGlove®, which is no longer manufactured, is instructive. This interface device, which was derived from the Data Glove®, was marketed for video games as an intuitive control device to replace the joysticks and fire buttons. But it proved fatiguing since it required the users to keep their hands held in the air for extended periods, and yet, since no special-purpose software was ever written to exploit its unique control capacities, provided no particular advantage to its user. It was thus marketed for a pure novelty value which soon wore off. A successful virtual environment product will have to find a real communications need to fill for it to be successful in the long term. Interaction metaphors that are merely amusing will need to be separated from those that are actually useful. One way to ensure that a real communications need is met is to imbed the use of virtual displays in a thematic context as has been done in the Battle Tech video-game parlours and as is planned by Paramount for a similar Star Trek attraction. Thus, virtual environments may become a unique medium for storytelling through participation.

2.4.5 Future mass markets

It is difficult to foretell the future practical mass-market applications for virtual environments. Like three-dimensional movies, the technology could be only a periodic infatuation of visionary technophiles waxing and waning over the years, but the situation is more likely analogous to the introduction of the first personal computer, the Altair. At its introduction, the practical uses for which small computers like it have since become essential, such as word processing, databases and spreadsheets, seemed well beyond its reach. In fact, spreadsheet programs like VISICALC had not even been conceived! Possibly, once the world is densely criss-crossed with high bandwidth, public access, fibre-optic 'information highways' (Gore, 1990), mass demand will materialize for convenient, virtual environment displays of high-resolution imagery.

2.4.6 Social and psychological impact

As in the flight simulators which were their predecessors, extended time in virtual environments can produce nausea and altered visual and visuomotor coordination. These after-effects can interfere with automobile driving and other aspects of normal life in the physical environment to which all users must ultimately return.

Life in virtual environments may, indeed, also have social after-effects, especially if the high level of violence in existing video games is transferred into this new medium. There are, for example, anecdotal instances of aggressive driving behaviour transferring from arcade auto-simulator video games to real driving (J. Wharton, Personal communication, 1992). Consequently, the design of virtual environments may provide not only technical, but also social and possibly political challenges in a manner related to the impact of video-game technology (see section 10.10).

As video-game technology has matured, the innocuous table tennis game of *Pong* has evolved to more and more realistic martial arts simulations with a level of violence that would satisfy Rambo. Though this cartoon violence is still abstract and somewhat like killing a chessman, parents may find it prudent to filter the video games that get into their house. The wisdom of giving children practice in being a deity, unleashing natural disasters, and declaring Armageddon while massacring uncounted millions in increasingly sophisticated graphic simulations, remains unclear. Many of these games appeal to an infantile, omnipotent egocentrism hardly conducive to good citizenship and their translation into a virtual environment in which the violence is literally 'acted out' in a more and more realistic synthetic visual and acoustic settings can only be expected to introduce or worsen untoward behavioural side-effects.

The existence of these negative side-effects is now an open issue, but it is not hard to suppose that as the visual and interactive fidelity of the simulation improves, the synthetic violence will become harder to distinguish from the real thing and may transfer from the virtual to the real world. One day legal theories, seemingly as bizarre as 'the-backwards-subliminal-message-in-the-music-made-me-do-it', may have to be faced. They may in fact come to have a demonstrable basis as virtual environments more closely approximate the physical world.

Some eager developers, such as Jaron Lanier of VPL Research, have been optimistic in suggesting that the virtual environments could provide an outlet for violence and have a beneficial social effects. This issue remains as open as the worry about negative consequences. Both need investigation (section 10.13 suggests ways of doing this). Virtual environments, like all powerful new technologies, can have light as well as dark sides which will have to be addressed by policy makers interested in 'ponography' as well as pornography.[4]

Echoing Plato's lament that the rise of literacy would drive the art of memory into obsolescence, those reflecting on the future social impact of virtual environment technology might be tempted to worry that their rise will herald the ruin of the capacity of human imagination. This ruin seems to me unlikely. The type of technologically driven intellectual change at issue is best seen as analogous to the transposition of a melody into a new key: the detailed old, low-level habits become obsolete, but provided the transition is skilfully done, the range of human expression is conserved, and possibly extended.

There should be no question that new forms of expression and control will be possible through the maturing of virtual environment technology. In fact, however, the future uses for virtual environments are hard to foresee since the technology could provide a new general medium for communication. Since the situation is probably parallel to the introduction of the first personal computer, the ultimate, most prominent application of virtual environments is possibly today still unknown.

Notes

1. Earlier versions of some parts of this chapter appeared as 'Nature and origin of virtual environments: a bibliographical essay,' in *Computer Systems in Engineering*, **2** (4), 321–46, 1991, and as 'What are virtual environments?' in *IEEE Computer Graphics and Applications*, **14** (1), 17–22, 1994.
2. Higher-dimensional displays have also been described. See Inselberg (1985) or Feiner and Beshers (1990) for alternative approaches.
3. This 'knowledge' should not be thought of as the conscious, abstract knowledge that is acquired in school. It rather takes the form of tacit acceptance of specific constraints on the possibilities of change such as are reflected in Gestalt Laws, e.g. common fate or good continuation. Its origin may be sought in the phylogenetic history of a species, shaped by the process of natural selection and physical law, and documented by the history of the Earth's biosphere.
4. Ponography: the writing that causes or is associated with pain (from Greek *ponos*).

References

Adelstein, B. D. and Ellis, S. R., 1993, Effect of head-slaved visual image roll on spatial situation awareness, in *Proceedings of the 37th Meeting of the Human Factors and Ergonomics Society*, 11–15 October, 1993, pp. 1350–54, Santa Monica, CA: Human Factors and Ergonomics Society.
Adelstein, B. D. and Rosen, M. J., 1991, A high performance two degree of freedom kinesthetic interface, in *Human Machine Interfaces for Teleoperators and Virtual Environments*, pp. 108–13, Santa Barbara, CA: NASA CP 91035, NASA Ames Research Center, Moffett Field, CA.

44 *Simulated and virtual realities*

Adelstein, B. D. and Rosen, M. J., 1992, Design and Implementation of a Force Reflecting Manipulandum for Manual Control Research, in *Proceedings of the Annual Meeting of ASME*, Anaheim, CA, pp. 1–2, New York: American Society of Mechanical Engineers.
Adelstein, B. D., Johnston, E. R. and Ellis, S. R., 1992, A test-bed for characterizing the response of virtual environment spatial sensors, in *The 5th Annual ACM Symposium on User Interface Software and Technology*, pp. 15–20, Monterey, CA: ACM.
AGARD, 1988, Motion cues in flight simulation and simulator induced sickness, in *AGARD Conference Proceedings No. 433* (AGARD CP 433), Springfield, VA: NTIS.
Airey, J. M., Rohlf, J. H. and Brooks Jr, 1990, Towards image realism with interactive update rates in complex virtual building environments, *Computer Graphics*, **24** (2), 41–50.
Apple Computer Co., 1992, *Newton Technology: An Overview of a New Technology From Apple*, Apple Computer Co., 20525 Mariani Ave, Cupertino, CA 95014.
Ascension Technology Corp., 1990, Product description, Ascension Technology Corporation, Burlington, VT 05402.
Bahill, A. T., Adler, D. and Stark, L., 1975, Most naturally occurring human saccades have magnitudes of 15 degrees or less, *Investigative Ophthalmology*, **14**, 468–9.
Barfield, W. and Kim, Y., 1991, Effect of geometric parameters of perspective on judgments of spatial information, *Perceptual and Motor Skills*, **73** (2), 619–23.
Barnes, J., 1992, Acoustic 6 dof sensor, Internal Report, Logitech, 6505 Kaiser Dr., Fremont, CA 94555.
Barrette, R., Dunkley, R., Kruk, R., Kurtz, D., Marshall, S., Williams, T., Weissman, P. and Antos, S., 1990, Flight simulation advanced wide FOV helmet mounted infinity display, USAF ASD Report AFHRL-TR-89-36, Air Force Human Resources Laboratory.
Basalla, G., 1988, *The Evolution of Technology*, New York: Cambridge University Press.
Bassett, B., 1992, Virtual reality head-mounted displays, Internal Report, Virtual Research, 1313 Socorro Ave, Sunnyvale, CA, 94089.
BASYS, 1990, Product description, Basys Gesellschaft für Anwender und Systemsoftware mbH, Nürnberg, Germany.
Bejczy, A. K., 1980, Sensor controls and man–machine interface for teleoperation, *Science*, **208**, 1327–35.
Bejczy, A. K. and Salisbury Jr, K. S., 1980, Kinesthetic coupling between operator and remote manipulator, in *Advances in Computer Technology, Proceedings ASME International Computer Technology Conference*, San Francisco, CA, pp. 197–211, New York: American Society of Mechanical Engineers.
Bejczy, A. K., Kim, W. S. and Venema, S. C., 1990, The phantom robot: predictive displays for teleoperation with time delay, in *Proceedings of the IEEE International Conference on Robotics and Automation*, 13–18 May 1990, San Francisco, CA, pp. 546–51, New York: IEEE.
Bertin, J., 1967 / 1983, *Semiology of Graphics: Diagrams, Networks, Maps*, Madison, WI: University of Wisconsin Press.
Bishop, P. O., 1987, Binocular vision, in Moses, R. A. and Hart, W. M. Jr (Eds), *Adler's Physiology of the Eye*, pp. 619–89, Washington, DC: Mosby.
Blauert, J., 1983, *Spatial Hearing*, Cambridge, MA: MIT Press.
Blinn, J. F., 1987, The mechanical universe: an integrated view of a large animation project (Course Notes: Course #6), in *Proceedings of the 14th Annual Conference on Computer Graphics and Interactive Techniques*, Anaheim, CA, ACM SIGGRAPH and IEEE Technical Committee on Computer Graphics.
Blinn, J. F., 1991, The making of the mechanical universe, in Ellis, S. R., Kaiser, M. K. and Grunwald, A. J. (Eds), *Pictorial Communication in Virtual and Real Environments*, pp. 138–55, London: Taylor & Francis.
Boff, K. R., Kaufman, L. and Thomas, J. P., 1986, *Handbook of Perception and Human Performance*, New York: John Wiley.
Borah, J., Young, L. R. and Curry, R. E., 1978, Sensory mechanism modeling, USAF ASD Report AFHRL TR 78-83, Air Force Human Resources Laboratory.
Brand, S., 1987, *The Media Lab: Inventing the Future at MIT*, New York: Viking.
Brehde, D., 1991, CeBIT: Cyberspace-Vorstoss in eine andere Welt (Breakthrough into another world), *Stern*, **44** (12), 130–42.
Brooks, F. Jr, 1988, Grasping reality through illusion-interactive graphics serving science, in *Proceedings of CHI 1988*, 15–19 May 1988, Washington, DC, pp. 1–12, New York: ACM.

Brooks, T. L. and Bejczy, A. K., 1986, Hand controllers for teleoperation, NASA CR 175890, JPL Publication 85-11.

Bussolari, S. R., Young, L. R. and Lee, A. T., 1988, The use of vestibular models for design and evaluation of flight simulation motion, in *AGARD Conference Proceedings No. 433: Motion Cues in Flight Simulation and Simulator Induced Sickness (AGARD CP 433)*, Springfield, VA: NTIS.

CAE Electronics, 1991, Product literature, CAE Electronics, Montreal, Canada.

Cardullo, F., 1993, Flight Simulation Update 1993, Internal Report, Binghamton, New York: Watson School of Continuing Education, SUNY Binghamton.

Carroll, L., 1883, *Through the Looking-Glass and What Alice Found There*, London: Macmillan and Co.

Cellier, F., 1991, *Modeling Continuous Systems*, New York: Springer-Verlag.

Clark, J. H., 1980, A VLSI geometry processor for graphics, *IEEE Computer*, **12**, 7.

Clark, J. H., 1982, The geometry engine: a VLSI geometry system for graphics, *Computer Graphics*, **16** (3), 127-33.

Collewijn, H. and Erkelens, C. J., 1990, Binocular eye movements and the perception of depth, in Kowler, E. (Ed.), *Eye Movements and their Role in Visual and Cognitive Processes*, pp. 213-62, Amsterdam: Elsevier Science Publishers.

Comeau, C. P. and Bryan, J. S., 1961, Headsight television system provides remote surveillance, Electronics, 10 November, 86-90.

Cooper, G. E. and Harper, R. P. Jr, 1969, The use of pilot ratings in the evaluation of aircraft handling qualities, NASA TN D 5153, Moffett Field, CA: NASA Ames Research Center.

Cotter, C. H., 1966, *The Astronomical and Mathematical Foundations of Geography*, New York: Elsevier.

Cowdry, D. A., 1986, Advanced visuals in mission simulators, in *AGARD Flight Simulation*, pp. 3.1-10, Springfield, VA: NTIS.

Crampton, G. H., 1990, *Motion and Space Sickness*, Boca Raton, FL: CRC Press.

Cruz-Neira, C., Sandin, D. J., DeFanti, T. A., Kenyon, R. V. and Hart, J. C., 1992, The cave: audio visual experience automatic virtual environment, *Communications of the ACM*, **35** (6), 65-72.

Curry, R. E., Hoffman, W. C. and Young, L. R., 1976, Pilot modeling for manned simulation, AFFDL-TR-76-124, Air Force Flight Dynamics Laboratory Publication, December 1.

Daley, B., 1982, *.Tron / a novel by Brian Daley based on the screenplay by S. Listberger*, New York: Ballantine Books.

D'Arcy, J., 1990, Re-creating reality, *MacCleans*, **103** (6), 36-41.

Deering, M., 1992, High resolution virtual reality, *Computer Graphics*, **26** (2), 195-201.

Durlach, N. I., Sheridan, T. B. and Ellis, S. R., 1991, Human machine interfaces for teleoperators and virtual environments, NASA CP91035, NASA Ames Research Center.

Elkind, J. I., Card, S. K., Hochberg, J. and Huey, B. M., 1989, *Human Performance Models for Computer-Aided Engineering*, Washington, DC: National Academy Press.

Ellis, S. R., 1990, Pictorial communication, *Leonardo*, **23**, 81-6.

Ellis, S. R., 1991, Prologue, in Ellis, S. R., Kaiser, M. K. and Grunwald, A. J. (Eds), *Pictorial Communication in Virtual and Real Environments*, pp. 3-11, London: Taylor & Francis.

Ellis, S. R., 1994, What are virtual environments?, *IEEE Computer Graphics and Applications*, **14** (1), 17-22.

Ellis, S. R. and Bucher, U. J., 1994, Distance perception of stereoscopically presented virtual objects superimposed by a head mounted see through display, in *Proceedings of the 38th Annual Meeting of the Human Factors Society*, pp. 1300-5, Santa Monica, CA: Human Factors Society.

Ellis, S. R. and Grunwald, A. J., 1989a, Visions of visualization aids: design philosophy and observations, in *Proceedings of the SPIE OE/LASE '89, 1083, Symposium on Three-Dimensional Visualization of Scientific Data*, Los Angeles, CA, pp. 220-7, Bellingham, WA: SPIE.

Ellis, S. R. and Grunwald, A. J., 1989b, The dynamics of orbital maneuvering: design and evaluation of a visual display aid for human controllers, in *AGARD Space Vehicle and Flight Mechanics Symposium CP 489*, pp. 29.1-13, Springfield, VA: NTIS.

Ellis, S. R. and Hacisalihzade, S. S., 1990, Symbolic enhancement of perspective displays, in *Proceedings of the 34th Annual Meeting of the Human Factors Society*, pp. 1465-9, Santa Monica, CA: Human Factors Society.

Ellis, S. R. and Hitchcock, R. J., 1986, Emergence of Zipf's Law: spontaneous encoding optimization by users of a command language, *IEEE Transactions on Systems Man and Cybernetics*, **SMC-16**, 423–27.

Ellis, S. R., McGreevy, M. W. and Hitchcock, R., 1987, Perspective traffic display format and airline pilot traffic avoidance, *Human Factors*, **29**, 371–82.

Ellis, S. R., Kaiser, M. K. and Grunwald, A. J., 1993, *Pictorial Communication in Virtual and Real Environments*, 2nd Edn, London: Taylor & Francis.

Erkelens, C. J. and Collewijn, H., 1985a, Eye movements and stereopsis during dichoptic viewing of moving random dot stereograms, *Vision Research*, **25**, 1689–700.

Erkelens, C. J. and Collewijn, H., 1985b, Motion perception during dichoptic viewing of moving random dot stereograms, *Vision Research*, **25**, 583–8.

Exos, 1990, Product literature, Exos Inc., 8 Blanchard Rd, Burlington, MA.

Fagan, B. M., 1985, *The Adventures of Archaeology*, Washington, DC: National Geographic Society.

Feiner, S. and Beshers, C., 1990, Worlds within worlds: metaphors for exploring n-dimensional virtual worlds, in *Proceedings of 3rd Annual Symposium on User Interface Technology*, Snowbird, Utah, 3–5 October 1990, ACM 429902, New York: ACM.

Feldon, S. E. and Burda, R. A., 1987, The extraocular muscles: Section 2, The oculomotor system, in Moses, R. A and Hart, W. M. Jr. (Eds), *Adler's Physiology of the Eye*, pp. 122–68, Washington, DC: Mosby.

Fisher, P., Daniel, R. and Siva, K. V., 1990, Specification of input devices for teleoperation, in *IEEE International Conference on Robotics and Automation*, Cincinnati, Ohio, pp. 540–5, New York: IEEE.

Fisher, S. S., McGreevy, M., Humphries, J. and Robinett, W., 1986, Virtual Environment Display System, in *ACM 1986 Workshop on 3D Interactive Graphics*, Chapel Hill, North Carolina, 23–24 October 1986, ACM.

Foley, J. D., 1987, Interfaces for advanced computing, *Scientific American*, **257** (4), 126–35.

Foley, J. M., 1980, Binocular distance perception, *Psychological Review*, **87**, 411–34.

Foley, J. M., 1985, Binocular distance perception: Egocentric distance tasks, *Journal of Experimental Psychology: Human Perception and Performance*, **11**, 133–49.

Furness, T. A., 1986, The supercockpit and its human factors challenges, in *Proceedings of the 30th Annual Meeting of the Human Factors Society*, Dayton, OH, pp. 48–52, Santa Monica, CA: Human Factors Society.

Furness, T. A., 1987, Designing in virtual space, in Rouse, W. B. and Boff, K. R. (Eds), *System Design*, Amsterdam: North-Holland.

Gibson, J. J., 1950, *The Perception of the Visual World*, Boston: Houghton Mifflin.

Gibson, W., 1984, *Neuromancer*, New York: Ace Books.

Goertz, R. C., 1964, Manipulator system development at ANL, in *Proceedings of the 12th Remote Systems Technology Conference, Argonne National Laboratory*, pp. 117–36, ANS.

Goertz, R. C., Mingesz, S., Potts, C. and Lindberg, J., 1965, An experimental head-controlled television to provide viewing for a manipulator operator, in *Proceedings of the 13th Remote Systems Technology Conference*, pp. 57–60, Washington, DC: American Nuclear Society.

Goodale, M. A., 1990, *Vision and Action: The Control of Grasping*, Norwood, NJ: Ablex.

Gore, A., 1990, Networking the future, *Washington Post*, 15 July, B3.

Green, P., Satava, R., Hill, J. and Simon, I., 1992, Telepresence: advanced teleoperator technology for minimally invasive surgery, *Surgical Endoscopy*, **6**, 62–7.

Greenberg, A. and Woodhead, R., 1980, Wizardry: proving ground of the mad overload, Sir-Tech Software Inc., 6 Main Street, Ogdenburg, NY 13669.

Greenberg, D. P., 1991, Computers and architecture, *Scientific American*, **264** (2), 104–9.

Gregory, R. L., 1968, Perceptual illusions and brain models, *Proceedings of the Royal Society*, *B*, **171**, 278–96.

Gregory, R. L., 1980, Perceptions as hypotheses, *Philosophical Transactions of the Royal Society*, *B*, **290**, 181–97.

Gregory, R. L., 1981, *Mind in Science*, London: Weidenfeld & Nicolson.

Gregory, R. L. and Wallace, J. G., 1974, Recovery from early blindness: a case study, in R. L. Gregory (Ed.), *Concepts and Mechanisms of Perception*, pp. 65–129, London: Methuen.

Grudin, J. and Norman, D., 1993, Language evolution and human–computer interaction, (manuscript submitted for publication).

Grunwald, A. J. and Ellis, S. R., 1988, Interactive Orbital Proximity Operations planning system, NASA TP 2839, NASA Ames Research Center.

Grunwald, A. J. and Ellis, S. R., 1991, Design and evaluation of a visual display aid for orbital maneuvering, in Ellis, S. R., Kaiser, M. K. and Grunwald, A. J. (Eds), *Pictorial Communication in Virtual and Real Environments*, pp. 207–31, London: Taylor & Francis.

Grunwald, A. J. and Ellis, S. R., 1993, A visual display aid for orbital maneuvering: experimental evaluation, *AIAA Journal of Guidance and Control*, **16** (1), 145–50.

Hannaford, B., 1989, A design framework for teleoperators with kinesthetic feedback, *IEEE Transactions on Robotics and Automation*, **5** (4), 426–34.

Hart, S. G. and Staveland, L. E., 1988, Development of the NASA TLX, Task Load Index: results of empirical and theoretical research, in Hancock, P. A. and Meshkati, N. (Eds), *Human Mental Workolad*, New York: North-Holland.

Hashimoto, T., Sheridan, T. B. and Noyes, M. V., 1986, Effects of predictive information in teleoperation with time delay, *Japanese Journal of Ergonomics*, **22**, 2.

Hatada, T., Sakata, H. and Kusaka, H., 1980, Psychophysical analysis of the sensation or reality induced by a visual wide-field display, *Society of Motion Pictures Technology and Engineering Journal*, **89**, 560–9.

Heeger, D. J., 1989, Visual perception of three-dimensional motion, *Neural Computation*, **2**, 127–35.

Heilig, M. L., 1955, El cine del futuro (The cinema of the future), *Espacios*, No. 23–24, January–June, Apartado Postal Num. 20449, Mexico: Espacios SA.

Held, R. and Durlach, N., 1991, Telepresence, time delay and adaptation, in Ellis, S. R., Kaiser, M. K. and Grunwald, A. J. (Eds), *Pictorial Communication in Virtual and Real Environments*, pp. 232–46, London: Taylor & Francis.

Held, R., Efstathiou, A. and Greene, M., 1966, Adaptation to displaced and delayed visual feedback from the hand, *Journal of Experimental Psychology*, **72**, 887–91.

Hess, R. A., 1987, Feedback control models, in Salvendy, G. (Ed.), *Handbook of Human Factors*, New York: John Wiley.

Hirose, M., Kijima, R., Sato, Y. and Ishii, T., 1990, A study for modification of actual environment by see-through HMD, in *Proceedings of the Human Interface Symposium*, Tokyo, October.

Hirose, M., Hirota, K. and Kijma, R., 1992, Human behavior in virtual environments, in *Symposium on Electronic Imaging Science and Technology*, pp. 548–59, San José, CA: SPIE.

Hitchner, L. E., 1992, Virtual planetary exploration: a very large virtual environment, course notes, in *SIGGRAPH '92*, Chicago, IL, pp. 6.1–16., New York: ACM.

Hitchner, Lewis E. and McGreevy, Michael W., 1993, Methods for user-based reduction of model complexity for virtual planetary exploration, in *Human Vision, Visual Processing and Digital Display IV, Proceedings SPIE 913*, San José, CA, pp. 622–36, Bellingham, WA: SPIE.

Hochberg, J., 1986, Representation of motion and space in video and cinematic displays, in Boff, K. R., Kaufman, L. and Thomas, J. P. (Eds), *Handbook of Perception and Human Performance*, *1*, pp. 22:1–63, New York: John Wiley.

Howard, I., 1982, *Human Visual Orientation*, New York: John Wiley.

Howlett, E. M., 1991, Product literature, Leep Systems, 241 Crescent Street, Waltham, MA.

Hung, G., Semlow, J. L. and Cuiffreda, K. J., 1984, The near response: modeling, instrumentation and clinical applications, *IEEE Transactions in Biomedical Engineering*, **31**, 910–19.

Hussey, K. J., 1990, *Mars the Movie* (video), Pasadena, CA: JPL Audiovisual Services.

Huxley, A., 1932, *Brave New World: A Novel*, London: Chatto & Windus.

Inselberg, A., 1985, The plane with parallel coordinates, *The Visual Computer*, **1**, 69–91.

Jacobson, S.C., Knutti, D.F., Biggers, K.B., Iversen, E.K. and Woods, J.E., 1984, The Utah/MIT dexterous hand: Work in progress, International Journal of Robotics Research, 3 (4), 21–50.

Jacobson, S. C., Iversen, E. K., Knutti, D. F., Johnson, R. T. and Biggers, K. B., 1986, Design of the Utah/MIT Dexterous hand, in *IEEE International Conference on Robotics and Automation*, pp. 1520–32, San Francisco, CA: IEEE.

Jacobus, H. N., 1992, Force reflecting joysticks, CYBERNET Systems Corporation Imaging and Robotics, 1919 Green Road, Suite B-101, Ann Arbor, MI 48105.

Jacobus, H. N., Riggs, A. J., Jacobus, C. J. and Weinstein, Y., 1992, Implementation issues for telerobotic hand controllers: human–robot ergonomics, in Rahmini, M. and Karwowski, W. (Eds), *Human–Robot Interaction*, pp. 284–314, London: Taylor & Francis.

Janin, A. L., Mizell, D. W. and Caudell, T. P., 1993, Calibration of head-mounted displays for augmented reality applications, in *Proceedings of the IEEE VRAIS '93*, Seattle, WA, pp. 246–55, New York: IEEE.

Jenkins, C. L. and Tanimoto, S. I., 1980, Oct-trees and their use in representing three-dimensional objects, *Computer Graphics and Image Processing*, **14** (3), 249–70.

Jex, H. R., McDonnell, J. D. and Phatak, A. V., 1966, A critical tracking task for man–machine research related to the operators effective delay time, NASA CR 616, NASA.

Jones, G. M., Berthoz, A. and Segal, B., 1984, Adaptive modification of the vestibulo-ocular reflex by mental effort in darkness, *Brain Research*, **56**, 149–53.

Jones, R. K. and Hagen, M. A., 1980, A perspective on cross cultural picture perception, in Hagen, M. A. (Ed.), *The Perception of Pictures*, pp. 193–226, New York: Academic Press.

Kaiser Electronics, 1990, Product literature, Kaiser Electronics, San José, CA 95134

Kaiser, M. K., MacFee, E. and Proffitt, D. R., 1990, Seeing beyond the obvious: Understanding perception in everyday and novel environments, NASA Ames Research Center, Moffett Field, CA.

Kalawksy, R., 1993, *The Science of Virtual Reality and Virtual Environments*, Reading, MA: Addison-Wesley.

Kalman, R. E., 1960, Contributions to the theory of optimal control, *Boletin de la Sociedad Matematico Mexicana*, **5**, 102–19.

Kim, W. S., Takeda, M. and Stark, L., 1988, On-the-screen visual enhancements for a telerobotic vision system, *Proceedings of the 1988 International Conference on Systems, Man and Cybernetics*, Beijing, 8–12 August 1988, pp. 126–30, New York: IEEE.

Kleinman, D. L., Baron, S. and Levison, W. H., 1970, An optimal control model of human response, Part I: Theory and validation; Part II: Prediction of human performance in a complex task, *Automatica*, **6**, 357–69.

Koenderink, J. J. and van Doorn, A. J., 1977, How an ambulant observer can construct a model of the environment from the geometrical structure of the visual inflow, in Hauske, G. and Butenandt, E. (Eds), *Kybernetik*, Munich: Oldenbourg.

Kramer, J., 1992, Company literature on head-mounted displays, Virtex/Virtual Technologies, PO Box 5984, Stanford, CA 94309.

Krueger, M. W., 1977, Responsive environments, in *NCC Proceedings*, pp. 375–85.

Krueger, M. W., 1983, *Artificial Reality*, Reading, MA: Addison-Wesley.

Krueger, M. W., 1985, VIDEOPLACE – An artificial reality, in *SIGCHI 85 Proceedings*, April, pp. 35–40, New York: ACM.

Lackner, J. R., 1988, Some proprioceptive influences on the perceptual representations of body shape and orientation, *Brain*, **111**, 281–97.

Larimer, J., Prevost, M., Arditi, A., Bergen, J., Azueta, S. and Lubin, J., 1991, Human visual performance model for crew-station design, in *Proceedings of the 1991 SPIE*, February, pp. 196–210, San José, CA: SPIE.

Laural, B., 1991, *Computers as Theatre*, Reading, MA: Addison-Wesley.

Levine, M., 1984, The placement and misplacement of you-are-here maps, *Environment and Behavior*, **16**, 139–57.

Levit, C. and Bryson, S., 1991, A virtual environment for the exploration of three dimensional steady flows, *SPIE*, 1457, pp. 161–8, Bellingham, WA: SPIE.

Licklider, J. C. R., Taylor, R. and Herbert, E., 1978, The computer as a communication device, *International Science and Technology*, April, 21–31.

Lippman, A., 1980, Movie maps: an application of optical video disks to computer graphics, *Computer Graphics*, **14** (3), 32–42.

Lipton, L., 1982, *Foundations of Stereoscopic Cinema*, New York: Van Nostrand.

Lypaczewski, P. A., Jones, A. D. and Vorhees, M. J. W., 1986, Simulation of an advanced scout attack helicopter for crew station studies, in *Proceedings of the 8th Interservice/Industry Training Systems Conference*, Salt Lake City, Utah., pp. 18–23.

Mandelbrot, B., 1982, *The Fractal Geometry of Nature*, San Francisco: Freeman.

Marcus, O. B., 1991, Personal communication, Exos, 8 Blanchard Rd, Burlington, MA.

Martin, L. and Fox, C. R., 1989, Visually perceived eye-level and perceived elevation of objects: linearly additive influences from visual field pitch and from gravity, *Vision Research*, **29**, 315–24.

McDowall, I. E., Bolas, M., Pieper, S., Fisher, S. S. and Humphries, J., 1990, Implementation and integration of a counterbalanced CRT-base stereoscopic display for interactive viewpoint control in virtual environment applications, in *Stereoscopic Displays and Applications II*, San José, CA, Bellingham, WA: SPIE.

McGreevy, M. W., 1993, Virtual reality and planetary exploration, in Wexelblat, A. (Ed.), *Virtual Reality Applications: Software*, pp. 163–97, New York: Academic Press.

McGreevy, M. W. and Ellis, S. R., 1986, The effect of perspective geometry on judged direction in spatial information instruments, *Human Factors*, **28**, 439–56.

McKinnon, G. M. and Kruk, R., 1991, Multiaxis control of telemanipulators, in Ellis, S. R., Kaiser, M. K. and Grunwald, A. J. (Eds), *Pictorial Communication in Virtual and Real Environments*, pp. 247–64, London: Taylor & Francis.

McRuer, D. T. and Weir, D. H., 1969, Theory of manual vehicular control, *Ergonomics*, **12** (4), 599–633.

Meagher, D., 1984, A new mathematics for solids processing, *Computer Graphics World*, November, 75–88.

Monheit, G. and Badler, N.I., 1990, A Kinematic Model of the Human Spine and Torso, Technical Report MS-CIS-90-77, August 29, University of Pennsylvania, Philadelphia.

Monmonier, M., 1991, *How to Lie with Maps*, Chicago: University of Chicago Press.

Myers, T. H. and Sutherland, I. E., 1968, On the design of display processors, *Communications of the ACM*, **11** (6), 410–14.

NASA, 1985, *Rendezvous/Proximity Operations Workbook, RNDZ 2102*, Lyndon B. Johnson Space Center, NASA Mission Operations Directorate Training Division.

NASA, 1990, Computerized reality comes of age, *NASA Tech Briefs*, **14** (8), 10–12.

Nemire, K. and Ellis, S. R., 1991, Optic bias of perceived eye level depends on structure of the pitched optic array, paper presented at the 32nd Annual Meeting of the Psychonomic Society, San Francisco, CA, November.

Netrovali, A. N. and Haskell, Barry G., 1988, *Digital Pictures: Representation and Compression*, New York: Plenum Press.

Nomura, J., Ohata, H., Imamura, K. and Schultz, R. J., 1992, Virtual space decision support system and its application to consumer showrooms, in T. L. Kunii (Ed.), *Visual Computing*, pp. 183–96, Tokyo: Springer Verlag.

Octree Corporation, 1991, Product literature, Octree Corporation, Cupertino, CA 95014.

Oman, C. M., 1991, Sensory conflict in motion sickness: an observer theory approach, in Ellis, S. R., Kaiser, M. K. and Grunwald, A. J. (Eds), *Pictorial Communication in Virtual and Real Environments*, pp. 362–76, London: Taylor & Francis.

Oman, C. M., Lichtenberg, B. K., Money, K. E. and McCoy, R. K., 1986, MIT/Canada Vestibular Experiment on the SpaceLab 1- Mission: 4 Space motion sickness: systems, stimuli and predictability, *Experimental Brain Research*, **64**, 316–34.

Open University and BBC, 1991, *Components of Reality* (Video #5. 2 for Course T363: Computer Aided Design), Walton Hall, Milton Keynes, England MK7 6AA.

Ouh-young, M., Beard, D. and Brooks Jr, F., 1989, Force display performs better than visual display in a simple 6D docking task, in *Proceedings of the IEEE Robotics and Automation Conference*, May, pp. 1462–6, New York: IEEE.

Pedotti, A., Krishnan, V. V. and Stark, L., 1978, Optimization of muscle force sequencing in human locomotion, *Mathematical Bioscience*, **38**, 57–76.

Phillips, C., Zhao, J. and Badler, N. I., 1990, Interactive real-time articulated figure manipulation using multiple kinematic constraints, *Computer Graphics*, **24** (2), 245–50.

Polhemus Navigation Systems, 1990, Product Description, Polhemus Navigation Systems, Colchester, VT 05446.

Pollack, A., 1989, What is artificial reality? Wear a computer and see, *New York Times*, 10 April, A1L.

Poulton, E. C., 1974, *Tracking Skill and Manual Control*, New York: Academic Press.

Pritsker, A. A. B., 1986, *Introduction to Simulation and SLAM II*, 3rd Edn, New York: John Wiley.

Raab, F. H., Blood, E. B., Steiner, T. O. and Jones, H. R., 1979, Magnetic position and orientation tracking system, *IEEE Transactions on Aerospace and Electronic Systems*, **AES-15** (5), 709–18.

Regan, D. and Beverley, K. I., 1979, Visually guided locomotion: psychophysical evidence for a neural mechanism sensitive to flow patterns, *Science*, **205**, 311–13.

Robinett, W., 1982, *Rocky's Boots*, Fremont, CA: The Learning Company.

Robinson, A. H., Sale, R. D., Morrison, J. L. and Muehrcke, P. C., 1984, *Elements of Cartography*, 5th Edn, New York: John Wiley.

Rolfe, J. M. and Staples, K. J., 1986, *Flight Simulation*, London: Cambridge University Press.

Rolland, J., 1995, Towards quantifying depth and size perception in virtual environments, *Presence*, (in press).

SAGES, 1991, Panel on future trends in clinical surgery, *American Surgeon*, March.

Satava, R. M. and Ellis, S. R., 1994, Human interface technology: an essential tool for the modern surgeon, *Surgical Endoscopy*, **8**, 817–20.

Schuffel, H., 1987, Simulation: an interface between theory and practice elucidated with a ship's controllability study, in Bernotat, R., Gärtner, K.-P. and Widdel, H. (Eds), *Spektrum der Anthropotechnik*, pp. 117–28, Wachtberg-Werthoven, Germany: Forschungsinstitut für Anthropotechnik.

Senden, M. V., 1932, *Raum und Gestaltauffassung bei operierten Blindgeborenen vor und nach Operation*, Leipzig: Barth.

Sheridan, T. B., 1992, *Telerobotics, Automation and Human Supervisory Control*, Cambridge, MA: MIT Press.

Silicon Graphics, 1993, Product literature, Silicon Graphics Inc., Mountain View, CA.

Smith, D. C., Irby, C., Kimball, R. and Harslem, E., 1982, The Star User Interface: an overview, in *Office Systems Technology*, pp. 1–14, El Segundo, CA: Xerox Corp.

Spatial Systems, 1990, Spaceball product description, Spatial Systems Inc., Concord, MA 01742.

Stewart, D., 1991, Through the looking glass into an artificial world – via computer, *Smithsonian Magazine*, January, 36–45.

Stone, R. J., 1991a, Advanced human–system interfaces for telerobotics using virtual reality and telepresence technologies, in *Fifth International Conference on Advanced Robotics*, Pisa, Italy, pp. 168–73, IEEE.

Stone, R. J., 1991b, personal communication, National Advanced Robotics Research Laboratory, Salford, UK.

Stritzke, J., 1991, 'Automobile Simulator', Daimler-Benz AG, Abt FGF/FS, Daimlerstr. 123, 1000, Berlin 48, Germany.

Sutherland, I. E., 1965, The ultimate display, *International Federation of Information Processing*, **2**, 506.

Sutherland, I. E., 1970, Computer Displays, *Scientific American*, **222** (6), 56–81.

Tachi, S., Tanie, K., Komoriya, K. and Kaneko, M.,1984, Tele-existence (I): design and evaluation of a visual display with sensation of presence, in *Proceedings of the 5th International Symposium on Theory and Practice of Robots and Manipulators*, Udine, Italy, 26–29 June 1984, [CISM-IFToMM-Ro Man Sy '84], pp. 245–53.

Tachi, S., Hirohiko, A. and Maeda, T., 1989, Development of anthropomorphic tele-existence slave robot, in *Proceedings of the International Conference on Advanced Mechatronics*, 21–24 May 1989, pp. 385–90, Tokyo.

Takeda, T., Fukui, Y. and Lida, T., 1986, Three dimensional optometer, *Applied Optics*, **27** (12), 2595–602.

Taylor, R. H., Paul, H. A., Mittelstadt, B. D., Hanson, W., Kazanzides, P., Zuhars, J., Glassman, E., Musits, B. L., Williamson, B. and Bargar, W. L., 1990, An image-directed robotic system for hip replacement surgery, *Japanese Remote Systems Journal*, **8** (5), 111–16.

Tobler, W. R., 1963, Geographic area and map projections, *The Geographical Review*, **53**, 59–78.

Tobler, W. R., 1976, The geometry of mental maps, in Golledge, R. G. and Rushton, G. (Eds), *Spatial Choice and Spatial Behavior*, Columbus, OH: The Ohio State University Press.

Tomovic, R., and Boni, G., 1962, An adaptive artificial hand, in *IRE Transactions on Automatic Control*, **AC-7**, 3–10 April.

Tufte, E. R., 1983, *The Visual Display of Quantitative Information*, Cheshire, CO: Graphics Press.

Tufte, E. R., 1990, *Envisioning Information*, Cheshire, CO: Graphics Press.

UCSD Medical School, 1994, *Abstracts of the Interactive Technologies in Medicine II: Medicine Meets Virtual Reality*, 27–30 January 1994, San Diego, CA: UCSD Medical School.

Veldhuyzen, W. and Stassen, H. G., 1977, The internal model concept: an application to modeling human control of large ships, *Human Factors*, **19**, 367–80.

Vertut, J. and Coiffet, P., 1986, *Robot Technology: Teleoperations and Robotics: Evolution and Development*, Vol. 3A, and *Applications and Technology*, Vol. 3B (English Translation), Englewood Cliffs, NJ: Prentice Hall.

Wang, J.-F., Chi, V. and Fuchs, H., 1990, A real-time optical 3D tracker for head-mounted display systems, *Computer Graphics*, **24**, 205–15.

Weintraub, D. J. and Ensing, M., 1992, *Human Factors Issues in Head-up Display Design: The Book of HUDs*, Wright-Patterson AFB, OH: CSERIAC.

Welch, R. B., 1978, *Perceptual Modification: Adapting to Altered Sensory Environments*, New York: Academic Press.

Wells, M. J. and Venturino, M., 1990, Performance and head movements using a helmet-mounted display with different sized fields-of-view, *Optical Engineering*, **29**, 810–77.

Wenzel, E. M., 1991, Localization in virtual acoustic displays, *Presence*, **1** (1), 80–107.

Wenzel, E. M., Wightman, F. L. and Foster, S. H., 1988, A virtual display system for conveying three-dimensional acoustic information, in *Proceedings of the 32nd Meeting of the Human Factors Society*, Anaheim, 22–24 October 1988, pp. 86–90, Santa Monica, CA: Human Factors Society.

Wharton, John, 1992, personal communication, Microprocessor Report, Palo Alto, CA.

White, K. D., Post, R. B. and Leibowitz, H. W, 1980, Saccadic eye movements and body sway, *Science*, **208**, 621–23.

Wickens, C. D., 1986, The effects of control dynamics on performance, in Boff, K. R., Kaufman, L. and Thomas, J. P. (Eds), *Handbook of Perception and Human Performance*, pp. 39–60, New York: John Wiley.

Wightman, F. L. and Kistler, D. J., 1989a, Headphone simulation of free-field listening I: stimulus synthesis, *Journal of the Acoustical Society of America*, **85**, 858–67.

Wightman, F. L. and Kistler, D. J., 1989b, Headphone simulation of free-field listening II: psychophysical validation, *Journal of the Acoustical Society of America*, **85**, 868–78.

W Industries, 1991, Product literature, ITEC House, 26–28 Chancery St, Leicester LE1 5WD, UK.

Witkin, A., Fleisher, K. and Barr, A., 1987, Energy constraints on parameterized models, *Computer Graphics*, **21** (4), 225–32.

Witkin, A., Gleicher, M. and Welch, W., 1990, Interactive dynamics, *Computer Graphics*, **24** (2), 11–22.

Zangemeister, W. H., 1991, Voluntary presetting of the vestibular ocular reflex permits gaze stabilization despite perturbation of fast head movement, in Ellis, S. R., Kaiser, M. K. and Grunwald, A. J. (Eds), *Pictorial Communication in Virtual and Real Environments*, pp. 404–16, London: Taylor & Francis.

Zangemeister, W. H. and Hansen, H. C., 1985, Fixation suppression of the vestibular ocular reflex and head movement correlated EEG potentials, in O'Reagan, J. K. and Levy-Schoen, A. (Eds), *Eye Movements: From Physiology to Cognition*, pp. 247–56, Amsterdam: Elsevier.

Zeltzer, D. and Johnson, 1991, Motor planning: specifying the behavior and control of autonomous animated agents, *Journal of Visualization and Computer Animation*, **2**, 2.

Zimmerman, T., Lanier, J., Blanchard, C., Bryson, S. and Harvil, Y., 1987, A hand gesture interface device, in *Proceedings of the CHI and GI*, 5–7 APril 1987, Toronto, Canada, pp. 189–92, New York: ACM.

Zipf, G. K., 1949, *Human Behavior and the Principle of Least Effort*, Cambridge, MA: Addison-Wesley.

3

Visual realism and virtual reality: a psychological perspective

Chris Christou and Andrew Parker

3.1 Introduction: what is visual realism?

We experience the real world through the operation of our senses of sight, touch, smell, audition and taste, all of which convey information to the brain. The senses are complex systems but their basic mode of operation involves being stimulated by external energies such as light or sound. This stimulation is transformed into a form usable by the brain and provides information regarding the form and content of the external world and of the perceiver's position within this world. In normal sensory perception the various senses work together in providing a stable, unified and consistent 'picture' of reality. This stability is essential if the subject is to be able to use sensory information to guide actions and make decisions (see sections 7.4 and 7.5).

A virtual reality, in general terms, exists in the sense that it is arranged to have a certain character and structure determined by the (artificial) stimulation of the observer's various senses. Any sense of reality is, of course, illusory because the 'objects' of the artificial world are non-existent. The artificial world is simulated or synthesized by the appropriate stimulation of the observer; since so much of our experience and knowledge is directly derived from the senses, it is possible to fool the perceiver by making it difficult for them to discern that the world they are experiencing is artificial.

Realism in this sense can be equated with how closely the artificial world resembles a corresponding possible real world: that is, how similar is the sensory stimulation originating from the artificial environment to that originating from a equivalent real environment. For many reasons the level of realism attainable through the use of artificial computer-generated realities is now, and may always be, incomparable to the intricate and complex structure of the real world. In view of this we must therefore employ a more operational criterion for determining the level of realism portrayed in a virtual scene. Such a criterion is derived by considering the function of sensory information in guiding actions to achieve a particular goal. In these terms realism means providing the kinds of sensory information necessary in the performance of a particular task or in the achievement of a particular goal within the real environment. Furthermore, the more information that is provided, the greater the experienced sense of reality (see also the discussion of redundancy in section 7.4.3). In this respect it is important that information from various sources is globally consistent. Only if these basic requirements are satisfied will the resulting sense of realism allow the observer to forget that what they are experiencing is actually an imaginary world.

Visual realism, in particular, requires the stimulation of the visual system in a manner consistent with natural image formation, so that the visual system can extract the same kinds of information from the artificial stimulus that are available in the real world. Vision provides information regarding relative bodily posture, self-motion, the location and character of external objects and the overall structure of the three-dimensional environment that surrounds the observer. Vision researchers now have a good idea of what kinds of stimulation the visual system finds informative in particular tasks. For instance, the stereo images of the two eyes and motion parallax are known to provide extremely useful information regarding the relative distances of various objects from the observer. The detection of relative motion is also useful for this purpose. Other forms of visual stimulation provide information regarding solidity and structure and allow for identification of objects under varying environmental conditions.

Given that we have some idea of (i) the information to which the visual system is sensitive, (ii) the ecological significance of this information and (iii) limitations on the use of information that are peculiar to our own visual systems, it is possible to attempt the design of artificial environments by stimulating the visual system in the way to which it is accustomed in the real world. There are two factors whose careful consideration can simplify this process. Firstly, the visual system has limitations in the kinds and magnitudes of visual stimulation that it is sensitive to. It would be pointless to expend considerable time on a particular attribute of virtual imagery if the visual system is incapable of utilizing such attributes effectively. Secondly, the nature of the visual stimulus is determined in most cases by the form of the external world. The visual stimulus, light, is structured by its interaction with a solid and well-structured environment. The character of light reflects this structure and indeed in many cases humans use the structure in light as information. Therefore, by analysing how light behaves in the real world we can learn how to create more convincing imagery.

In this chapter we attempt to show that visual realism in the portrayal of three-dimensional virtual scenes can be improved by a better understanding both of the aspects of the real world that are visually informative and also of the sensitivities and limitations of the visual system itself. Some reference is made to the efforts of artists and painters, who have attempted to capture the essential character of light that makes it capable of depicting three-dimensional structure. The chapter begins with a brief introduction to the history and philosophy of visual psychology. We see from this history that the debate between the philosophical positions of empiricism and nativism has shaped our modern-day conception of what visual perception involves. From this debate arose an awareness of the importance of phenomenology, the content of visual experience, and an awareness of the informative richness of light as a visual stimulus. The fact that the image formation process, the process by which a three-dimensional environment is projected onto a two-dimensional surface, obeys physical laws means that images can be simulated or synthesized. Some notable advances in representational art have taken place through an improved awareness of the rules involved in the image formation process and also of the particular sensitivities of vision. For instance, Leonardo da Vinci studied the physics of light and image formation and consequently devised many rules regarding how the painter should use light and colour to create particular impressions of depth and solidity.

The kind of image synthesis used in modern virtual reality technologies utilizes computer graphics. Computer graphics uses rules of image formation for capturing natural image features, such as diffuse light and shade, specularities or highlights and textures of surfaces.

The approach begins by looking at the physics of how light interacts with surfaces to generate an image of a scene, when the scene is viewed from a particular vantage point. Understanding image formation leads to the formulation of rules or models for these effects and subsequently the implementation of these models on a computer. Naturally, the closer a particular model approximates the real process of image formation, the more realistic the final effect should appear. This accuracy of modelling is therefore closely linked with visual perception. The main requirement in making effective virtual reality systems therefore involves not only providing adequate kinds of visual information but also making this information as naturalistic as possible.

The final part of this chapter offers a demonstration of how a better understanding and modelling of natural light–surface interaction can result in images which are perceptually more appealing, stable and informative. There is an introduction to the radiosity method, which has been developed over the last ten years to accurately simulate how light bounces between surfaces in natural scenes. To illustrate the benefits of more accurate models of image formation, such as the radiosity method, we provide three examples where using radiosity improves visual performance.

3.2   Historical overview of visual psychology

3.2.1   Empiricism

A central question in visual psychology is how we are able to perceive three dimensions when all we have at our disposal are the two-dimensional images that are formed on the retina of each eye. Current conceptions of visual perception derive from philosophical arguments concerning the acquisition of knowledge. The empiricist tradition has been the predominant force behind the development of much of current psychology and is exemplified by John Locke's *Essay Concerning Human Understanding* (1690). Seeking to refute the nativist position that knowledge of the world is innate he wrote:

> Let us suppose the mind to be, as we say, white paper, void of all characters, without any ideas; how comes it to be furnished? Whence comes it by that vast store, which the busy and boundless fancy of man has painted on it with an almost endless variety? Whence has it all the materials of reason and knowledge? To this I answer in one word, from experience: in that all our knowledge is founded, and from that it ultimately derives itself. (Book II, chapter i, sec. 2).

Knowledge is derived from experience by way of the senses which produce in us simple ideas regarding the qualities of objects that we perceive. Simple ideas are those of colour, taste and smell for example. These simple ideas, according to Locke, are combined by the imagination to form complex ideas. Thus, the idea of a horse is a derivative of the various simple ideas that are associated with it through experience. George Berkeley (1709) extended Locke's theory of knowledge-by-association by insisting that no simple idea can be a derivative of more than one sense modality. Simple ideas of sight (colours, shades etc.) and touch (hardness, softness, roundedness, etc.) are distinct by their very nature, although they can be combined by experience.

With regard to vision, Locke and Berkeley and many empiricist psychologists who followed them contended that the ideas of, for example, solidity and depth are necessarily complex ideas derived from simple ideas of touch such as firmness, roundedness and extension. The retinal image is two-dimensional and consists only of distinct patches or points of colour. Thus, it cannot provide sufficient information on its own to induce in us ideas of three dimensions. Only through our interaction with the world (through touching objects, and manoeuvring) do we learn to see the conglomeration of points that make up the retinal image as objects with form and extension in three-dimensions. In this respect the retinal image serves only as a cue to the various properties of the world and we must reconstruct reality in our minds using inference and reasoning from past experience.

3.2.2 Nativism and the beginnings of phenomenology

There are several problems with the empiricist view of perception but two main criticisms are its conception of sensation as distinct from perception and the belief that the basic elements of perception are just meaningless points of colour transferred to the mind for interpretation. This opinion is thought to result in a philosophical scepticism; reality, if there is such a thing, is only perceived through a shroud or veil of 'sense data' (i.e. points of colour) which only provide us with cues as to its true nature. Perception is said to be indirect and mediated by sense data that serve as cues for inference. Nativism, another popular philosophical position of the 18th century, was opposed to this view and considered that the immediacy of perception was proof that the ideas of form and depth are inborn and also that the visual stimulus is sufficient to arouse these ideas in us. Nativist philosophers and psychologists included Immanuel Kant, Johannes Muller and Ewald Hering. For instance, Kant's view of *a priori* intuitions of space and time as prerequisites for perception opened the way for a nativist view of vision. Kant believed that we cannot perceive, say, solidity unless we have a conception of space and this concept cannot be an abstraction from experience (because we require the concept of space in order to perceive it as such), so it must be innate (Kant, 1891). Later, Johannes Muller (1826, pp. 71–90; see Boring, 1942, p. 29) suggested that the spatial relations between objects is given directly by the spatial organization of receptors in the eye. Elaborating on this idea of local signs, Hering (1861) argued that retinal points produce three components, two of which define the point's retinal location, and a third which gives a measure of relative depth of points in the world. Since the brain registers the output of nerves from the eyes and because these nerves are spatially organized across the retina, then the brain can sense spatial form and extension in depth directly from the visual stimulus without the need for learning and inference from past experience.

3.2.3 Elements of perception – Gestalt psychology

The main argument between empiricists and nativists amounted, partly, to a disagreement about what are the primary elements of perception. The former claimed that, in the extreme, only points of colour are given to the eyes. The latter believed that the perceptions of extension, form and even depth are evoked directly by visual sensations. Gestalt psychology, which superseded nativist psychology in 20th-century Germany, was greatly influenced by nativist doctrine. Gestalt theory stresses the importance of the spatial and temporal relationships between the various parts of the visual stimulus. The theory states that the brain uses organizational principles to interpret visual appearance. These principles act on the stimulus and determine both how sub-regions are grouped together in the perception of form and also how objects are segregated from each other and from their backgrounds. The perception of two-dimensional form and even three-dimensional space is derived from the effects of these internalized organizational forces. One aim of the Gestalt psychologists was to discover the nature of the organizational principles used by the visual system (see for instance, Koffka, 1935).

3.2.4 Direct perception: removing the veil of sense data

The Gestalt movement, with its emphasis on phenomenology and introspection, stimulated a lot of research into perceptual saliency of image features. Unfortunately, the movement could only provide a handful of 'rules' or principles, which could be used to indicate that one particular pattern is more likely to be perceived in a stimulus than another. These rules, however, did not tell us anything about how the brain uses information in the visual stimulus or the link between the form of the visual stimulus and the perception of three dimensions.

The American psychologist James Gibson also argued for an alternative to the analysis and perceptual inference of the constructivist empiricists (Gibson, 1950, 1966). Gibson, like the Gestaltists, believed that the distinction between sensation and perception was unwarranted. He praised the Gestalt movement for shifting the focus from the subjective to what is phenomenally available in the visual stimulus. Gibson's main achievement, however, was to attempt an explanation of what it is about the 'world-out-there' that can provide information concerning three-dimensional structure without the requirement of associating sight with other sense modalities such as touch. He directed the attention of the visual psychologist to the higher-order properties of the visual stimulus, such as gradients of size, shading and texture, and ratios of stimulus energies. These higher-order image variables, when perceived, could give direct visual information regarding the structure of the world because they are visual correlates of physical states of affairs (Gibson, 1950). Thus the smooth change or gradient in texture on a surface as we look at it is the direct result of the relative orientation of the surface with respect to our line of sight and therefore such gradients can provide direct information regarding spatial orientation (see section 9.3.2).

Gibson's views are held by many theorists in high esteem and have inspired research aimed at investigating the informative nature of spatial variations in light. Just as the empiricists over-emphasized the role of learning in perception, however, Gibson over-emphasizes the phenomena of perception. Modern physiology has revealed that the brain cells which are involved in the early stages of vision concentrate on processing information in small regions of the retinal image known as their receptive fields. These cells have particular 'preferred' characteristics of the image to which they respond maximally such as lines or edges in particular orientations and directions of motion. In this sense the early stages of vision are very much transformed versions of the retinal image. An adequate theory of how the global structure of objects can be extracted from these early representations must explain how representations are formed and how they interact. This concept of early vision as a transformation of representations is captured by computational models of visual processing (Marr, 1976, 1982).

3.2.5 Visual psychophysics

Human perception is difficult to study because it is controlled and determined by the brain whose functional operation is not open to inspection. We can, however, use the subjective content of experience to determine human sensitivities to visual stimuli. The desire to make sensory psychology more scientific led the German physicist and philosopher G. T. Fechner (1860) to establish the methods and measures of psychophysics: the physical measurement of sensations. Psychophysics is based on the concepts of absolute and differential thresholds. An absolute threshold is the limiting value on a perceptual scale; for instance, it could be the dimmest light that is detectable or, in auditory psychophysics, it could mark the lowest audible frequency. A differential threshold, on the other hand, is the smallest difference between two points on a scale that is perceptible. Thus, in visual terms we may talk about the just noticeable difference in intensity as the smallest difference in intensity between two target light sources that is discernible by a human subject.

Fechner established a number of psychophysical methods to measure these thresholds and these methods have subsequently been improved and developed by sensory psychophysicists (Newell Jones, 1974). Because some experimental methods are more prone to experimental and subjective error than others, part of the psychophysicist's job involves devising an adequate task for a particular sensitivity to be measured and deciding which particular psychophysical method is the most appropriate given the required level of accuracy. In general, measuring subjects' thresholds using psychophysical methods not only allows us to objectively determine the limits of human perceptual abilities but also allows us to decide between competing theories of visual function, both within psychology and visual physiology.

3.3   Perceiving three dimensions

The ecological, or functional, basis of vision is that it allows us to judge spatial relations and make decisions and plan actions in a three-dimensional world. What is it about the composition of the visual system that provides this functionality? Berkeley (1709) noted that depth cues could be provided by the convergence and accommodation of the eyes in fixating objects variously located in depth. Convergence occurs when the lines of sight[1] meet at a point in space. Accommodation results from the increase in curvature, and therefore refractive power, of the eye's lens in order that an image may be brought to focus on the retina. (The physiology of accommodation is described by Edgar and Bex in section 4.3.1).

Wheatstone added a third, so-called physiological, cue to this list by demonstrating that binocular disparity results in a very striking impression of depth (Wheatstone, 1838). When the eyes fixate a point in space all image features resulting from objects at different depths from the fixation point will be horizontally shifted on the retina. This shift is different for the two eyes. It can be shown that the difference in the magnitude of shift, the disparity, between the two eyes is proportional to the depth difference between the fixation point and other objects in space (Mayhew, 1982; Mayhew and Longuet-Higgins, 1982; Prazdny, 1983). Thus, the visual system appears to use this disparity to form the impression of depth.

These physiological 'cues' have been contrasted with so-called secondary, or pictorial, cues which arise from the image-formation process: the projection of three dimensions onto two. These are the image features used by artists to depict three-dimensionality. In line with Gibson we say that these features are the stimulus correlates of the physical state of affairs that brought them about. Pictorial cues include the following:

1. Linear perspective – compression and foreshortening.
2. Retinal size – changes of size with distance.
3. Interposition – occlusion of distant objects by near objects.
4. Aerial perspective – change of colour with distance.
5. Height in the field – increase in image height with distance.
6. Shading – gradients in image intensity.

The physical basis of image attributes has resulted in vision research geared towards determining how these attributes can be used to obtain an impression of the three-dimensional environment. Pictorial cues of occlusion or interposition have been studied by Ratoosh (1949), and Chapanis and McCleary (1953). Retinal size as a relative depth cue has been studied by Gogel (1964, 1965). Linear perspective has been studied with respect to changes in shape by Clark *et al.* (1955), and in terms of size gradients by Freeman (1966). The perceptual importance of texture gradients is reported in Gibson (1950). If we consider the visual stimulus as potential information, then one of the goals of vision can be considered to be the identification, extraction and transformation of this information into a usable form (Attneave, 1954). The remainder of this chapter will be concerned with how information is extracted from patterns of shading and with the use of realistic computer graphics to enhance and facilitate the pick-up of visual information from shading.

3.4 Natural images

3.4.1 Perspective projection

Real images (such as those produced in the eye or in a camera) are formed when light is projected onto a flat surface. The basic property of light that determines this projection is that, for present purposes, light travels in straight lines. Euclid's optics (*c*. 300 BC) set out the physical rules which govern this process and are used in its explanation. Rays of light from different spatial locations converge[2] at the optical centre, or nodal point, of the eye where they form the apex of a pyramid (see Figure 3.1). The location of the apex is called the 'centre of projection'. The size of the angle at the centre of projection, known as the 'visual angle', varies with viewing distance. As the distance between the observer and the object of regard is increased, the size of the visual angle is, approximately, linearly reduced. Objects laid out in front of an observer with increasing distance will therefore form increasingly smaller visual angles and will disappear at the 'vanishing point' (*p* in Figure 3.1). The relation between the visual angles formed by different objects at different distances from an observer at a particular vantage point is called natural perspective. Perspective projection, which occurs in the eye, is the result of the intersection between a two-dimensional surface, called the projection or picture plane, and the pyramid of light. There are three visual effects that result from the perspective projection of light:

1. Objects at increasingly greater distances from the observer make smaller visual angles at the eye.
2. Parallel lines which are not parallel to the line joining the two eyes appear to converge at the horizon or vanishing point.
3. The outlines of objects appear to diminish, or flatten, in the direction of the vanishing point, causing what is known as perspective foreshortening.

3.4.2 The 'discovery' of perspective

Because the physical world is three-dimensional there are an infinite number of vantage points from which it may be viewed and therefore an infinite number of possible perspective projections. Visually, the diminution of size and foreshortening of surfaces provide powerful impressions of three dimensions as demonstrated by Renaissance paintings of the 14th and 15th centuries. Although Roman and Greek art dating back to the 8th century BC utilized perspective (Richter, 1970), the rules of linear perspective, interestingly, do not appear to have been formulated in rigid terms until the time of Renaissance Florence, i.e. until the architects Filippo Brunelleschi (1377–1446) and Leon Battista Alberti (1404–1472) and Florentine painters such as Paolo Uccello (1397–1475), Masaccio (1401–1428), and Andrea Mantegna (1431–1506). These painters used the rules of linear or artificial perspective, based on optical principles, developed by Brunelleschi and Alberti.

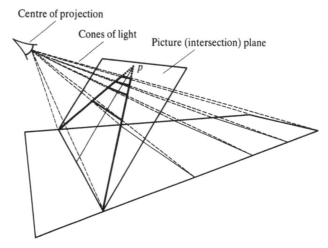

Figure 3.1 Perspective projection occurs when the pyramid of rays from a scene (in this case, equally spaced horizontal segments) are intersected by a two-dimensional surface. The projected size of each rectangular segment varies with the distance from the projection plane and its shape is altered according to convergence of parallel sides and foreshortening of its longitudinal dimension.

Because these rules were based on principles of physical optics that determine the projection of light in the eye, they resulted in imagery which was astoundingly realistic compared to contemporary art. This is because the projective process could be represented accurately for any spatial structure. By studying the optics of light and the phenomenal impression of projected light, these artists where therefore able to devise methods for producing a greater sense of visual depth and therefore for enhancing the visual realism of their pictures.

3.4.3 Physics of light–surface interaction

3.4.3.1 The transport of light

We see the objects around us because their surfaces emit, reflect or refract (transmit) light and this light then propagates to the eye. Light can travel through some media and not others and light is reflected and absorbed by some surfaces more than others. There are four types of light–surface interaction: diffuse reflection, diffuse transmission, specular reflection and specular transmission (see Figure 3.2). It is the accurate portrayal of such light–surface interactions both in representational art and in computer graphics that ultimately determines the level of realism attained.

3.4.3.2 Diffuse reflection

Diffusely reflected light is produced to varying degrees at dull or matt surfaces, including some forms of vegetation, soil, stone, fabrics and paper. Incident light is absorbed by the material and then re-emitted in any direction above the surface. The effect is that the amplitude of reflected light is constant and, to a certain extent, independent of the position from which an observer views the surface. An ideal diffuse, or Lambertian, surface is a special case where the reflected light is totally independent of viewing angle and depends only on the relative orientation of the surface with respect to the incident illumination. The relative orientation is defined in terms of the angle produced between the surface normal

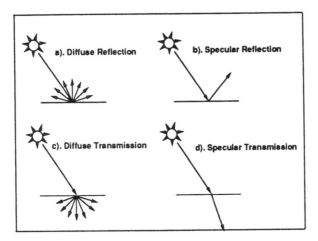

Figure 3.2   Four mechanisms of light transport.

(normal to the tangent plane at point of incidence) and the direction of incident light. One example of an approximately ideal diffuse material is chalk.

3.4.3.3 Diffuse transmission

Diffuse transmission is analogous to diffuse reflection in that light incident at a surface is dispersed in all directions but in this case it is dispersed within the body of the second material. The amplitude of the transmitted light is again only dependent on the angle the incident light makes with the surface normal. Diffusely transmitting materials include milky substances and some translucent plastics which contain suspended pigmentation.

3.4.3.4 Specular reflection

Specular reflection occurs at very shiny interfaces such as polished stone or metal. Ideally, the incident light is not absorbed at all by the illuminated material but bounces off the surface immediately. The surface microstructure on such materials is relatively smooth and ensures that most incident light is reflected with little absorption. Thus, the reflected light is not affected by the spectral properties of the surface and, in most circumstances, continues to have the colour of the light source. In ideal, or perfect, specular reflection light is reflected in a single direction determined by the angle of incidence of incoming light; thus, specular reflection is view-dependent. Mirrors or highly polished surfaces (such as polished metal) exhibit specular reflection of this form and hence reflections in mirrors move as the observer moves. Most real shiny surfaces, however, reflect incident light in a 'pear-drop' of directions, with a maximal amplitude in one particular direction and a gradual drop-off as the angle between the ideal specular (or mirror) direction and the line of sight increases. Visually, this produces fuzzy highlights on directly illuminated surfaces of objects.

3.4.3.5 Specular transmission

Materials such as transparent glass allow light energy to be specularly transmitted or refracted through them. Unlike diffuse transmission, however, light incident at a specularly transmitting surface is refracted or bent along one single direction (ideal specular transmission) or a pear-drop-shaped envelope of directions determined by the angle of incidence. This bending occurs as a result of the different velocity at which light travels in different media. The amount of refraction is also dependent on wavelength and this accounts for the dispersion of colours observed as white light is refracted by thick glass.

3.5 Artificial images and requirements for optimal presentation

3.5.1 Artificial images

This section is concerned with the links between requirements for the accurate depiction of scenes on various types of display system and the performance characteristics of human observers. Clearly, it is a waste of computational effort to spend resources on the calculation of visual features that are either invisible to or insignificant for the human observer. Generally, pictures generated by computers will be displayed on cathode-ray tube (CRT) monitors, although the use of liquid-crystal displays and other technologies is increasing and there is the further possibility of some form of printed output for static scenes.

An artificial image consists of a rectangular grid of discrete picture elements, or pixels. Each pixel is assigned an intensity value and a colour. Depending on the dimensions of the pixels, their combination forms an almost continuous image of the scene portrayed. A basic grasp of the link between the display technology and the computer software that generates the pictures can be developed by considering how well the display technology would reproduce video sequences of the natural world. Even ordinary broadcast television achieves a level of 'realism' that is typically way beyond the present-day limitations of computer-generated images. This highlights the point that much more effort is needed on the software side and its methodology before the true potential of developments in display hardware can be exploited.

At the lowest level of description, an image (or sequence of images) is simply a set of pixel intensity and colour values that are a function of 2-D spatial position and time. For considering the relationships between display technology and the images that it is desired to display, this level of description is generally the most appropriate. Even this low level, however, has interactions with human perceptual capabilities, particularly in the links between image intensities and perceived brightness and colour. Moreover there are circumstances where the combined effects of two or more of the basic variables need to be considered, such as the case of visual motion where space and time are combined. Finally, there are special extensions to the basic scheme that are linked with specific human capabilities. The clearest example of this is the use of binocular stereoscopic displays to generate a sense of depth.

3.5.2 Accurate representation of brightness values

It should be acknowledged immediately that almost all of the present display technologies have a very limited range of intensity values compared with those experienced by human observers as they move around in the natural world. On a typical CRT screen, the effective range of intensity values might be 100:1 (perhaps a little better on some of the brightest displays). By comparison, humans often experience simultaneously differences of 1000:1 or more (when bright sunshine enters a dim room), can maintain good discrimination of differences in relative intensity over a wide range of absolute intensities and can maintain some form of visual function over a total range of 10^9:1. The limitations of present-day

CRT screens in this respect are shared by photographic prints and by classic oil or water-colour painting. Yet, artists using these media have been able to convey a sense of the range of dark and light values. Helmholtz (1893) argued that this was because human observers interpret a scene in which the brightness values are all low and shifted to the blue end of the spectrum as being a night-time scene. Again, television and cinema can convey these differences without needing to recreate the fully-extended range of intensities that would be received from the natural world.

Clearly, then, some form of compression must be used. The ideal form for this compression is less obvious, since this could well depend upon the range of scenes that needs to be depicted. In practice, CRT screens tend to have a region of more or less linear response, bounded at the upper end by factors such as the saturation of the electron beam currents and at the lower end by the threshold for emission (or, more trivially, the ambient light in the room falling on the face of the CRT screen can hide very low light levels from the screen). The quasi-linear region can be adjusted to have a smoothly nonlinear response, not unlike that of some photographic materials. For some purposes, it is important to know exactly what type of characteristic is used by a particular display system. This would be essential for implementing some forms of anti-aliasing techniques (see below) and for matching the characteristics of the display system to human perceptual discriminability. This latter case is especially important in 'filmless' radiographic systems, where the medical diagnosis can often depend upon the appreciation of quite subtle differences in grey-level in the final image. The presence of noise in the display system in such cases would be highly deleterious.

The same points about the restricted range of output luminances apply equally strongly to the final representation of coloured regions. Typically, each separate colour channel has the restrictions described above and these will operate independently from one another. Since some coloured phosphors are capable of overall higher brightness output than others (blue phosphors are typically quite dim) and no phosphor is perfectly monochromatic, then the range of colours that can be displayed is smaller than the overall range that can be perceived. This is best illustrated by referring to the CIE diagram (Wysecki and Stiles, 1967, p. 276), which reveals that the range of visible colours available from a particular display screen is limited to selections from within the triangle of colours defined by the CIE coordinates of the three colour phosphors. (This limitation has nothing to do with the 'palette' of colours (4096 or 16 million are common values) as quoted for the video electronics of a graphics board. The palette of colours only limits the density of sampling that can be achieved within the bounding triangle on the CIE diagram.)

For the psychophysical work described later in this chapter, the main aim was to achieve a faithful rendition of the grey-levels specified by the radiosity-based computer graphics algorithm. In this case, the display screens were checked carefully for linearity of operation and any deviations from linearity were compensated for by use of the 'look-up' tables on the video output stage of the graphics display device (Watson *et al.*, 1986a). Again, to keep matters as simple as possible, only monochromatic (white light) display screens were used, although the radiosity approach to computer graphics can be applied to coloured scenes as well.

3.5.3 Smooth representations of surfaces

One of the features of human vision that regularly defeats simplistic approaches in computer graphics is the extraordinary sensitivity of humans to small differences in local brightness, contour alignment and other local surface properties. Accordingly, work in computer graphics needs to be planned with this in mind from the beginning. For example, many computer graphics systems approximate the surfaces of curved objects with a set of connected planar patches. When the final image is produced by shading this patch-wise representation, the boundaries between the patches often remain visible, owing to the high sensitivity of human observers to the local contrast across a step-edge of brightness. A further example arises in the use of binocular stereo to convey a sense of depth. Often quite significant changes of depth are geometrically signalled by very small angular differences between the right and left images. If a conventional CRT display system is used, the size of angular difference that really needs to be displayed is often much smaller than the size of the pixels on the display screen and the step-change in disparity given by shifting a feature through a whole pixel is generally well supra-threshold. Unless anti-aliasing techniques are used, a surface that should vary smoothly in depth (such as an Egyptian pyramid viewed from above the apex) often takes on a step-wise appearance (like a Babylonian or Assyrian ziggurat).

Most of the problems relating to the representation of smooth surfaces can be eliminated with some advance planning and, typically, with the application of additional computing power to implement anti-aliasing methods. This needs to be thought about well before the final stage of image rendering. A low-level, hardware-based, anti-aliasing system may be useful in that it allows lines to be placed at arbitrary positions and orientations in the final display without introducing unwanted variations in brightness when a line is parallel to the raster of the display as compared with an oblique line. Nonetheless, anti-aliasing at this level is no use for removing either of the features described above, since it serves in the case of the unwanted features (along the edges of surface patches or at a step-change in binocular disparity) only to render those unwanted features even more precisely!

3.5.4 Requirements determined by human factors

In most cases when a computer display device of sufficiently high luminance is inspected by a human observer, the main limit on the visual resolution of the human observer will be the quality of the optical image falling on the fovea of the retina. The work of Campbell and Green (1965) demonstrated that, with optimal pupil size, careful correction of refractive errors and paralysis of the lens accommodation system, the best performance from the human optical system is quite closely matched to the best performance from the neural apparatus that receives information from the optics. It will be immediately understood that these optimal circumstances are rarely achieved in practice and so the more usual situation will be one in which the optical performance falls short of that potentially available. Therefore, considerable attention should be given to the need for careful optical correction of observers, particularly with helmet-mounted displays where it is difficult for the observer to wear a standard pair of spectacles. Normally, accommodation will not be a problem, although prolonged periods of near-field viewing may be tiring, especially for presbyopic subjects (Edgar and Bex describe other possible accommodation problems; see section 4.3).

The combined performance of the optical and neural systems can provide spatial acuity good enough to resolve high contrast sinusoidal grating patterns whose spatial frequency is as much as 60 cycles/degree. This means that a single dark or light bar in such a grating is about the size of a single cone photoreceptor in the human fovea. But this value is only achieved with bright displays (> 100 cd m^{-2} average luminance). For dimmer displays, the achievable performance will be lower (van Nes *et al.*, 1967). Similarly, if the contrast of the display is lower, then the acuity limit can be predicted from the spatial contrast sensitivity function. These comments relate to the luminance contrast in the display. If the display contains chromatic information, then the effective luminance contrast needs to be considered because the resolution of spatial variations on the basis of colour alone (i.e. with isoluminant patterns) is typically four or five times poorer than for luminance.

A different kind of limit applies to the spatial precision with which a stimulus feature can be located relative to others. For example, in the classic Vernier acuity configuration, two lines of the same orientation are placed end-to-end, either with or without a small gap separating the end-points of the lines. When one line is shifted orthogonal to its long axis relative to the other line, it is found that human observers are exquisitely sensitive to these small shifts in position. Thresholds as low as 10 seconds of arc are found for this configuration. When this visual angle is projected onto the retina, it occupies a distance considerably smaller than the diameter of a single cone photoreceptor in the human fovea. Consequently, thresholds at this level have been termed 'hyperacuity'. (Even lower thresholds have been reported but these are for very high contrast thin lines on otherwise empty backgrounds, a situation unlike most of those found in computer graphics displays of naturalistic scenes.) Performance at the hyperacuity level is achieved for a wide variety of spatial configurations of the target (Westheimer, 1981).

It should be understood that the achievement of thresholds in the hyperacuity range does not mean that the fundamental spatial resolution limit of 60 cycles/degree has somehow been mysteriously evaded. Indeed, there is evidence that hyperacuity performance can be sustained by spatial frequencies considerably lower than the grating resolution limit. Figure 3.3 shows two sampled one-dimensional distributions: each sample bin is equivalent to a single photoreceptor and the sample size in each bin is equivalent to the light received by the photoreceptor. It can be appreciated that statistical mean of this 1-D distribution can be recovered with greater precision than the width of any one sample bin, provided that it can be assumed that the samples represent a single continuous distribution. If this principle is extended to the 2-D case as a better model of the photoreceptor mosaic on the retina, it can be seen that the spatial shape of the distribution, such as its elongation or orientation, could be exploited in addition to the positional information that is available from the 1-D case. There are, of course, close similarities between this principle and the techniques for anti-aliasing used in computer graphics systems.

Rather similar points apply to displays that are temporally varying. The flicker sensitivity of human observers varies with the stimulus size, illumination level and position within the visual field. Again, the predominant variable is the luminance variation since the flicker sensitivity to pure chromatic variation is weak (de Lange, 1958; Kelly and van Norren, 1977). Studies that combine temporal and spatial variation of luminance contrast demonstrate that the sensitivity to high spatial frequencies is considerably diminished by temporal variation, either by flickering or by means of directed motion. At the earliest stages of processing, stimuli with a specific direction of motion are limited in much the same way as stationary, flickering stimuli. For a photoreceptor, which has no direction selectivity, it does not matter whether the temporal variation in its signal is created by flicker or by a moving pattern that passes across it. At later stages, neurons in the visual system are specifically sensitive to the direction of motion and their sensitivity to spatial patterns and their temporal variations are specifically coupled together. This has some important consequences for CRT displays, which are periodically refreshed, since in these displays a moving target is displayed as a set of spatial discrete samples at specific temporal intervals. This is much like the classical apparent motion display. Whether or not apparent motion will appear smooth depends on the spatial and temporal characteristics of the motion (Morgan, 1979; Watson *et al.*, 1986b, Edgar and Bex, section 4.2). It is also possible to obtain a form of motion hyperacuity in which a moving target can be assigned to a

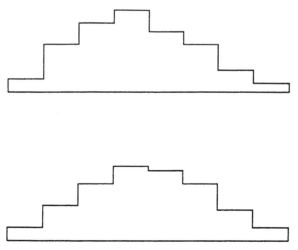

Figure 3.3 Imagine a single row of photoreceptors upon which a distribution of light is falling. The above histograms represent the quantal catch in each photoreceptor for two slightly different spatial distributions of the incident light. The mean of each spatial distribution can be calculated more accurately than the limit set by the photoreceptor size (the bin size in the illustrated histograms). This principle can be used in computer graphics for the accurate positioning of features on a pixel display system and is used by biological visual systems for making fine-grain spatial discriminations.

spatial position with a precision not limited by the spatial intervals of the sampled motion.

Some special considerations apply to the recovery of 3-D structure from binocular stereopsis and from motion parallax. For both these cases, the sensitivity of the human observer to depth variation depends upon the 3-D structure of the surface that is being displayed. For example, observers are relatively insensitive to shallow gradients of depth variation but highly sensitive to abrupt changes in depth. There is evidence that the sensitivity to abrupt changes in depth can dominate the whole percept as demonstrated by the existence in the depth domain (Anstis *et al.*, 1978) of an equivalent to the Craik–Cornsweet–O'Brien illusion induced by AC-filtered step changes in luminance (Cornsweet, 1970). For both binocular stereo and motion parallax, it is possible to define a disparity-modulation limit as a function of the spatial frequency of depth variation. This function summarizes most of the aspects of depth sensitivity mentioned here. In addition, these measurements show that human observers cannot detect very closely spaced corrugations in depth: anything above about 5 cycles/degree of depth variation is invisible.

In summary, an important concept for the display of complex visual scenes on a computer graphics screen is to consider the 'window of visibility' available to a human observer. Clearly, subtle aspects of 3-D structure cannot be appreciated if the information necessary to define them is outside the limits of visibility for the early stages of visual processing. A great deal of information on the capabilities of the early stages of human visual processing is available for the user of computer graphics systems to make some reasoned choices about the display system that is needed (see Watson, 1993). It must, however, be acknowledged that whilst many variables have been studied individually or in pairs, there are relatively few studies that have investigated systematically the possibility of multiple interactions between the relevant variables. Where this has been carried out, there are some indications that a simple analysis in terms of the visibility of individual components of a complex pattern will not always be successful (Heeley and Buchanan-Smith, 1994).

Translating these requirements into a specification for a particular display system for the portrayal of naturalistic images depends inevitably on some choices outside of questions about human performance. Also relevant are questions about what needs to be displayed and for what purpose. Many display systems at the moment are refreshed at about 60–70 Hz, which is just on the limit of temporal sensitivity. Two or three times this value would be better. In particular, for fast-moving stimuli, it could be worth sacrificing some spatial detail, particularly if additional temporal resolution can be achieved as a consequence. Alternatively, for many aspects of display performance in virtual reality systems, true 'photographic' quality is a desired goal. In this case, displays of high brightness (ideally > 100 cd m^2) of high resolution (good enough to resolve 60 cycles/degree, so as to bear inspection by foveal vision) are needed. Since observers will scan such a display to inspect different features in the picture it needs to have high resolution wherever the observer might be looking. This can be achieved by making the display have high resolution all over, but there is growing interest in displays that are dynamically altered as the observer scans around the scene. These would leave high resolution vision only where the observer's fovea is pointing (see also sections 2.4.3 and 6.3.2).

The main computational requirement for improving the low-level features of the display to allow the accurate depiction of high-level aspects such as 3-D shape is to ensure that anti-aliasing techniques are appropriately applied. A shape such as a sphere, ellipsoid or face is always going to be unconvincing if it is evidently made up from a connected set of flat plates rather than being smoothly curved. At a lower level, the best precision that can be achieved is needed to depict features such as binocular stereo disparities and motion disparities. Both of these depth cues depend critically upon the quality of the image as actually presented on the display screen. Here the correct use of anti-aliasing can achieve a smooth disparity field where inaccurate use will lead to a jagged or piecewise planar representation of what should be a smooth shape. Clearly, coordination is also necessary between various aspects of anti-aliasing, such that both shading and stereo disparity depict the same shape.

Edgar and Bex (Chapter 4) and So and Griffin (Chapter 5) discuss in detail three human factors issues which determine requirements for virtual reality displays.

3.6 Computer graphics

3.6.1 Simulation and representation

The aim of three-dimensional computer graphics is, in principle, the same as that of representational art. Both are concerned with the accurate portrayal of the real world. If the visual environment is represented accurately, the overall level of realism is increased because we are providing the visual system with the kinds of information expected of the real world. Although the aims of pictorial art and computer graphics may be similar, the means of arriving at the final image are very different. In pictorial art, the brush strokes and markings are placed on the canvas in order to mimic the visual product of the projection of light. In three-dimensional computer graphics, the light projection process itself is simulated; yielding the features of natural images as a consequence. In this sense, realism is improved by accurate modelling of the natural image formation process.

The starting point in image generation is a specification of the artificial three-dimensional world to be represented. This artificial world consists of objects at various locations and orientations. Objects are collections of surfaces which are represented mathematically according to the operations which are performed on them (for example, see Watt, 1989). One important calculation at a surface is that of its brightness or intensity. Images are formed by calculating and projecting intensity values onto a projection plane (the screen).

In creating computer images there are, therefore, two problems that need to be addressed: how to determine the intensity of light at a point on a surface in the artificial three-dimensional world and how to project this value onto the screen for viewing.

3.6.2 Illumination models

The method used to calculate the intensity of reflected or transmitted light at a point on a surface is known as the illumination model. Essentially, the illumination model is used to capture the physical characteristics of the four types of light–surface interaction described above. The light reflected at a diffusely reflecting surface depends on the intensity of the light source and whether the light from the source strikes the surface directly, or at some oblique angle. This relationship is modelled by Lambert's law, which has been used to calculate diffusely reflected light:

$$I = \rho I_0 \cos(\theta) \tag{3.1}$$

This says that the intensity I of a surface is a function of the intensity of the light source I_0 and the cosine of the angle of incidence (see Figure 3.4). The value ρ is called the surface albedo or reflectance and is a constant fractional quantity that determines the amount of the incident light that is reflected. Thus, white surfaces have a high reflectance (approximately 80% or 0.8) and reflect a large amount of incident light, and black surfaces have a low reflectance (around 30% or 0.3) and therefore reflect very little.

This model can be extended by including factors that determine the amounts of diffuse transmission and specular reflection and specular transmission (Bui-Tuong, 1975; see Rogers, 1985, p. 314). For simplicity, only diffuse reflection will be discussed here.

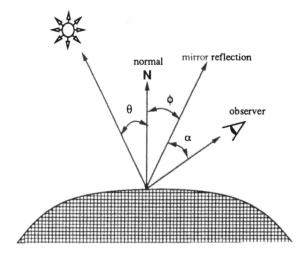

Figure 3.4   Schematic diagram of the reflection process. The intensity of light reflected from a surface may be determined from the intensity of the light source and the relative orientation of the surface with respect to the light source. This relative orientation is determined by the angle of incidence, θ, which is the angle between the incoming light and the surface normal (the direction that is perpendicular to the surface at a point). The angle of reflection, φ, is the angle between the normal and the direction of the reflected light. For diffuse surfaces there is no particular reflection direction as light energy is absorbed and re-emitted in random directions. In ideal specular reflection the light is reflected in one direction only, known as the mirror direction, and in this case the angle of reflection is equal to the angle of incidence.

3.6.3 Making images

Given that we can calculate the shading across a surface, how do we assign intensity values to pixels in order to create an image? The initial method used in computer graphics has involved variants of the scan conversion or polygon rendering method (see Watt, 1989, p. 97; Rogers, 1985). Surfaces are represented or approximated by a number of planar (polygonal) patches. The polygons are shaded (according to the illumination model) and projected onto the projection plane to form an image. One problem that must be considered in scan conversion is the determination of visible surfaces from the viewing position. In complex scenes many polygons are used to represent the scene and for each pixel only the relevant polygon with the smallest distance between it and the viewer can be projected as the visible surface. Another problem is the determination of cast shadows in complex scenes. Additional visibility routines have to be used to determine portions of each polygon that are in shadow before projection (see Foley *et al.*, 1990, p. 746). Despite such problems, scan conversion algorithms are computationally inexpensive and many hardware implementations exist. For this reason, most, if not all, virtual reality image generation is produced by such methods.

Another means of projecting surface intensity values onto an image is to use visible surface ray tracing (Foley *et al.*, 1990; Whitted, 1980; see Glassner, 1989). This method is analogous to the natural image formation process; straight line rays are shot from a vantage point (eye point) through each pixel of a declared image plane (see Figure 3.5). The rays pass through the world space containing the objects constituting the scene. Intersection tests between the rays and the surfaces of objects determine the visible surface with respect to the viewer and ensure hidden surface removal as a natural consequence. Intensity calculations at intersection points are then used to calculate the intensity or colour to be attributed to the appropriate pixel through which the ray has passed.

Figure 3.6 was generated using this method of ray tracing in conjunction with an illumination model based on the simple Lambertian relationship defined above. Ray tracing has the advantage that no additional work has to be carried out to obtain accurate perspective projections of surfaces because the method resembles the natural light projection process. Furthermore, the method ensures that accurate cast shadows are also obtained simply by shooting additional rays from each intersection point to the light sources. If the paths of these, so-called, shadow rays are blocked by other surfaces, the intersection point is in

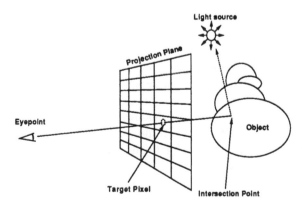

Figure 3.5 Visible surface ray tracing is a method of generating images closely related to real image generation. Unlike the real world, however, light rays are propagated backwards from the eye through each pixel and into the scene.

Figure 3.6   Computer-generated image of two rooms joined together by a connecting passageway. The image was generated by using a visible surface ray tracing algorithm that shoots rays from a vantage point and attempts to intersect them with the surfaces in the scene. At intersection points the intensity is calculated according to how much light energy from the visible light source is reflected. This is dependent on the surface orientation with respect to the light source, the intensity of the light source itself and, also, on the reflectance of the reflecting surface.

shadow and a constant ambient illumination factor is often used such that the resulting shadowed region is not unrealistically dark.

3.6.4 Problems with local illumination models

The Lambertian illumination model described above, or variants of it, have been in widespread use in computer graphics applications, such as computer-aided design and also in head-mounted virtual reality systems. Because these models determine image intensity only as a function of local surface intensity, which in turn depends only on direct illumination from light sources, they are known as local illumination models (at least, they are local as far as diffuse reflection is concerned). Although they can be used to generate some realistic effects, they do not fully capture the character of natural diffuse reflection. In nature, surface intensity is not only determined by energy arriving directly from light sources ('direct illumination'), but is also dependent on 'indirect illumination' reaching the surface by inter-reflections between non-emitting surfaces. That is, in a real scene, light that comes from a light source bounces around between surfaces before it reaches the eye of an observer (see Figure 3.7). One consequence is that shadowed regions are not completely void of light. Moreover, we often see the world not through direct lighting from light sources, such as the sun or room lights but from light that is diffusely reflected between surfaces as, for instance, on a cloudy day.

In an attempt to handle ambient indirect illumination, an arbitrary ambient component has been added to the Lambertian model. The Lambertian relationship for diffuse reflection then becomes:

$$I = \rho I_0 \cos(\theta) + I_a \qquad (3.2)$$

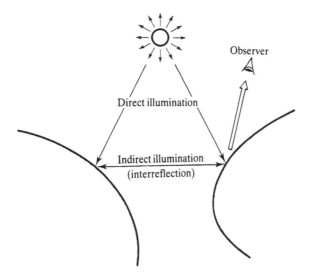

Figure 3.7 The intensity of a diffusely reflecting surface in a complex environment is the result of direct and indirect illumination. Direct illumination is light from an emitting object (a light source). Indirect illumination is light energy that is bounced between proximal surfaces. This bouncing of light energy is known as mutual illumination or mutual inter-reflection. The two components of direct and indirect illumination constitute global diffuse illumination.

where I_a is the constant ambient component. This, however, is insufficient because, in reality, indirect illumination is not constant but varies according to the proximity and relative orientation of nearby surfaces. In some cases mutual inter-reflection of light produces remarkable highlights where two inclined surfaces meet (Forsyth and Zisserman, 1990). Some of the perceptual consequences of these effects will be described later. In general, images generated by standard ray tracing methods using local illumination models exhibit a characteristic 'plastic' quality and lack the smooth variations in luminance apparent in real scenes (see Figure 3.6). The graphics community, aware of this lack of realism, has made efforts over the past ten years to derive more realistic illumination models and consequently this has led to illumination models with a stronger grounding in the physics of natural illumination.

3.6.5 Physics of global diffuse illumination

Once light energy is introduced into an environment of surfaces it bounces around between them. Assuming that we only have diffusely reflecting monochromatic surfaces in a given configuration, an equation is formed that specifies how much energy is reflected from each surface. This equation is then said to model the light transport between the surfaces. The light energy radiated from a surface is determined by the total illumination that arrives from all emitting or reflecting surfaces (assuming they are visible). If the surface in question is also an emitter (light source), then the radiated energy is also dependent on how much energy it emits. A global diffuse illumination model therefore takes the form:

$$B = \epsilon + \rho E \qquad (3.3)$$

where B is the energy radiated into the environment by the surface, E is the total incoming illumination, or irradiance, at a surface and ρ is the surface reflectance and ϵ is the quantity of light energy emitted by the surface. More specifically, *B* is known as the radiosity of the surface and is measured in W m^{-2}, that is, it is the energy radiated into the environment per second per unit area. Once we can determine the radiosity at any point on a surface we can proceed to generate images of the surface in the usual way.

There are, however, three difficult obstacles in calculating radiosity from this relationship. Firstly, the incident illumination, *E*, must be calculated from all surfaces not just light sources. Because surfaces are extended entities, this involves mathematical integration of equation (3.3) across all visible surfaces. Secondly, equation (3.3) only tells us what the radiosity is at a single point on a surface. Calculating the total radiosity across the whole of a surface involves further integration across the target surface. Lastly, the radiosity at one point on a surface is determined by the radiosity of other surfaces. But, the radiosity of these surfaces is equally dependent on the radiosity of the target surface. This interdependence means that equations of the form expressed in equation (3.3) must be solved simultaneously for all surfaces in the environment. Fortunately, mathematical techniques exist for overcoming these problems and these are incorporated in the radiosity method described below.

3.6.6 The radiosity method

The radiosity method, introduced to the field of computer graphics by Goral *et al.* (1984), allows an approximate calculation of global light energy exchange, as described above. It was originally used by thermal engineers to calculate heat exchange between surfaces (Hottel and Sarofim, 1967; Siegel and Howell, 1981). The method involves a finite element approximation to the radiosity distribution across surfaces within a given scene. All surfaces are subdivided or tessellated into a finite number of polygonal elements, which serve as sites for radiating and receiving energy from each other. By assuming that the radiosity across each element is constant, light energy exchange between the defined surfaces is approximated by calculating the energy exchanged between these elements. For each element, *i*, in a scene consisting of *N* elements, the total incoming light energy is found by summing over the radiosities of every other element, *j*:

$$E = \sum_N B_j F_{ij} \tag{3.4}$$

where B_j is the radiosity of the *j*th element and F_{ij} is a constant that determines what proportion of light leaving one element *j*, actually arrives at the target element *i*. This constant, known as the form factor, is dependent only on surface geometry but is extremely difficult to calculate. Methods for its calculation are, fortunately, well established (Howell, 1982; see also Cohen and Wallace, 1993). Given that we have a relationship for the total incoming energy at each element, each element can therefore be assigned a corresponding radiosity equation:

$$B_i = \epsilon_i + \rho_i \sum_N B_j F_{ij} \tag{3.5}$$

Equation 3.5 says that the radiosity of surface *i* is equal to its emission, ϵ_i, plus the proportion of the total incident illumination which is reflected back into the environment. If each element has such an equation, then all equations of this form must be solved simultaneously for a global solution to be obtained. Further details of such solutions, and of methods for calculating form factors, can be found in Cohen and Wallace (1993).

The radiosity method is an object-based analysis of light exchange and is therefore view-independent. That is, the radiosity solution is calculated for the whole scene independently of any viewpoint. Once the radiosity value for each element is calculated, an image can be generated from any viewpoint by converting radiosity values into image intensities. Early implementations used polygon scan conversion of the actual scene tessellation for this purpose. More recent implementations have shown that sharp shadow boundaries and other illumination effects are best produced by using a method of visible surface ray tracing (Jones *et al.*, 1993). Figure 3.8 shows the same scene depicted in Figure 3.6 but rendered using ray tracing in conjunction with a radiosity solution for the whole scene. Notice that the walls of the room no longer appear flat-shaded but have smooth gradients of intensity across them. This is because the intensity is determined partly by direct lighting and partly by light bouncing from nearby surfaces. Moreover, the intensity of reflected light across the walls does not change rapidly with change in surface orientation, as it does in the locally illuminated scene in Figure 3.6. All surfaces clearly have the same colour and the only sharp variations in image intensity occur at the reflectance changes observed on the chequered floor. Comparing these two images demonstrates the remarkable increase in realism that can be achieved simply by using a more natural illumination model.

3.7 Global illumination and virtual reality

3.7.1 Applying the radiosity method

The view-independence of the radiosity method makes it ideal for generating simulated computer graphics 'walkthroughs' where the rendered image must change continually as the viewer progresses through a static environment. As long as the geometry of the environment remains constant, a single radiosity solution is sufficient for the generation of any number of views. In head-mounted virtual reality systems this could be very advantageous, allowing the viewer to move around a more realistic setting generated from only a single radiosity solution. But if the observer actively changes the geometry of the

Figure 3.8 Computer-generated image of two rooms with the same geometry as shown in Figure 3.6 but using a global diffuse illumination model to calculate surface intensity.

artificial environment, say by moving an object, then a new radiosity solution would be required. Radiosity calculations are still comparatively expensive (computationally). One possibility would be to generate an initial global illumination solution which can be locally adjusted according to any changes that subsequently occur in surface geometry. Here, we are only interested in assessing the importance of global illumination on visual realism. Much research is currently under way to reduce the computational overheads involved in the global illumination model. We now turn to the question of what makes radiosity images more realistic and how modelling of natural illumination can be beneficial for virtual reality.

3.7.2 Shape from shading

Shading, the use of gradients of light intensity, has featured in many forms of pictorial art. The power of shading in revealing surface solidity and relief was stressed by Leonardo da Vinci who wrote of it in his books on painting:

> Shadows and lights are the most certain means by which the shape of any body comes to be known, because a colour of equal lightness or darkness will not display any relief but gives the effect of a flat surface which, with all its parts at equal distance, will seem equally distant from the brightness that illuminates it. ('Treatise on painting' by Leonardo da Vinci, edited and translated by A. P. McMahon (1956))

More formally, shape from shading is the ability to extract some representation of three-dimensional surface shape from smooth gradients in image intensity. Because the image intensity of a surface depends, in part, on the orientation of the surface with respect to the light source, it has been assumed that visual mechanisms can extract orientation measures from gradients in image intensity. Many computational algorithms have therefore been proposed for reconstructing shape from shading (Horn, 1975; Pentland, 1989; see also Brooks and Horn, 1989).

The saliency of shading in vision is demonstrated in the animal kingdom where various animals have evolved reversed shading patterns on their skins (dark on the top and light on the bottom), presumably in order to reduce their appearance of solidity – either to evade predators or to minimize the possibility of detection by their prey. A number of psychophysical studies of human sensitivity to shading have shown that they assume direct illumination is from directly above the shaded surface (see Figure 3.9) when interpreting surface shape (Benson and Yonas, 1973; Yonas *et al.*, 1979; Kleffner and Ramachandran, 1992). This powerful assumption is probably related to ecological development of a visual system that is used to the sun as a major illuminant (Ramachandran, 1988, 1990). If human visual systems make such an assumption, then it is likely that it is made by many other animals. Thus, reverse shading is a counteractive measure, like camouflage, intended to reduce the likelihood of the animal's shape being separated from the immediate background. Reverse shading of this kind is found in many animals such as deer, antelope, tigers, leopards and most species of fish.

What are the benefits for shape perception when using a global illumination model as opposed to a local illumination model? Because global diffuse inter-reflection varies with the geometry and proximity of surfaces, it may be that it aids the visual discrimination of shape by increasing the appearance of surface relief with respect to the immediate background. Consider Figure 3.10 which shows two elliptical bumps on a planar background. The scene on the left was generated using a local illumination model and so we see the bump in terms of the reflected light coming directly from the light source (which in this case is above and to the left of the shape). The scene on the right was generated using the global illumination model where light exchange between the various portions of the figure that are visible to each other means that areas which do not receive light directly from the source are still illuminated by indirect illumination.

Clearly, indirect illumination makes a significant difference to the phenomenological appearance of the scenes which, in geometrical terms, are identical. Indeed, psychophysical studies using such images have revealed that a greater impression of depth or relief is

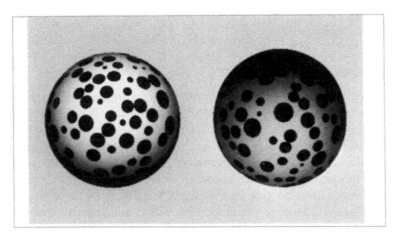

Figure 3.9   The hemispherical shapes depicted both lie on the same background surface. The shape on the left is, in fact, concave (receding from the viewer) and that on the right is convex (protruding towards the viewer). The light source is below the two shapes, although most people assume it is from above and hence the concave shape on the left appears more convex than that on the right (there also appears to be a preference for convexity in the absence of information indicating the contrary). If the picture is turned upside down, the light from above preference now causes the truly convex shape to appear rather more convex than the truly concave shape.

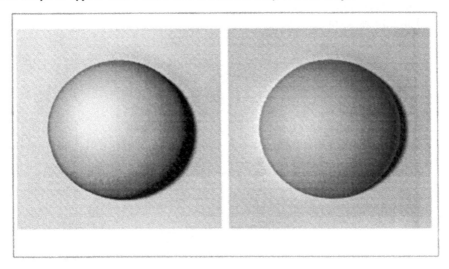

Figure 3.10   Example image used to test whether indirect illumination (mutual illumination) affects perceived depth in shaded images. The image on the left was generated using a local illumination model, as employed in computational studies of shape from shading. The image on the right was generated using the global illumination model. Both images are of the same 3·5 cm hemisphere resting on a white plane. The use of global illumination appears to enhance the appearance of solidity as demonstrated by comparing the apparent depth of the two shapes.

obtained from images portraying indirect illumination (Christou, 1994). In these experiments subjects were shown similar shapes which had varying degrees of elongation in depth (towards the viewer). They had to judge whether the elongation in depth of these shapes was as great as their half-height (see Figure 3.11). The subjects' results in this monocular experiment indicated that, although underestimations of shape were made in both conditions, the level of underestimation was significantly reduced when indirect illumination was utilized. This suggests that the impression of solidity in a scene can be enhanced by global illumination and this has important implications for graphics systems attempting to portray accurate three-dimensional structure.

3.7.3 Reducing structural ambiguity

The importance of providing sufficient information for accurate perception is demonstrated by the human tendency to reverse or invert figures in depth. Notable examples include the Necker cube and Schroeder's staircase (see Figure 3.12). In both cases, the perceptual data remain constant but the observer can 'decide' how the data are to be interpreted. Under normal conditions, however, our percepts of the world are stable, consistent and determinate. Only on those rare occasions when we are confronted with little visual detail, or when information from different sources is contradictory, is the perceived spatial layout of a scene ambiguous. Much of the observed stability in human perception is probably due to the wealth of consistent visual information. Furthermore, information deriving from expectations and past experience may also be responsible for much of this stability. An example which demonstrates the influence of past experience is the Ames room designed by Adelbert Ames (see Gregory, 1972). The Ames room is distorted in depth but the surface textures and perspective of the room are proportioned such that at a certain vantage point the room appears normal and rectangular. The 'desire' to see the room as normal and conforming to past experience is so strong, however, that when familiar objects are placed in different corners of the room (which are in fact at different depths), they are grossly distorted in size as a consequence.

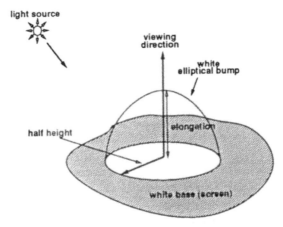

Figure 3.11   Schematic diagram of the geometry used to determine whether indirect illumination affects our perception of depth and solidity. The psychophysical task involved subjects judging whether the elongation of the elliptic bump (prolate spheroid) was as great as its half-height. The half-height is simply the radius of the circle formed by the intersection of the shape with the background (which was coincident with the screen).

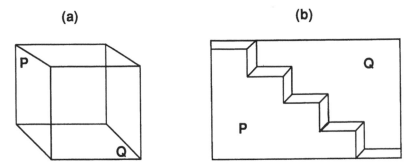

Figure 3.12   Examples of figures that are reversible in depth. (a) The Necker cube. The orientation of the cube changes depending on whether point P or point Q is visualized as closer to the viewer. (b) The Schroeder staircase. The 'stairs' reverse their orientation if the area Q is visualized as being in front of area P.

Can the use of a more natural model of illumination reduce the possibility of ambiguity in complex scenes? Consider Figure 3.13. In Figure 3.13(a), there are three possible three-dimensional interpretations of the two-dimensional image, a variant of the Mach book illusion (Wade, 1982). Firstly, the figure in the middle of the image might appear as a series of coplanar black and white parallelograms forming a chevron pattern on a uniform grey background. Secondly, and more likely, we may see the central figure as approximately perpendicular to the line of sight and consisting of inclined rectangular facets joined together to form a three-dimensional zigzag corrugation. The alternating light and dark appearance of the surface would in this case be attributable to a light source placed to the left of the figure. Lastly, the figure may again appear as a three-dimensional corrugation but with the direct light now coming from the right of the figure. In this case convex corners appear as concave corners, and vice versa, and the whole figure appears to be inclined to the line of sight. In fact, the actual modelled structure of Figure 3.13(a) is consistent with the first interpretation. The scene in this case consists of coplanar light and dark parallelograms. But our predisposition to perceive such patterns as three-dimensional results in one of the other two interpretations.

Figure 3.13(b) on the other hand was modelled as a three-dimensional 'corrugated' surface as described above. The image in this case was produced using the local illumination model and exhibits no mutual inter-reflection. The geometry in this case consists of a horizontal ground plane with the corrugated surface resting perpendicular to it. Notice how similar the images in Figure 3.13(a) and Figure 3.13(b) appear, even though the geometries that produced these two images were entirely different. Figure 3.13(c) on the other hand, is the same geometry as in Figure 3.13(b), but rendered using the more natural global illumination model. Indirect illumination in this case produces a dramatic change in the appearance of the depicted structure. There are two differences. Firstly, the increase in surface intensity within the concave regions of the figure appear to increase the overall impression of three-dimensionality. These increases in intensity are mutual illumination effects and would not have occurred if the facets of the surface were coplanar. Secondly, and more importantly, the inter-reflection of light between the corrugated surface and the ground plane at the point of contact causes highlights on both surfaces. The effect of this is to increase the impression of connectedness between the two figures and also reveals the relative orientation of the figure to the ground plane. The image of this scene now appears more like the modelled three-dimensional structure.

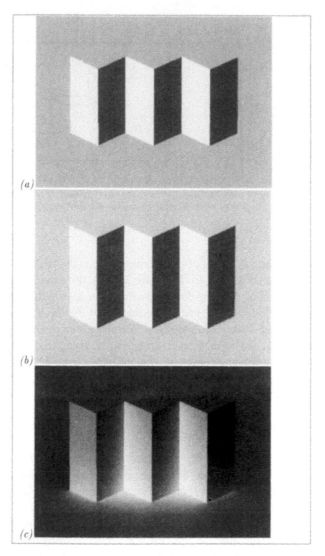

Figure 3.13   Demonstration of the effects of indirect illumination on the perception of spatial structure and relative orientation. (a) A series of alternating light and dark parallelograms that appear as a three-dimensional surface. (b) Directly illuminated three-dimensional geometry consisting of rectangular polygons inclined at 45 degrees to each other. The whole figure in this case is almost perpendicular to the viewing direction and the background is a horizontal surface. (c) Same geometry as in (b) but rendered using the global illumination model. The indirect illumination makes the structure of the scene more apparent than in (b).

3.7.4   Discriminations of reflectance and illumination

Another area that is liable to be affected by the use of global illumination is in the psychological discrimination between changes in reflectance and changes in illumination. The amount of indirect light exchanged between surfaces is, of course, dependent on the surfaces' reflectance as well as the geometry: a room full of black surfaces will exhibit very little indirect illumination. In such a room the majority of indirect illumination is absorbed after the first or second inter-reflection of light. A room containing mostly white surfaces, on the other hand, will contain a great deal of indirect illumination because a large proportion of direct light is reflected back into the scene. Since most environments we encounter contain mainly light surfaces, we must be acquainted with indirect light in visual perception. What are the visual effects of indirect illumination with respect to revealing the reflectance of surfaces in complex scenes?

It was noted previously, when comparing Figures 3.6 and 3.8, that indirect illumination seems to make it easier to discriminate between reflectance changes and rapid changes in illumination. The ability to correctly judge surface reflectance under variations in illumination is known as lightness constancy; lightness is the psychological correlate of reflectance. A piece of paper that is perceived as white in sunlight will still be judged to be white under normal room illumination and even in moonlight. Lightness constancy is important for two reasons. Firstly, the ability to gauge lightness under varying illumination means that objects can be recognized easily even though the light that illuminates them changes significantly. Secondly, the ability to discriminate between changes in illumination and changes in reflectance is necessary for making decisions about the structure of the visual world. A change in the luminance of a surface may be caused by incidental changes in incident illumination, which in turn occur either when the orientation of surface changes with respect to the illuminant or when the incident light is intercepted by another object, resulting in a cast shadow. Surface reflectance on the other hand is an enduring property of a surface and remains constant under variations in illumination. Therefore, whereas illumination changes imply some variations in the spatial structure of the world, changes in reflectance occur as variations in the material properties of surfaces alone. Discriminating between the two is vital for the correct interpretation of what we see. The fact that these two sources of variation in image intensity can in principle be confused, especially when inaccurate illumination models are used, is demonstrated both by Figure 3.6 and by Figure 3.13. In the latter case the pattern of reflectances in Figure 3.13(a) was found to be easily confused with illumination changes on a totally different structure. If it were not for other geometrical information from cues such as stereopsis and motion parallax such mistakes would occur quite often.

The contribution of indirect illumination in providing a better impression of surface lightness is demonstrated by Figures 3.14 (a) and (b). Here, the geometry consists of a corner of a room with a modelled table and chair. The walls, floor and all other surfaces have a constant reflectance of 40% (middle grey) apart from the wall on the left which has a reflectance of 85% (average white). Figure 3.14(a) shows the image resulting from a local illumination model whereas Figure 3.14(b) shows the result from the radiosity method. Because a single light source has been placed to the left of this scene, we notice from the cast shadows that in both cases the left wall should be in relative shadow compared to the wall on the right. In Figure 3.14(a), the brightest wall is that on the right which is directly illuminated by the light source. In the instance, most people will not really be able to judge that the wall on the left has a higher reflectance than the right-hand wall. In the globally illuminated room, Figure 3.14(b), the brightest wall is the left-hand wall. Clearly we can tell from this image that the reflectance of the left-hand wall is greater than that of the right. This information is provided primarily because the brightness of the left hand wall is increased as a result of the light inter-reflections between the various surfaces in the scene. Again this shows that using a realistic and natural model of light reflection makes significant contributions to the accurate perception of surface properties.

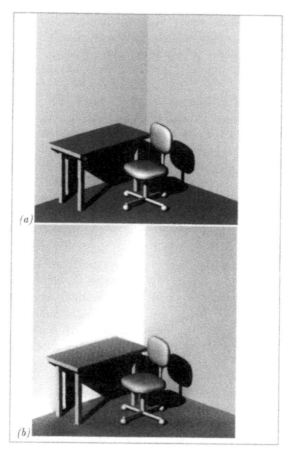

Figure 3.14   The contribution of indirect illumination in revealing true surface reflectance is demonstrated by these two computer-generated images. (See text for explanation.)

3.8   Conclusion

In this chapter, the aim has been to introduce the reader to the study of vision from a psychological standpoint and to show how an understanding both of the kinds of information the visual system utilizes and is sensitive to, and of the physics of natural image formation, can combine to improve visual realism in computer-generated imagery. From the history of art we see that most steps towards visual realism have taken place in conjunction with a greater awareness of how images are formed. A good example of this was the development by Florentine artists of the rules of perspective. The ensuing imagery which this artistic revolution produced were considered astounding compared to the earlier art of the Medieval and Byzantine periods (Gombrich, 1972). Computer graphics methods can depict three-dimensional environments by simulating the natural image formation process. As in representational art, realism is attained with the accurate representation of the natural structure of such environments, that is, their extension in depth, solidity, opacity, etc. These features are represented by simulation of the naturally occurring forms of light–surface interaction. Because we have evolved to use the spatial character of light as visual information, our impressions of extension and solidity, etc., are derived from these pictorial cues and also from the cues of convergence, accommodation, stereopsis and motion parallax. Thus, synthetic environments can be made more realistic, more believable, by incorporating more and more of the information normally available to the visual system in real scenes. The amount of information that is provided depends, to a certain extent, on the task of the observer. In general, however, the greater the redundancy in visual information the less likelihood there is of ambiguity and illusion. The particular advance in realism we have concerned ourselves with here is the progression from an idealized model of diffuse reflection to a more naturalistic one. The more natural illumination model captured by the radiosity method produces images of greater realism. This is because we are providing the visual system with the information it expects of natural environments. The typical characteristics of local shading methods used prior to the simulation of global diffuse illumination were of unnaturally sharp discontinuities in intensity across surfaces. This in turn conveyed a harsh, rather plastic, appearance. Globally illuminated scenes, however, exhibit more subtle gradients of intensity.

The realistic appearance of imagery generated using the radiosity method is not just a matter of aesthetics. We have attempted to show that important visual information is conveyed by the natural inter-reflection of light between surfaces of relatively high reflectance. Firstly, inter-reflection between surfaces can accentuate the shape and relief of the depicted objects. Shading can be used as a cue for judging surface relief and also it can be used to segment a curved surface from its background. In this respect accurate portrayal of global shading effects can be very important for vision. Secondly, inter-reflection between surfaces can reveal spatial structure. In locally illuminated scenes, the connectedness and adjacency of surfaces is sometimes not established by available information and this results in the appearance that objects are floating above the ground. Determining the relative spatial location of objects with respect to each other, however, is important for the perception of object size (Gibson, 1950). Surfaces which exhibit light inter-reflection, on the other hand, appear more connected and the use of global illumination in this respect aids the process of determining relative location and relative orientation between surfaces. Lastly, we have shown that light inter-reflection can also reduce the possibility of confusing changes in surface reflectance with changes in scene illumination. In computer-generated environments, where visual information may be scarce compared to natural scenes, this confusion becomes a real possibility. Changes in illumination signal changes in surface structure. Changes in surface reflectance, on the other hand, have no such geometrical significance. Distinguishing between the two is therefore a significant problem, if a scene is to be correctly interpreted. In conclusion, we hope to have provided some guidance as to which aspects of artificial imagery and artificial scenes are of particular relevance in visual perception. A knowledge of the sensitivities and limitations of the human visual system is necessary, both for reducing the amount of superfluous effort involved in constructing virtual worlds and also in order to maximize the level of reality and utility of the resultant imagery.

Notes

1. The lines originating near the fovea and passing through the optical centre of the eye.
2. Euclid believed in the reverse of this process. Rays of light being shot out from the eye into the world to be intersected with surfaces. The geometrical interpretation is, however, unaffected in either case.

References

Anstis, S. M., Howard, I. P. and Rogers, B.J., 1978, A Craik–Cornsweet–O'Brien illusion for visual depth, *Vision Research*, **18**, 213–17.

Attneave, F., 1954, Some informational aspects of visual perception, *Psychological Review*, **61**, 183–93.

Benson, C. and Yonas, A., 1973, Development of sensitivity to static pictorial depth information, *Perception and Psychophysics*, **13** (3), 361–6.

Berkeley, G., 1709, *Towards a New Theory of Vision and Other Writings*, New York: Dutton (1992).

Boring, E. W., 1942, *Sensation and Perception in the History of Experimental Psychology*, New York: Appleton-Century.

Brooks, M. J. and Horn, B. K. P., 1989, *Shape from Shading*, Cambridge, MA: MIT Press.

Bui-Tuong, P., 1975, Illumination for computer generated images, unpublished PhD thesis, University of Utah, also in *Communications of the ACM*, **18** (6), 311–17.

Campbell, F. W. and Green, D. C., 1965, Optical and retinal factors affecting visual resolution, *Journal of Physiology (Lond.)*, **181**, 576–93.

Chapanis, A. and McCleary, R. A., 1953, Interposition as a cue for the perception of relative distance, *Journal of General Psychology*, **48**, 113–33.

Christou, C.G., 1994, Human vision and the physics of natural images, unpublished PhD thesis, University of Oxford.

Clark, W. C., Smith, A. H and Rabe, A., 1955, Retinal gradient of outline as a stimulus for slant, *Canadian Journal of Psychology*, **9**, 247–53.

Cohen, M. F. and Wallace, J. R., 1993, *Radiosity and Realistic Image Synthesis*, New York: Academic Press.

Cornsweet, T. N., 1970, *Visual Perception*, New York: Academic Press.

Foley, J. D., van Dam, A., Feiner, S. K. and Hughes, J. F., 1990, *Computer Graphics: Principles and Practice*, Reading, MA: Addison-Wesley.

Forsyth, D. and Zisserman, A. P., 1990, Shape from shading in the light of mutual illumination, *Image and Vision Computing*, **8**, 42–9.

Forsyth, D. and Zisserman, A. P., 1991, Reflections on shading, *IEEE Transactions on Pattern Analysis and Machine Intelligence*, **13** (7), 671–9.

Freeman, R. A., 1966, Effect of size on visual slant, *Journal of Experimental Psychology*, **71**, 96–103.

Gibson, J. J., 1950, *The Perception of the Visual World*, Boston, MA: Houghton Mifflin.

Gibson, J. J., 1966, *The Senses Considered as Perceptual Systems*, Boston, MA: Houghton Mifflin.

Glassner, A., 1989, *An Introduction to Ray Tracing*, London: Academic Press.

Gogel, W. C., 1964, Size cue to visually perceived distance, *Psychological Bulletin*, **62**, 217–35.

Gogel, W. C., 1965, Size cues and the adjacency principle, *Journal of Experimental Psychology*, **70**, 289–93.

Gombrich, E. H., 1972, *The Story of Art*, London: Phaidon.

Goral, C. M., Torrance, K. E. and Greenberg, D. P., 1984, Modelling the interaction of light between diffuse surfaces, *ACM Computer Graphics—SIGGRAPH*, **18** (3), 213–22.

Gregory, R. L., 1972, *Eye and Brain, The Psychology of Seeing*, World University Library, London: Weidenfeld & Nicolson.

Heeley, D. W. and Buchanan-Smith, H., 1994, Changes in the perceived direction of drifting plaids, induced by asymmetrical changes in the spatio-temporal structure of the underlying components, *Vision Research*, **34**, 775–97.

Helmholtz, H. von, 1893 (translated by Atkinson, E.), On the relation of optics to painting. II. Shade; III. Colour, in *Popular Lectures on Scientific Subjects*, Second Series, Vol. II. pp. 94–123. London: Longmans, Green and Co.

Hering, E., 1861 (translated by Hurvich, L. M. and Jameson, D. from 1874 edition), *Outlines of a Theory of the Light Sense*, Cambridge, MA: Harvard University Press (1964).

Horn, B. K. P., 1975, Obtaining shape from shading information, in Winston, P. H. (Ed.), *The Psychology of Computer Vision*, pp. 115–55, New York: McGraw-Hill.

Hottel, H. C. and Sarofim, A. F., 1967, *Radiative Transfer*, New York: McGraw-Hill.

Howell, J. R., 1982, *A Catalog of Radiation Configuration Factors*, New York: McGraw-Hill.

Jones, G. J., Christou, C. G., Cumming, B. G., Zisserman, A. and Parker, A. J., 1993, Accurate rendering of curved shadows and interreflections, in Cohen, M., Puech, C. and Sillion, F.

(Eds, on behalf of the European Association for Computer Graphics), *Proceedings of the Fourth Eurographics Workshop on Rendering*, Paris, 14–16 June 1993, pp. 337–47, Aire-la-Ville, Switzerland: Eurographics Association.

Kant, I., 1781 (translated by Kemp Smith, N. from 1781 edition), *Critique of Pure Reason*, London: Macmillan (1934).

Kelly, D. and van Norren, D., 1977, Two-band model of heterochromatic flicker, *Journal of the Optical Society of America*, **67**, 1081–91.

Kleffner, D. A. and Ramachandran, V. S., 1992, On the perception of shape from shading, *Perception and Psychophysics*, **52** (1), 18–36.

Koffka, K., 1935, *Principles of Gestalt Psychology*, New York: Harcourt-Brace.

de Lange, H., 1958, Research into the dynamic nature of the human fovea-cortex systems with intermittent and modulated light. I. Attenuation characteristics with white and coloured light. *Journal of the Optical Society of America*, **48**, 777–84.

Marr, D., 1976, Early processing of visual information, *Philosophical Transactions of the Royal Society of London, Series B*, **275**, 483–524.

Marr, D., 1982, *A Computational Investigation into Human Representation and Processing of Visual Information*, San Francisco, CA: W. H. Freeman.

Mayhew, J. E. W., 1982, The interpretation of stereo disparity information: the computation of surface orientation and depth, *Perception*, **11**, 387–403.

Mayhew, J. E. W. and Longuet-Higgins, H. C., 1982, A computational model of binocular depth perception, *Nature*, **297**, 376–9.

McMahon, A. P. (Ed. and Trs.), 1956, *Treatise on Painting (Codex Urbinas Latinus 1270) by Leonardo da Vinci*, Princeton, NJ: Princeton University Press.

Morgan, M. J., 1979, Perception of continuity in stroboscopic motion: a temporal frequency analysis, *Vision Research*, **19**, 491–500.

Muller, J., 1826, *Zur vergleichenden Physiologie des Gesichtssinnes*, Leipzig.

van Nes, F. L., Koenderink, J. J., Nas, H. and Bouman, M. A., 1967, Spatiotemporal modulation transfer in the human eye, *Journal of the Optical Society of America*, **57**, 1082–8.

Newell Jones, F., 1974, History of psychophysics and judgement, in Carterette, E. C. and Friedman, M. P. (Eds), *Handbook of Perception*, Volume II: *Psychophysical Judgement and Measurement*, pp. 1–22, New York: Academic Press.

Pentland, A. P., 1989, A possible neural mechanism for computing shape from shading, *Neural Computation*, **1** (2), 208–17.

Prazdny, K., 1983, Stereoscopic matching, eye position and absolute depth, *Perception*, **12**, 151–60.

Ramachandran, V. S., 1988, Perception of shape from shading, *Nature*, **331**, 163–6.

Ramachandran, V. S., 1990, Visual perception in people and machines, in Blake, A. and Troscianko, T. (Eds), *AI and the Eye*, pp. 21–77, New York: John Wiley.

Ratoosh, P., 1949, On interposition as a cue for the perception of distance, *Proceedings of the National Academy of Science, USA*, **35**, 257–9.

Richter, G. M. A., 1970, *Perspective in Greek and Roman Art. A Study of its Development from about 800 B.C. to A.D. 400*, London: Phaidon.

Rogers, D. F., 1985, *Procedural Elements of Computer Graphics*, New York: McGraw-Hill.

Siegel, R. and Howell, J. R., 1981, *Thermal Radiation Heat Transfer*, Washington, DC: Hemisphere Publishing Corp.

Wade, N., 1982, *The Art and Science of Visual Illusions*, London: Routledge & Kegan Paul.

Watson, A. B. (Ed.), 1993, *Digital Images and Human Vision*, Cambridge, MA: MIT Press–Bradford Books.

Watson, A. B., Nielsen, K. R. K., Poirson, A. B, Fitzhugh, A., Bilson, A., Nguyen, K. and Ahumada, A. J. Jr, 1986a, Use of a raster framebuffer in vision research, *Behavior Research Methods, Instrumentation and Computers*, **18**, 587–94.

Watson, A. B., Ahumada, A. and Farrell, J. E., 1986b, Window of visibility: a psychophysical theory of fidelity in time-sampled visual-motion displays, *Journal of Optical Society of America*, *A, **3***, 300–7.

Watt, A., 1989, *Fundamentals of Three-Dimensional Computer Graphics*, Wokingham, UK: Addison-Wesley.

Westheimer, G. 1981, Hyperacuity, in Antrum, H., Ottoson, D., Perl, E. R. and Schmidt, R. F. (Eds), *Progress in Sensory Physiology 1*, pp 1–30, Berlin: Springer Verlag.

Wheatstone, C., 1838, Contributions to the physiology of vision. I: On some remarkable, and hitherto unobserved phenomena of binocular vision, *Philosophical Transactions of the Royal Society of London, Series B*, **128**, 371-94.

Whitted, T., 1980, An improved illumination model for shaded display, *Communications of the ACM*, **23** (6), 343-9.

Wysecki, G. and Stiles, W. S., 1967, *Color Science*, New York: John Wiley.

Yonas, A., Kuskowski, M. and Sternfels, S., 1979, The role of frames of reference in the development of responsiveness to shading information, *Child Development*, **50**, 495-500.

4

Vision and displays

Graham K. Edgar and Peter J. Bex

4.1. The 'real' and the 'virtual'

A working definition of the word 'virtual' may be found in any dictionary, and will be something along the lines of 'having the essence or effect, but not the appearance or form of'. Extending this definition to virtual reality (VR) seems quite appropriate. VR systems, as described in this book, are often designed to emulate, or indeed replace, the 'real world'. To this end, the visual image provided by VR systems will, in many ways, be similar to that provided by the 'real world'. Limitations in the technology, however, mean that there will be certain detectable differences between a VR image and a 'real world' one. These differences are important. The differences (some of which may be very difficult to eradicate) affect both how 'real' the virtual world feels and also how easy and comfortable it is to interact with the virtual world. Using the definition given above, *virtual* reality has already been achieved; whether it will ever be indistinguishable from reality will depend on solutions being found to many of the problems discussed below. The optical definition of 'virtual' is also appropriate to most types of virtual reality, as shown in section 2.2.1.

This chapter will consider in detail two ways in which a virtual world may differ from the real one. Some differences may have greater impact on 'realism' than others, but, potentially, any difference may be a problem for perception. It is beyond the scope of this chapter to cover all the differences in detail (that would require a book in itself!), but a brief review of several differences has been given by Christou and Parker (section 3.5), and So and Griffin in Chapter 5 discuss a futher aspect of display technology and perception. Different factors will be more or less important, depending on the configuration of the system (e.g. 'immersive' or 'desk-top' virtual reality).

Temporal aliasing and misaccommodation are contrasting examples of current knowledge about the effects of visual display technology on perception. Both problems occur because the VR system cannot faithfully mimic the real visual world. Temporal aliasing has, however, received a considerable amount of attention (mainly in the film and computer graphics industries) and there are several techniques designed to ameliorate the problem. Misaccommodation with displays has, in contrast, received relatively little attention and, at present, there appear to be no real solutions to the problem. Also, it is easy to see how improvements in technology (such as higher display update rates) might reduce the perceptual effects of temporal aliasing; it is not so easy to see how the improvements in technology could reduce the likelihood and effects of misaccommodation.

4.2   Temporal aliasing

4.2.1   Sampled and continuous motion

The images used to simulate motion in virtual environments paradoxically do not usually contain any motion. Cathode ray oscilloscopes (CROs) and liquid crystal displays (LCDs) on which virtual images are usually displayed are updated at a finite rate. This means that motion may only be simulated by presenting a sequence of static images at spatially discrete locations. Under certain viewing conditions, this sampled motion is indistinguishable from real motion by a human observer (Sperling, 1976; Fahle and Poggio, 1981; Burr *et al.*, 1986a; Watson *et al.*, 1986). Occasionally, however, the coherence of sampled motion may be lost and a number of perceptual phenomena may occur, including the appearance of reversals in perceived direction, motion seeming jerky, and multiple images trailing behind the target. This class of perceptual loss is called temporal aliasing because the problems are principally associated with a coarse temporal sampling regime. When the sampling rate of a motion sequence is low or the simulated velocity is high, fewer samples along the trajectory of the target are possible (see Figure 4.1). Under these conditions, large spatial displacements are required between image updates in the apparent motion sequence and each image is static at each location for a longer duration: both factors may affect motion perception. Additionally, the same perceptual losses may arise if the spatial resolution of the display is so coarse that images may only be presented at a few positions in its motion path. This means that fewer displacements are possible and each displacement may be larger. Consequently, some aspects of low spatial resolution are equivalent to a low image sampling rate.

Within the computer programming and animation literature, the problem of temporal aliasing has long been recognized; cartoonists have employed several anti-aliasing techniques over many years. Several computer programmers have observed that blurred images alias less frequently than sharp images. This has led several researchers to propose

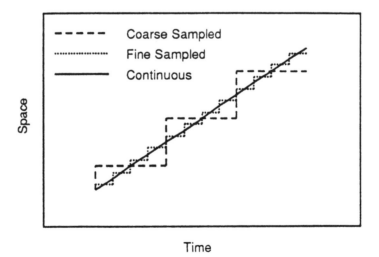

Figure 4.1   The effects of temporal frequency on apparent motion. This figure illustrates the spatio-temporal distortions which are caused by different sampling frequencies. When the sampling rate is very low, spatial displacements are greater and the duration of each static image is longer. Increases in the sampling rate produce closer approximations to continuous motion.

algorithms for adding blur to each sampled image (Norton *et al.*, 1982; Korein and Badler, 1983; Potmesil and Chakravarty, 1981, 1983; Dippe and Wold, 1985; Grant, 1985; Max and Lerner, 1985; Mitchell and Netravali, 1988) and spatio-temporal filters for video animation (Amanatides and Mitchell, 1990). In the latter example, the concept of spatio-temporal filtering was drawn from the observation that stills in cinematographic sequences depicting motion were blurred. Before considering the different types of temporal aliasing that may occur, it is useful to consider some of the current theories of human motion detection.

4.2.2 Current approaches to motion detection

The extraction of motion from sequences of static images can be considered the result of cognitive processes (Kolers, 1972; Johansson, 1975). The occurrence of motion may be inferred by tracking the sequential appearance of the same target through its successive positions. A model of feature tracking has been proposed by Ullman (1979). Ullman's minimal mapping theory provides algorithms for computing possible correspondences between elements of a motion sequence. Thus, an affinity measure is established for each possible match between frames and motion is seen between elements with the highest affinity.

Alternatively, sampled motion may be perceived by low-level motion detectors which operate on delay and compare principles (Reichardt, 1957) (Figure 4.2). Basically, Reichardt detectors are direction selective units. Each unit consists of a pair of receptors which sample the visual field at two different points and a second stage where the output of the receptors is compared. A moving object activates each receptor sequentially. If the output of the first receptor is delayed before reaching the second stage, but the output of the second receptor passes directly, then the output of both receptors will reach the second stage at approximately the same time. Motion will be signalled when the correlation (by multiplication) between the outputs of the two receptors is high. The signal will indicate that motion has occurred in the direction to which the detector is sensitive. When the external delay of the motion closely matches the internal delay between the two receptors, the correlation is high and the sub-unit signals motion. It can readily be seen that sampled and continuous motion will appear indistinguishable to such motion detectors under the appropriate spatial and temporal conditions. This basic principle forms the basis of several recent models of motion detection (Adelson and Bergen, 1985; van Santen and Sperling, 1985; Watson and Ahumada, 1985). In the more recent models, however, the receptors have been elaborated such that they perform a local spatio-temporal frequency (Fourier) analysis of the stimulus. Each receptor is sensitive to only a limited spatial frequency range (pass-band). The output of the receptors is again compared at a second stage. Rather than computing the correlation between luminance points which have been displaced, these models analyse temporal variations in spatial frequency energy within a limited pass-band of spatial frequencies. The responses of a number of such units responsive to a broad range of different pass-bands can be combined to signal the direction of movement of objects with a broad spatial frequency spectrum.

4.2.3 Perceptual consequences of temporal aliasing

Motion reversals

The appearance of motion reversals may readily be explained by either motion energy detection or feature matching systems. Consider the 'wagon wheel illusion', where the

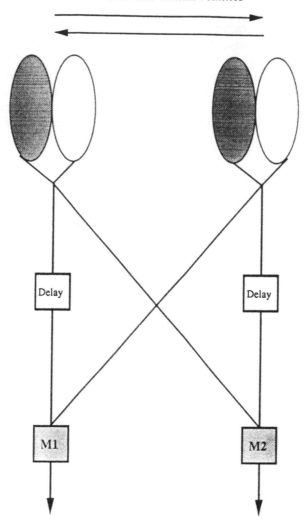

Figure 4.2 The basic scheme of a Reichardt motion detector. Signals from two receptive fields are multiplied at units M1 and M2. The signal from one unit is delayed relative to that of the other unit. If the velocity of a moving pattern is such that its time to move between the two receptive fields corresponds to the delay, the output of the multiplier will be optimal. M1 will respond optimally to rightward motion, M2 to leftward motion. Subtraction of the responses of M1 and M2 would convert the difference between the two M units' responses to a directional response.

direction of movement of the spokes of a wagon wheel change direction as the rotation speed of the wheel increases. In a sampled motion sequence, each spoke in one frame must be matched to a spoke in the next frame. Between two successive frames, the spoke may be displaced by up to half the angle between the spokes so that the nearest match in the next frame is itself (displaced to its new position). But when the velocity of the wheel is such that the distance travelled between frames is greater than half the angle between spokes, then the nearest match is the spoke behind it on the wheel (displaced to its new position). The apparent reversal in the direction of movement is clearly understood in terms of motion detection by feature matching, or for energy displacements detectable by any of the motion energy detectors outlined above. The normal cinematographic rate is 24 frames per second (although each frame is presented at three times this rate for reasons which will be discussed below) and therefore it is possible to calculate the rotation speed at which this form of temporal aliasing will occur if the angle between spokes is known. It can easily be seen that reversals of motion may be avoided by increasing the sampling rate of the moving image. Increasing the update rate will permit more representations of moving targets along the trajectory and increase the proximity between successive images. The increased sampling rate means that the maximum angle between image updates may occur more over a given time period. Therefore, greater velocities may be simulated before mismatches are probable.

Jerky motion

In terms of the spatial–temporal structure of moving images, sampling preserves motion energy at the fundamental frequency of the image, but it also introduces artefactual components of motion, symmetrically placed about integral multiples of the frequency of sampling. Morgan (1979), Burr *et al.* (1986a) and Watson *et al.* (1986) argue that stroboscopic motion is discriminable from real motion when the additional temporal and spatial frequencies added by sampling are detectable. Thus, in addition to the motion energy of the target, artefactual motion energy will be present and, under certain conditions, will have sufficiently high contrast to be above threshold. Smooth motion will be perceived when the additional spatial and temporal frequencies are higher than the resolution limits of the visual system. When the sampling artefacts are visible, however, sampled motion will be betrayed and temporal aliasing may occur. The higher the sampling rate, the higher the temporal frequency of the artefacts. Since the visual system is less sensitive to high temporal frequencies, the temporal artefacts of a high sampling rate are less likely to be visible.

Burr *et al.* (1986a) demonstrated that the perception of smooth motion varies with spatial frequency, temporal frequency and contrast. At high contrast and low temporal frequencies, the artefactual components are visible and may be in apparent motion both in the same and opposite direction to the target. A comparison of the calculated contrast of the artefact of highest contrast with the temporal sensitivity function reveals that motion appears smooth as the artefacts become invisible. Thus, smooth motion is perceived when sampling artefacts are below their individual thresholds. Burr *et al.* (1986b) suggest that motion detection, including sampled motion, is mediated by oriented spatio-temporal receptive fields. In a series of experiments, spatio-temporal fields were mapped producing receptive fields that are oriented in space and time. These fields correspond to the receptive fields of the motion energy detectors discussed above (Adelson and Bergen, 1985; van Santen and Sperling, 1985; Watson and Ahumada, 1985). As long as the sampled motion sequence passes through the oriented receptive field and there are sufficient samples, the receptor will respond identically to the smooth and sampled motion. If the strobe rate is too slow, motion energy detectors of the wrong orientation and spatio-temporal energy preference will respond to spurious temporal components, producing conflicting motion signals and the appearance of jerky motion.

Watson *et al.* (1986) observe that smooth and sampled moving stimuli will be subjected to spatio-temporal filtering by the visual system. The spatial and temporal limits of the visual system are relatively independent (Robson, 1966; Koenderink and van Doorn, 1979). Watson *et al.* argue that this arrangement of spatial and temporal filters results in a rectangular 'window of visibility' in terms of spatial and temporal frequency sensitivity. Spatial and temporal frequencies which fall outside the window are invisible. Thus, if the energy spectrum of smooth and sampled motion after this filtering is identical, then the two will be indiscriminable. Sampling may result in a temporally square wave image but the neural representation of a sampled moving image may be indistinguishable from that of a smoothly moving one after the spatio-temporal filtering.

The above analysis reveals that temporal aliasing may be understood in terms of the spatio-temporal structure of sampled motion sequences and the spatio-temporal resolution limits of the human visual system. Again, a high sampling frequency reduces the probability of temporal aliasing. In addition to a high sampling frequency, the visibility of sampling artefacts may be reduced by lowering the contrast of an image and this might be a more appropriate anti-aliasing technique in certain conditions. Contrast, however, has been shown to have no effect on multiple imaging (Bex *et al.*, 1993) and there is evidence that increasing the update rate may result in more multiple imaging (Allport, 1970; Farrell, 1984).

Multiple imaging

Investigations involving the appearance of multiple images have concentrated on the time course of the persistent images once they appear. It has been assumed that the appearance of multiple images is related to the duration of visual persistence. Visual persistence has been estimated by a large number of different methods in addition to multiple imaging. Estimates of the duration of visual persistence are extremely varied: from as little as five milliseconds (Burr, 1980) to several seconds (Sakitt, 1976) (for a review, see for example Nisly and Wasserman, 1989). This approach indicates that a reduction in the appearance of multiple images might be achieved by employing conditions which are related to shorter measures of visual persistence. The conditions associated with lower visual persistence, which might be employed to eliminate multiple imaging, are in direct contradiction to those prescribed above for the reduction of motion reversals and jerky motion. Multiple imaging has been associated with high temporal frequency. Allport (1970) measured visual persistence using a radius painted on a rotating disk. When viewed directly the radius was seen as a single rotating line. But when viewed illuminated by a stroboscope at low flash rates (approximately 10 Hz) a single line was seen adopting a rapid succession of radial positions around the disk. At high strobe rates it appeared as a sharply defined fan of radii rotating as a group and the perceived number of radii increased with flash rate. Farrell (1984) replicated these findings using a rotating bar displayed on a CRO with presentation rate adjusted by varying the image update rate. Subjects were required to report the number of lines simultaneously visible. She found that the perceived number of rotating lines increased linearly with the rate of stimulus presentation with a slope that was proportional to the spatial separation between the successive radii. The linear velocity varied along the length of a rotating radius. Farrell *et al.* (1990) avoided this problem by measuring persistence of bars in linear apparent motion. Stimulus lines were presented in successive positions moving in opposite directions above and below a fixation point. Subjects were required to indicate whether or not they perceived all lines simultaneously, thus seeing a grating (i.e. the final presentation occurred while the first and subsequent presentations persisted visually). Visual persistence was found to increase with spatial separation up to 0·24 degrees of visual angle. The appearance of multiple targets in these experiments is analogous to the multiple imaging which may occur in virtual environments. The results of these experiments suggest that multiple imaging will occur at high update rates and the visual persistence of the spurious images will last longer if large spatial displacements occur between image updates.

Farrell *et al.* (1990) suggested that multiple imaging may involve some form of gradual neural response decay (see Figure 4.3). Thus repeated presentation of a time-sampled

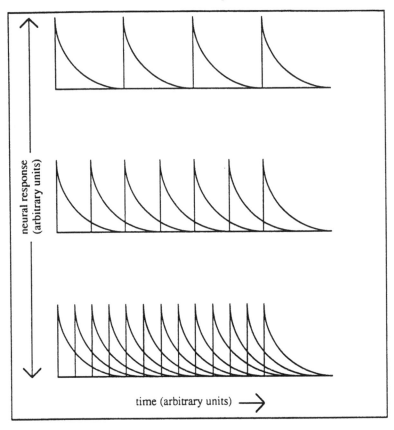

Figure 4.3 Model of gradual response decay as the determinant of multiple imaging (after Farrell et al., 1990). This figure illustrates the hypothesis that multiple imaging is caused by the gradual decay of neural representations of sampled moving targets. Thus, multiple images appear when successive presentations of a target occur before former presentations have fully decayed. In the upper panel, multiple images are not seen because each image appears after previous images have fully decayed. In the middle and lower panels multiple images appear and the number of visible images increases with update rate (Allport, 1970).

moving stimulus would result in multiple images if the second presentation occurred before the first had fully decayed. They suggested that the time course of visual persistence may be affected by a simple gain mechanism for modulating the duration of visual persistence as a function of the distance separating stimuli. Thus, the presence of adjacent stimuli reduces the time course of the neural response and the effect is greatest for the closest stimuli. Di Lollo and Hogben (1985, 1987) and Hogben and Di Lollo (1985) propose that the data are readily explained by a system of local inhibition (Breitmeyer and Ganz, 1976). Thus, a response to a stimulus inhibits those of nearby receptors and the strength of inhibition decreases with separation.

The underlying assumption of the above investigations is that two (or more) images will be seen when the neural representations of a single image presented in successive locations exist simultaneously. Usually, stimulus lines displaced between updates on a CRO appear to be in apparent motion and the larger the spatial displacement, the higher the velocity. It is surprising, given the relatively sluggish temporal response of the human visual system (Graham and Margaria, 1935; Barlow, 1958), that when an image is rapidly displaced, multiple images are not seen under most circumstances. Normally a single image is seen in apparent motion, which indicates active suppression of the multiple images that would otherwise be perceived, in accordance with local inhibition theories. The precise conditions under which multiple images may or may not appear have been explored by Bex *et al.* (1993, 1995). These researchers have shown that multiple images first appear when a constant spatial displacement is exceeded between image updates. It is argued that the value of the displacement corresponds to the spatio-temporal integration limit of energy-based motion detectors. Thus, multiple images appear when energy based motion detection fails between image updates and do not appear when motion energy is detected. This is analogous to the displacement limit (d_{max}) in random dot kinematograms at which there is a failure of motion detection. Bex *et al.* (1993, 1995) suggest that multiple neural representations of a target in apparent motion do exist because of the temporal integration of the visual system. When multiple representations of an image coexist, and if motion energy is detected between image updates, the presence of one drifting target is inferred. If motion energy is not detected, the presence of multiple targets is inferred. Bex *et al.* (1993, 1995) have shown that many of the manipulations of the spatial structure of random dot kinematograms which extend d_{max} result in an increase in the displacement between image updates (i.e. the simulation of higher velocity) before the appearance of multiple images. Thus, multiple images appear at a larger displacement for images viewed peripherally (cf. Baker and Braddick, 1985), for low-pass filtered images (cf. Chang and Julesz, 1983, 1985; Cleary and Braddick, 1990b) and band-pass filtered images (cf. Chang and Julesz, 1983, 1985; Bischof and Di Lollo, 1990, Cleary and Braddick, 1990a).

4.2.4 Anti-aliasing techniques

The association between the failure of motion energy detection and the appearance of multiple images, reported above, indicates that anti-aliasing may be achieved by stimulus manipulations which facilitate motion energy detection between image updates. There are several simple manipulations which may be used to this end. Increasing the image update rate means that the spatial displacements between image updates are smaller at a given velocity and are less likely to exceed the spatial displacement limit at which motion energy fails. This, however, may not be possible when the system is already at maximum update rate, or if the processing power is not available to compute complex images rapidly enough. Under these conditions, low-pass or band-pass filtering the images may increase the aliasing velocity. Similarly, images viewed peripherally alias at a higher velocity and this may be related to the poorer spatial resolution in the peripheral visual field (i.e. images viewed peripherally are effectively low-pass filtered). Low-pass, band-pass and peripheral images, however, appear blurred and the appearance of multiple images may be preferable to single, blurred targets. The particular anti-aliasing technique must, therefore, depend upon the requirements of the display user.

Analysis of sampled motion stimuli in the spatio-temporal domain and of the way in which the human visual system may extract motion signals from sequences of static images have demonstrated how several anti-aliasing techniques may operate. It can be seen that each of the anti-aliasing techniques operates by assisting the detection of motion energy at the appropriate spatial scale and reduces the visibility of artefactual motion energy signals. Thus, an elimination of some problems of temporal aliasing may be achieved by increasing image update rate, which reduces the spatial displacement between image updates and raises the temporal frequency of sampling artefacts above perceptual threshold. Anti-aliasing may also be achieved by blurring, which low-pass filters an image, allowing motion detection by motion energy detectors tuned to larger spatial scales and removing some of the high frequency sampling artefacts. Blurring is also achieved optically by viewing targets peripherally and this also assists anti-aliasing.

It can be seen that cinematographic films have a certain amount of anti-aliasing incorporated in them. Because of the low shutter speed, motion which occurs while the shutter is open will appear blurred and the faster the motion, the greater the blur (Burr, 1980). This means that faster moving images will be blurred more (i.e. a lower cut-off of the low-pass filter is simulated). Fortunately, additional blurring is required to eliminate temporal aliasing for more rapidly moving objects. Additionally, there is evidence that the blur which is present in the static stills of the motion sequence will not be detected by human observers. Although each image presented in isolation appears blurred, when presented as part of a motion sequence, the moving image may be seen in sharp focus (Ramachandran *et al.*, 1974). Consequently, the low-pass filtering which will eliminate certain temporal aliasing phenomena may not be visible to the observer. The temporal frequency at which cinematographic films are presented (24 Hz: a new frame every 42 ms) is, unfortunately, within the temporal resolution limits of the human visual system and flicker will be detected at this rate. This was a problem in the early film industry. It has been overcome, however, by presenting each image three times within the 42 ms frame time before advancing to the next frame. The resultant temporal frequency is 72 Hz, which is above the temporal resolution of foveal viewing.

Thus it can be seen that the factors that lead to temporal aliasing have been extensively studied, and there are several methods which may be used to reduce the salience of temporal aliasing. The next issue to be considered in some detail is likely to prove more intractable. The visual accommodation (focusing) response of the eye is governed by a complex interaction of factors and, indeed, may be influenced by many of the display characteristics mentioned by Christou and Parker in section 3.5, such as luminance, spatial and temporal resolution and field of view. Before considering the effect that a VR system may have on the accommodation response it is appropriate to understand some of the factors that influence the level of accommodation, and some of these are described below.

4.3 The accommodation response and virtual reality displays

4.3.1 The physiology of accommodation

The accommodation response refers to the change in the power of the lens in the eye. It was originally suggested that this was achieved by displacing the entire lens forward or backward (Kepler, 1611), as is the case in some fish. The actual mechanism, however, appears to be rather more elegant. The lens of the human eye is actually elastic and is held under tension by the ciliary muscles. Thus if the ciliary muscles contract, the lens is stretched and its focusing power drops. Conversely, if the ciliary muscles relax, the lens becomes thicker, and its focusing power increases. It is now generally accepted that the ciliary muscles are controlled by the action of the autonomic nervous system (ANS: see section 7.3.1). But there has been some debate about whether it is the sympathetic or parasympathetic branch of the ANS that innervates the ciliary muscle. Recent experiments suggest that in fact there is dual innervation of the ciliary muscles, with inputs from both the parasympathetic and sympathetic systems (see Gilmartin (1986) for a review). This dual innervation implies that the level of accommodation may be affected not only by physical factors (image blur, etc.), but also by psychological factors. Some of the physical and psychological factors that may affect the accommodation response are discussed briefly below.

4.3.2 Reflex accommodation

Reflex accommodation occurs as a result of some change in a particular property of the image being viewed. Several image properties have been implicated in the stimulation of reflex accommodation. Two which may be of relevance to the use of VR displays, stimulus blur and chromatic aberration, are of particular interest.

The strongest stimulus to reflex accommodation is almost certainly blur. If the eye is inappropriately accommodated to a particular object then the object will appear blurred. The eye will attempt to make the image as sharp as possible by changing its accommodative state (Fincham, 1951).

Another factor that has been considered important in the control of the accommodation response is chromatic aberration. Chromatic aberration refers to the fact that a simple lens system will tend to focus different wavelengths of light at different points. Thus, if a simple lens system (such as the human eye) is used to focus a point of white light, the point will tend to show coloured 'fringes' around the edges, which are due to chromatic aberration. Early work by Fincham (1951) and more recent work by, for example, Kruger and Pola (1986) suggest that chromatic aberration of the image may play an important role in the accommodation response, and that some individuals are more dependent on cues obtained from chromatic aberration than others.

4.3.3 Tonic accommodation

Tonic accommodation refers to the position at which the eye is focused when 'at rest'. It has traditionally been assumed that, in the absence of any visual stimulation (e.g. in darkness or when viewing a bright empty field), the eye will 'relax' and will be focused at infinity. There have been, however, a number of published reports of people becoming short-sighted in darkness: a phenomenon now referred to as 'night myopia' (see Knoll (1952) for an early review of this effect). A similar effect can occur when a person views a bright, unstructured field: this phenomenon is referred to as 'empty-field myopia'.

Following on from these early reports a large number of studies (e.g. Campbell, 1953; Fincham, 1951; Westheimer, 1957; Owens, 1984) have demonstrated that in the absence of visual stimulation, the focus of the eye usually lapses to a distance of between about 0·5 and 2·0 m (0·5–2·0 dioptres). This distance varies from person to person and is often referred to as the 'dark focus', the 'resting point of accommodation', or 'tonic accommodation'.

Another phenomenon of some concern is the Mandelbaum effect (Mandelbaum, 1960). Mandelbaum reported that when a scene was viewed through a wire-mesh screen, people often tended to focus on the screen rather than the scene beyond it, no matter how hard they tried to keep the scene in focus. Later studies (e.g. Owens, 1979) have found that the effect is greatest if an obstruction is placed at, or near, an individual's dark focus. Thus any object placed at, or near, an individual's dark focus may affect the level of accommodation.

4.3.4 Convergence accommodation

If a person is visually fixating an object situated at a particular distance and then changes fixation to another object at a different distance (or if the object being fixated moves closer or further away) not only will the person's level of accommodation change, but the angle of convergence of the two eyes will also alter (e.g. if the object moves closer the eyes will be angled inwards to a greater extent to align the images of the object in each eye). The convergence response, in a similar way to the accommodation response, is postulated to have several components. Maddox (1893) postulated that there are four components to the convergence response, similar to those proposed for accommodation. These are proximal convergence (convergence guided by the known position of an object), tonic convergence (the resting state of convergence), accommodative convergence (convergence linked to the accommodation response) and fusional or reflex convergence (convergence to align corresponding points on the retinae of each eye).

Given that accommodation seems to play a role in the convergence response and vice versa it seems likely that the two responses are linked. There is a large amount of evidence to suggest that this is the case, although the exact form of the relationship between the two is complex and beyond the scope of this chapter.

4.3.5 Proximal or psychic accommodation

Proximal accommodation refers to the effect that a person's knowledge of the proximity of an object will have upon accommodation. In relation to VR displays this is potentially the most important component of the accommodation response. Several studies (e.g. Malmstrom and Randle, 1976; Rosenfield and Cuiffreda, 1990; Jaschinski-Kruza, 1991) have demonstrated that knowledge of nearness of an object or even 'thinking near' or 'thinking far' can effect a person's level of accommodation; other studies have found that there is little or no proximal effect (e.g. Hennessy *et al.*, 1976).

Our understanding of convergence accommodation indicates that, in stereoscopic displays, a conflict could arise between the actual focusing distance (to the real distance of the image) and the perceived focusing distance (determined by the apparent position of objects in the display). Given that psychological factors can influence the accommodation response, the perceived distance of objects and the cues from stimulus blur will tend to conflict.

4.3.6 Mental effort

Some studies have suggested that mental effort or 'cognitive demand' may affect the level of accommodation (Jaschinski-Kruza and Toenies, 1988; Winn *et al.*, 1991; Iwasaki, 1993; Edgar *et al.*, 1994a, 1994b). The role of this factor in VR displays will be considered below.

4.3.7 Factors affecting the accommodation response in VR displays

It is now appropriate to consider the consequences of these factors when using a VR display. How well VR systems can mimic reality will depend, to some extent, on how well they are able provide reliable cues to accommodation.

Reflex accommodation

It is probable that the image in a VR display will not be as sharp as the equivalent 'real world' image, as a result of the intervening lens system, the poor spatial resolution of the display, or chromatic aberrations introduced by the lens system used. The generally poor quality of the optics used in VR displays may mean that the usefulness of chromatic aberration as a cue is reduced. Indeed, the optics usually introduce further chromatic aberration into the image, with different aberrations according to the characteristics of the lens.

The blur and chromatic aberration cues to accommodation may thus differ in a virtual reality system as compared to the real world.

Tonic accommodation

One phenomenon already known, and which is of obvious relevance to VR displays is that usually referred to as 'instrument myopia' (for reviews see Hennessy, 1975; Wesner and Miller, 1986). Instrument myopia describes the effect that a number of people, when using optical instruments, tend to over-accommodate. They will tend to adjust the focus of the instrument as though they were myopic (short-sighted). Instrument myopia is certainly not a recently discovered phenomenon; the astronomer Maskelyne, working about 200 years ago, found that he was better able to make night observations using an additional negative lens, indicating that he was becoming rather more short-sighted when using the telescope (Levene, 1965).

There are many factors which can contribute to instrument myopia, as with all accommodation phenomena. One possibility is that the extent to which individuals suffer from instrument myopia is related to their tonic level of accommodation. As discussed above, if the stimulus viewed is impoverished and, for instance, does not provide sufficient blur cues to allow correct accommodation, then the level of accommodation will tend to lapse towards an individual's tonic position. Thus, the poor spatial resolution of most VR displays may contribute to a form of instrument myopia. Most VR displays have limited resolution and, in some instances, the problem of visible pixellation is reduced by placing a diffuser in front of the displays which, of course, has the effect of blurring the image slightly. This makes the pixellation less salient but it also provides a poorer stimulus to accommodation, and it is possible that even a slight blurring could affect the accommodation response. Any lapse towards the dark focus is not an all-or-nothing response: some studies (e.g. Johnson, 1976) have shown shifts in accommodation towards the tonic position even when looking at quite well-structured images.

In most current VR displays, the image distance is typically less than 50 cm from the eye. In some ways, this should help to keep the image in sharp focus as, for some people, 50 cm may be quite close to the dark focus. Although this might have the effect of making it easier to keep the displays in focus, there will still be a mismatch between the actual level of accommodation and the level of accommodation approriate to the perceived distance of some objects in the display.

Luminance level also affects accommodation by causing it to lapse towards the resting position as the luminance drops (Alpern and David, 1958). Previous studies of instrument myopia have suggested that there may be an interaction between luminance and instrument myopia (see Wesner and Miller, 1986). Most immersive VR displays are binocular and this might help to reduce any instrument myopia. Some studies looking at microscope usage (e.g. Richards *et al.*, 1981; Schober *et al.*, 1970) found that there was less evidence of instrument myopia when subjects used a binocular microscope than when they used a monocular one. Unfortunately, there is usually very little adjustment of the lens-display configuration in VR displays and therefore the convergence response required to fuse the two images may be different from the accommodation response required to focus them (see below).

Convergence accommodation

Another factor that could affect the accommodation response in a VR display is the convergence response of the users' eyes. As discussed above, the accommodation and convergence responses are linked; if one changes then the other will tend to change also. In a VR system, the convergence and accommodation responses may be placed in conflict when objects in the virtual world are presented at different depths in stereoscopic images. Looking from one object to another, the convergence required to fixate each of the objects will vary – but the accommodation requirement will remain the same. Not only will this make the VR image different from the 'real world'; the conflict between the two systems may make the system uncomfortable to use for any length of time. Indeed a study by Mon-Williams *et al.* (1993) has suggested that this is the case, and Neary *et al.* (1994) have found that convergence can be less accurate with a conflicting focal distance. This visual conflict is probably less important for older people using VR, as the accommodation capability of the eye diminishes with age (Donders, 1864; Weale, 1989).

There is also some possibility that the inappropriate accommodation and/or inappropriate convergence will lead to misperceptions of distance. This suggestion is certainly not a new one. The effect of oculomotor (eye accommodation and convergence) responses on the perception of distance has been the subject of some debate for over three hundred years. Descartes (1677) postulated that oculomotor responses may provide cues to the distance of objects and Berkeley (1709) proposed a theory that oculomotor responses and perceived distance may become associated through experience. An early attempt at an experimental investigation of this problem was that of Hillebrand (1894). Since then, numerous investigators have found conflicting results when studying the role of the oculomotor response in the determination of perceived distance (e.g. Ittleson and Ames, 1950; Wallach and Norris, 1963; Heinemann and Nachmias, 1965; Holst, 1969). Whether or not oculomotor responses directly affect distance perception, they may degrade other cues to distance. For instance, if the image is more blurred (as a result of misaccommodation) then there will be a loss of contrast and texture: both important cues to distance. These cues tend to be fairly poor in VR systems at the moment, and effects such as those described above might well make things worse.

Proximal accommodation

Psychological factors are also likely to play a large part in the accommodation response when using a VR display. As discussed above, there may be a proximal accommodation response, driven by the knowledge of nearness of an object. No matter what optical distance the displays are imaged at, the user is fully aware that the actual displays may only be a few centimetres away from his or her eyes. Furthermore, the edge of the displays may well be visible (due to the limited field-of-view of most VR displays) as are the supports and casing of the VR headset. What effect these might have is debatable and will depend, to some extent, on how visible the surround is. Hennessy (1975) looked at the effects of various surrounds on the accommodation response and found that, although a chequerboard surround had an effect on accommodation, a dark surround did not.

Mental effort

Mental effort is perhaps the most difficult factor to even evaluate, let alone eliminate. A number of studies have found that mental effort affects the tonic level of accommodation (e.g. Post *et al.*, 1985; Bullimore and Gilmartin, 1987), or even the level of accommodation when looking at a 'real world' image (Iwasaki, 1993; Edgar *et al.*, 1994a, 1994b). The question that remains, of course, is whether this effect is likely to be more pronounced with a VR image than in the visual real world. If these two were indistinguishable, mental effort might affect the level of accommodation in a VR display—but the effect would be the same as in the real world and so would not affect the 'realism' of the VR system.

Some display characteristics could have indirect effects on accommodation. For instance, a lag between a head movement and the updating of the visual image may increase the amount of mental effort required to interact with the system in any kind of normal fashion, and thus have a secondary effect on accommodation. The same might be said of any differences between a VR system and the 'real world'. Potentially, any difference between a VR display and the 'real world' might increase the workload inherent in using the system, with a consequent undesirable effect on accommodation. The inappropriate change in accommodation will then also tend to increase the differences between the real world and virtual one.

Training

It may be possible to 'train' the accommodation system to try and reduce some of the problems mentioned above. There is some evidence that the accommodation response can be trained (Cornsweet and Crane, 1973; Trachtman, 1978; Randle and Malmstrom, 1982) and, indeed, there is evidence that instrument myopia decreases with experience (Wald and Griffin, 1947; Baker, 1966; Schober *et al.*, 1970). This abnormal 'trained' response would, in itself, distinguish interacting with a virtual world from interacting with the real one, making the VR experience less realistic, but possibly reducing some perceptual problems.

4.4 Conclusions

It has been shown that creating a perceptually 'realistic' virtual world is rather difficult. There are many differences between the images provided by a VR display and the real world. It is impossible to consider the factors in isolation; most (if not all) of the factors may interact and some factors that have little direct effect on the usability of the system could have quite strong indirect effects. The factors considered in detail in this chapter (temporal aliasing and accommodation) provide a good illustration of the difficulty of eliminating all differences. For instance, one way of reducing the saliency of temporal aliasing is to blur the image, but this then degrades the stimulus to accommodation. Reducing the effect of one problem can thus lead to aggravation of others, and a new set of difficulties! If the goal of VR is to create a perceptual experience that is indistinguishable from a real experience, then all the perceptual differences between the virtual world and the real world need to be eliminated. The aim may not be to make the virtual world itself indistinguishable from reality, as it might be useful to have a virtual world that does not contain all the features of the real one. It will still be important, however, that the virtual world is as easy to interact with as the real one, and some of the problems discussed above may inhibit this.

References

Adelson, E. H. and Bergen J. R., 1985, Spatiotemporal energy models for the perception of motion, *Journal of the Optical Society of America A*, **2**, 284–99.

Allport, D. A., 1970, Temporal summation and phenomenal simultaneity: experiments with the radius display, *Quarterly Journal of Experimental Psychology*, **22**, 686–701.

Alpern, M. and David, H., 1958, Effects of illuminance quantity on accommodation of the eyes, *Industrial Medicine and Surgery*, **27**, 551–5.

Amanatides, J. and Mitchell, D. P., 1990, Anti-aliasing of interlaced video animation, *Computer Graphics*, **24**, 77–85.

Baker, C. L. and Braddick, O. J., 1985, Eccentricity dependent scaling of the limits for short-range apparent motion, *Vision Research*, **25**, 803–12.

Baker, J. R., 1966, Experiments on the function of the eye in light microscopy, *Journal of the Royal Microscopical Society*, **85**, 231–54.

Barlow, H. B., 1958, Temporal and spatial summation in human vision at different background intensities, *Journal of Physiology, London*, **141**, 337–50.

Berkeley, G., 1709, An essay towards a new theory of vision, in Ayres, M. R. (Ed.) *Berkeley: Philosophical Works*, 4th Edn, 1837, London: Dent.

Bex, P. J., Edgar, G. K. and Smith, A. T., 1993, Temporal aliasing: investigating multiple imaging, *Ophthalmic and Physiological Optics*, **13**, 434.

Bex, P. J., Edgar, G. K. and Smith, A. T., 1994, Multiple images appear when motion energy detection fails, *Journal of Experimental Psychology: Human Perception and Performance*, **21**, 231–8.

Bischof, W. F. and Di Lollo, V., 1990, Perception of directional sampled motion in relation to displacement and spatial frequency: evidence for a unitary motion system, *Vision Research*, **30**, 1341–362.

Breitmeyer, B. G. and Ganz, L., 1976, Implications of sustained and transient models for theories of pattern masking, saccadic suppression and information processing, *Psychological Review*, **83**, 1–36.

Bullimore, M. A. and Gilmartin, B., 1987, Tonic accommodation, cognitive demand and ciliary muscle innervation, *American Journal of Optometry and Physiology*, **64**, 45–50.

Burr, D. C., 1980, Motion smear, *Nature*, **284**, 164–5.

Burr, D. C., Ross, J. and Morrone, M. C., 1986a, Smooth and sampled motion, *Vision Research*, **26**, 643–52.

Burr, D. C., Ross, J. and Morrone, M. C., 1986b, Seeing objects in motion, *Proceedings of the Royal Society of London*, **B227**, 249–65.

Campbell, F. W., 1953, Twilight myopia, *Journal of the Optical Society of America*, **43**, 925–6.

Chang, J. J. and Julesz, B., 1983, Displacement limits for spatial frequency filtered random dot kinematograms in apparent motion, *Vision Research*, **23**, 1379–85.

Chang, J. J. and Julesz, B., 1985, Cooperative and non-cooperative processes of apparent movement of random dot cinematograms, *Spatial Vision*, **1**, 39–45.

Cleary, R. and Braddick, O. J., 1990a, Direction discrimination for band-pass filtered random dot kinematograms, *Vision Research*, **30**, 303–16.

Cleary, R. and Braddick, O. J., 1990b, Masking of low frequency information in short range apparent motion, *Vision Research*, **30**, 317–27.

Cornsweet, T. N. and Crane, H. D., 1973, Training the visual accommodation system, *Vision Research*, **13**, 713–715.

Descartes, R., 1677, *Treatise of Man*, translated by Hall, T. S., 1972, Cambridge, MA: Harvard University Press.

Di Lollo, V. and Hogben, J., 1985, Suppression of visible persistence, *Journal of Experimental Psychology: Human Perception and Performance*, **11**, 304–16.

Di Lollo, V. and Hogben, J., 1987, Suppression of visible persistence as a function of spatial separation between inducing stimuli, *Perception and Psychophysics*, **41**, 345–54.

Dippe, M. A. Z. and Wold, E. R., 1985, Anti-aliasing through stochastic sampling, *Computer Graphics*, **19**, 69–78.

Donders, F. C., 1864, *On the Anomalies of Accommodation and Refraction of the Eye*, London: The New Sydenham Society.

Edgar, G. K., Reeves, C. A., Craig, I. and Pope, J. C. D., 1994a, The effect of mental processing on the closed-loop accommodation response, *Investigative Ophthalmology and Visual Science*, **35**, 1280.

Edgar, G. K., Pope, J. C. D. and Craig, I., 1994b, Visual accommodation problems with head-up and helmet-mounted displays?, *Displays*, **65**, 68–75.

Fahle, M. and Poggio, T., 1981, Visual hyperacuity: spatio-temporal integration in human vision, *Proceedings of the Royal Society of London*, **B213**, 451–77.

Farrell, J. E., 1984, Visible persistence of moving objects, *Journal of Experimental Psychology: Human Perception and Performance*, **10**, 502–11.

Farrell, J. E., Pavel, M., and Sperling, G., 1990, The visible persistence of stimuli in stroboscopic motion, *Vision Research*, **30**, 921–36.

Fincham, E. F., 1951, The accommodation reflex and its stimulus, *British Journal of Ophthalmology*, **35**, 381–93.

Gilmartin, G., 1986, A review of the role of sympathetic innervation of the ciliary muscle in ocular accommodation, *Ophthalmic and Physiological Optics*, **6**, 23–37.

Graham, C. H. and Margaria, R., 1935, Area and intensity-time relation in the peripheral retina, *American Journal of Physiology*, **113**, 299–305.

Grant, C. W., 1985, Integrated analytic spatial and temporal anti-aliasing for polyhedra in 4-space, *Computer Graphics*, **19**, 79–84.

Heinemann, E. G. and Nachmias, J., 1965, Accommodation as a cue to distance, *American Journal of Psychology*, **78**, 139–42.

Hennessy, R. T., 1975, Instrument myopia, *Journal of the Optical Society of America*, **65**, 1114–20.

Hennessy, R. T., Iida, T., Shina, K. and Leibowitz, H. W., 1976, The effect of pupil size on accommodation, *Vision Research*, **16**, 587–9.

Hillebrand, F., 1894, Das verhältnis von accommodation und konvergenz zur tiefenlokalisation, *Zeitschrift für Psychologie*, **7**, 97–151.

Hogben, J. H. and Di Lollo, V., 1985, Suppression of visible persistence in apparent motion, *Perception and Psychophysics*, **38**, 450–60.

Holst, E. von, 1969, The participation of convergence and accommodation in perceived size constancy, in *The Behavioural Physiology of Animal and Man* (translated by R. Martin), pp. 185–7, London: Butler & Tanner.

Ittleson, W. H. and Ames, A., 1950, Accommodation, convergence and their relation to apparent distance, *Journal of Psychology*, **30**, 43–62.

Iwasaki, T., 1993, Effects of a visual task with cognitive demand on dynamic and steady-state accommodation, *Ophthalmological and Physiological Optics*, **13**, 285–90.

Jaschinski-Kruza, W., 1991, On proximal effects in objective and subjective testing of dark accommodation, *Ophthalmological and Physiological Optics*, **11**, 328–34.

Jaschinski-Kruza, W. and Toenies, U., 1988, Effect of mental arithmetic on dark focus of accommodation, *Ophthalmological and Physiological Optics*, **8**, 432–7.

Johansson, G., 1975, Visual motion perception, *Scientific American*, **232**, 76–88.

Johnson, C. A., 1976, Effects of luminance and stimulus distance on accommodation and visual resolution, *Journal of the Optical Society of America*, **66**, 138–42.

Kepler, J., 1611, *Dioptrice*, Propos. 26, Augsburg.

Knoll, H. A., 1952, A brief history of nocturnal myopia, *American Journal of Optometry*, **29**, 69–81.

Koenderink, J. J. and van Doorn, A. J.,1979, Spatio-temporal contrast detection threshold is bimodal, *Optics Letters*, **4**, 32–4.

Kolers, P.A., 1972, *Aspects of Motion Perception*, Oxford: Pergamon Press.

Korein, J. and Badler, N., 1983, Temporal anti-aliasing in computer generated animation, *Computer Graphics*, **17**, 377–88.

Kruger, P. B. and Pola, J., 1986, Stimuli for accommodation: blur, chromatic aberration and size, *Vision Research*, **26**, 957–71.

Levene, J. R., 1965, Neil Maskelyne, F. R. S. and the discovery of night myopia, *Royal Society of London Notes and Reports*, **20**, 100–8.

Maddox, E. E., 1893, *The Clinical Use of Prisms*, 2nd Edn, Bristol: John Wright.

Malmstrom, F. V. and Randle, R. J., 1976, Effects of visual imagery on the accommodation response, *Perception and Psychophysics*, **19**, 450–3.

Mandelbaum, J., 1960, An accommodation phenomenon, *Archives of Ophthalmology*, **63**, 923–6.

Max, N. L. and Lerner, D. M., 1985, A two and a half-D motion-blur algorithm, *Computer Graphics*, **19**, 85–93.

Mitchell, D. P. and Netravali, A. N., 1988, Reconstruction filters in computer graphics, *Computer Graphics*, **22**, 221–8.

Mon-Williams, M., Wann, J. P. and Rushton, S., 1993, Binocular vision in a virtual world: visual deficits following the wearing of a head-mounted display, *Opthalmic and Physiological Optics*, **13**, 387–391.

Morgan, M. J., 1979, Perception of continuity in stroboscopic motion: a temporal frequency analysis, *Vision Research*, **19**, 491–500.

Neary, C., Fulford, K., Cook, M. and Williams, M., 1994, The effect of visuo-perceptual context on eye movements, *Investigative Ophthalmology and Visual Science*, **35**, 2032.

Nisly, S. J. and Wasserman, G. S., 1989, Intensity dependence of perceived duration, *Psychological Bulletin*, **106**, 483–96.

Norton, A., Rockwood, A. and Skolmoski, P., 1982, Clamping: a method of anti-aliasing textured surfaces by bandwidth limiting in object space, *Computer Graphics*, **16**, 1–8.

Owens, D. A., 1979, The Mandelbaum effect: evidence for an accommodative bias towards intermediate viewing distances, *Journal of the Optical Society of America*, **69**, 646–52.

Owens, D. A., 1984, The resting state of the eyes, *American Scientist*, **72**, 378–87.

Post, P. B., Johnson, C. A. and Owens, D. A., 1985, Does performance of tasks affect the resting focus of accommodation?, *American Journal of Optometry and Physiological Optics*, **62**, 533–7.

Potmesil, M. and Chakravarty, I., 1981, A lens and aperture camera model for synthetic image generation, *Computer Graphics*, **15**, 297–305.

Potmesil, M. and Chakravarty, I., 1983, Modelling motion blur in computer generated images, *Computer Graphics*, **17**, 389–99.

Ramachandran, V. S., Madhusudhan, V. and Vidyasagar, T. R., 1974, Sharpness constancy during movement perception, *Perception*, **3**, 97–8.

Randle, R. J. and Malmstrom, F. V., 1982, Visual field narrowing by non-visual factors, *Flying Safety*, **38**, 14–8.

Reichardt, W., 1957, Autokorrelationsauswertung als Funktionprinzip des Zentralnervensystems, *Zeitschrift für Naturforchung*, **12b**, 447–57.

Richards, O. W., Mathews, S. M. and Shaffer, S. M., 1981, Focused apparent image position with convergent, parallel binocular and monocular microscopes, *Journal of Microscopy*, **122**, 187–91.

Robson, J. G., 1966, Spatial and temporal contrast sensitivity function of the visual system, *Journal of the Optical Society of America*, **56**, 1141–2.

Rosenfield, M. and Cuiffreda, K. J., 1990, Proximal and cognitively induced accommodation, *Ophthalmological and Physiological Optics*, **10**, 252–6.

Sakitt, B., 1976, Iconic memory, *Psychological Review*, **83**, 257–76.

van Santen, J. P. H. and Sperling, G., 1984, A temporal covariance model of motion perception, *Journal of the Optical Society of America A*, **1**, 451–73.

van Santen, J. P. H. and Sperling, G., 1985, Elaborated Reichardt detectors, *Journal of the Optical Society of America A*, **2**, 300–21.

Schober, H., Dehler, H., and Kassel, R., 1970, Accommodation during observations with optical instruments, *Journal of the Optical Society of America*, **60**, 103–7.

Sperling, G., 1976, Movement perception in computer driven displays, *Behaviour Research Methods and Instrumentation*, **8**, 144–51.

Trachtman, J. N., 1978, Biofeedback of accommodation to reduce a functional myopia: a case report, *American Journal of Physiological Optics*, **55**, 400–6.

Ullman, S., 1979, *The Interpretation of Visual Motion*, Cambridge: MIT Press.

Wald, G. and Griffin, D. R., 1947, The change in the refractive power of the human eye in dim and bright light, *Journal of the Optical Society of America*, **37**, 321–36.

Wallach, H. and Norris, C. M., 1963, Accommodation as a distance cue, *Journal of Psychology*, **76**, 659–64.

Watson, A. B. and Ahumada, A. J., 1985, Model of human visual-motion sensing, *Journal of the Optical Society of America A*, **2**, 322–42.

Watson, A. B., Ahumada, A. J. and Farrell, J., 1986, Window of visibility: psychophysical theory of fidelity in time-sampled visual motion displays, *Journal of the Optical Society of America A*, **3**, 300–7.

Weale, R., 1989, Presbyopia towards the end of the 20th century, *Survey of Ophthalmology*, **34**, 15–30.

Wesner, M. F. and Miller, R. J., 1986, Instrument myopia conceptions, misconceptions and influencing factors, *Documenta Ophthalmologica*, **62**, 281–308.

Westheimer, G., 1957, Accommodation measurements in empty visual field, *Journal of the Optical Society of America*, **47**, 714–8.

Winn, B., Gilmartin, B., Mortimer, L. C. and Edwards, N. R., 1991, The effect of mental effort on open- and closed-loop accommodation, *Ophthalmic and Physiological Optics*, **11**, 335–9.

5

Head-coupled virtual environment with display lag

Richard H. Y. So and Michael J. Griffin

5.1 Introduction

5.1.1 Head-coupled VR systems and their benefits

A typical virtual reality (VR) system that can present head-slaved images will contain: (i) a helmet-mounted display, (ii) a head position sensor, and (iii) an image generator. The image generator will consist of a computer, or a head-slaved camera, or both a computer and a head-slaved camera (Figure 5.1). Such a system will be referred to as a 'head-coupled VR system' and the entire visual field will be called 'head-coupled virtual environment'.

Many benefits of such systems have already been mentioned in Chapter 2. For example, the ability to change the orientation of computer-generated images according to head position enables three-dimensional medical imagery to be viewed. Video captured by a head-slaved camera mounted on a remote controlled vehicle may provide the 'telepresence experience' (see section 2.2.4). In these applications, images may be presented in three-dimensional perspective and have constant coordinates with respect to the Earth. These images will be called 'space stationary images'.

Most images within a head-coupled environment will be space stationary. Two-dimensional text and symbols, however, can be projected along the operator's head-pointing angle. This enables information to be presented to an operator regardless of head orientation. Examples of such information include control parameters with 'telerobotics' applications and on-line instructions. In the field of aviation, pilots will be able to access flight control information through a head-coupled VR system without referring to conventional instrument panels (Gibson and Furness, 1987). This could be of great benefit during target searching or dog-fighting when the time during which the pilots take their eyes off targets is to be minimized.

5.1.2 Problems with head movements

5.1.2.1 Vibration problems and solutions (image blurring)

Head-coupled VR systems are subjected to movements, whether they are generated voluntarily or by external vibration. In applications where an image is presented at a fixed

Simulated and virtual realities

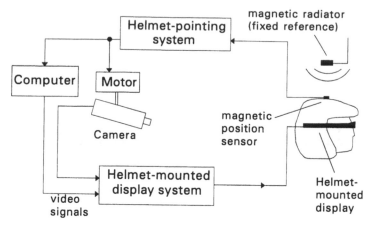

Figure 5.1 Configuration of a typical head-coupled VR system (adapted from So and Griffin (1992)).

position on a helmet-mounted display (e.g. on-line instructions), as the head moves, the image on the helmet-mounted display moves with the head. Because human eyes are space-stabilized by the 'vestibulo-ocular reflex' (see section 7.3.4), relative movements between the eyes and the moving image result in the blurring of the image (Figure 5.2). The vestibulo-ocular reflex has been shown to be active for movements from about 2–20 Hz vertically (Wells and Griffin, 1983) and at least 2–10 Hz horizontally (Benson and Barnes, 1978). The extent and causes of this vibration problem have been identified (Furness and Lewis, 1978). A solution has been developed and flight tested in a helicopter (Figure 5.3, Wells and Griffin, 1984, 1987a). The subjects read arrays of numerals presented on a helmet-mounted display during the flights. This space stabilization system utilized accelerometers to measure the movement of the head and electronically deflect the images in the opposite direction. An adaptive system was further developed to suppress the stabilization during voluntary head motions so that the images then followed the head (Lewis, 1984).

5.1.2.2 Lag problems (unwanted image oscillation)

A head-coupled VR system could utilize head position data to space-stabilize images. This, however, is not currently possible because of the lag between the moment at which the head position is sampled and the moment at which the image is displayed. The result is that in a head-coupled virtual environment, a space stationary image appears to 'swim' during the start and finish of a head motion (Allen and Hebb, 1983). This can be very disturbing to an operator and it is not the 'real world' experience (see section 2.2.2.).

Lags affect the dynamic response of the visual field of head-coupled VR systems to head movements. The lag has previously been called 'transmission lag' (Bryson and Fisher, 1990) and 'display lag' (So and Griffin, 1994a). It will hereafter be referred to as 'display lag'. In a computer-generated virtual environment, the display lag is pure and represents the time taken to measure the head orientation and to generate the corresponding graphics. In a virtual environment generated with images captured from a head-slaved camera, the display lag represents the response of the camera position to head movement. This lag is called exponential display lag.

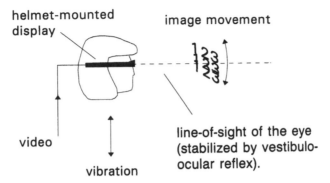

helmet-mounted
display

image movement

video

vibration

line-of-sight of the eye
(stabilized by vestibulo-
ocular reflex).

Figure 5.2 *Vibration-induced image movements, on a helmet-mounted display, relative to the operator line-of-sight.*

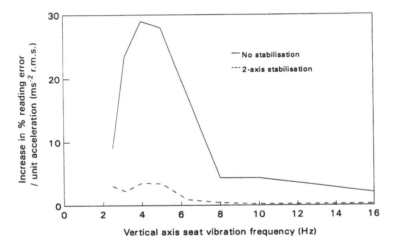

Figure 5.3 *Reading performance with a helmet-mounted display during exposure to whole-body vertical vibration (with and without space stabilization, adapted from Wells and Griffin (1984)).*

5.2 Effects of lags on head-coupled virtual environments

5.2.1 Overview

As discussed above, images in a head-coupled virtual environment are positioned incorrectly because of display lag. Such position error can be disturbing to an operator (Allen and Hebb, 1983). The error increases linearly with head movement velocity (Bryson and Fisher, 1990) and with display lag duration (So and Griffin, 1992). The discrepancies between head motion and the delayed response of the head-coupled VR system may cause motion sickness (Friedmann *et al.*, 1992; Oman, 1990). Regan and Price (1993a, 1993b) reported that with a 20-minute immersion in a head-coupled virtual environment and a 10-minute post-immersion period, 61% of the subjects reported symptoms of malaise. The symptoms ranged from headaches and eyestrain to severe nausea. A total of 150 subjects were used and 5% of the subjects had to withdraw from the experiment owing to severe dizziness. The system had a display lag in the order of 300 ms.

With teleoperation applications involving head-slaved cameras, the dynamic responses of the cameras will introduce exponential display lags. The time taken to search and recognize a symbol in a head-coupled virtual environment was found to increase with increasing exponential lag (Lewis, 1987). Grunwald *et al.* (1991) reported that exponential display lags constrained an operator from making fast head movements in a head-coupled virtual environment.

5.2.2 Specific results

5.2.2.1 *Head motion stimuli*

In a head-coupled virtual environment, the characteristics of head motion depend on the tasks. Examples of such tasks include visual search, target acquisition, tracking and aiming (see also section 2.2.3.). To study the effect of lag in a head-coupled virtual environment, a representative task is needed to stimulate head motion in a controlled manner. Two-axis continuous tracking tasks have been used for two reasons: (i) most head control tasks involve some tracking activity; (ii) the use of a continuous tracking task enables frequency domain analyses to be conducted.

5.2.2.2 *Duration of lag*

Head tracking error appears to be increased with pure display lags greater than, or equal to, 40 ms. The lags were imposed on a 40 ms system lag. Examples of head tracking error measurements are shown in Figure 5.4. As the lag increased, the error increased.

The effect of lags on head tracking responses has been modelled as the sum of two components: (i) an input-correlated response, and (ii) the remnant. The former represents the linear tracking response of the operator to the target and is expressed as the head tracking transfer function. The latter represents the nonlinear response of the operator. To determine how much of an operator tracking response was correlated with an input, a squared coherency function was measured between the two signals (Bendat and Piersol, 1986). Examples of such squared coherency functions are reproduced in Figure 5.5: as the lag increased, the squared coherency functions between the target motion and the head motion decreased. This is undesirable and a lag compensation technique is required.

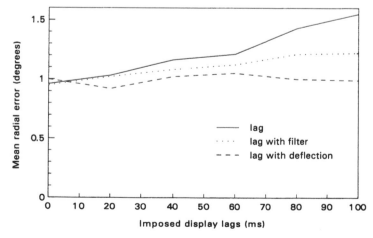

Figure 5.4 Mean radial tracking error with different display lags and three lag compensation conditions: no compensation, compensation with phase lead filter (48 ms lead) and compensation with image deflection (medians of six subjects, adapted from So and Griffin (1994b)).

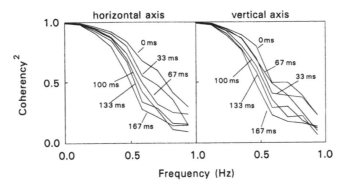

Figure 5.5 *Squared coherency functions between the target motions and the head motions during head tracking tasks with display lags (medians of six subjects).*

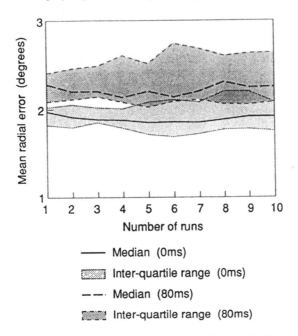

Figure 5.6 *Radial tracking error for 10 learning runs with 0 ms and 80 ms display lag imposed on a system display lag of 40 ms (medians of eight subjects, adapted from So and Griffin (1994a)).*

5.2.2.3 Effects of practice

Mean radial head tracking error was measured for ten consecutive runs with 0 ms and then with 80 ms pure display lag. With the 80 ms lag, no significant change in tracking performance with practice was found (Figure 5.6). This suggests that the problem with lags cannot be overcome with practice.

5.3 Lag compensation

5.3.1 Overview

The most natural solution to the lag problem will be to reduce the lags at source. But lags are inevitable in applications involving radio linkage, such as 'teleoperation', or with mechanical moving parts, such as head-slaved cameras. One possible solution to the display lag problem is to predict the future head position with a signal processing algorithm. For random movements of the head, prediction is, by definition, impossible. Fortunately, lags in the human operator restrict useful tracking to frequencies below about 1 Hz and prediction of head position becomes possible (Wells and Griffin, 1987b). List (1983) reported a simulation study with a simple nonlinear prediction algorithm. The objective was to compensate for lags during high-velocity step movements of the head. The algorithm used acceleration data and was implemented in a fibre-optic helmet-mounted display. Smith (1984) also conducted simulations to compensate for lags during step movements of the head. Although no human performance data were presented, it was reported that the predictor introduced jumps in the visual field which might be disturbing to an operator. An adaptive least-mean-square predictor was used to predict pilot head pointing direction (Albrecht, 1989). Simulation results showed that the predictor could predict input signals that change their characteristics linearly with time (e.g. a swept sine with decreasing amplitude). Improvement was necessary, however, to predict head movements whose characteristics change randomly with time. These studies investigated the nature of the predictors but human responses to the predictors in the presence of lags were not studied.

Another possible solution to the lag problem employs image deflection. With image deflection, an image in a head-coupled virtual environment with lags can be deflected to the correct angular position with respect to the Earth. Image deflection was originally developed to stabilize images projected on helmet-mounted displays exposed to vibration (see section 5.1.2.1). Applied to computer-generated images, the computational lag is measured and translated into horizontal and vertical position offsets. These offsets are then used to deflect the video image on the helmet-mounted display (So and Griffin, 1992). Image deflection can also be applied to images captured from a head-slaved camera. Allen and Hebb (1983) used a similar technique to restore the correct positions of images in a head-coupled virtual environment. No human experimentation was reported, however, and a helmet-mounted display was not used. The head-coupled virtual environment was projected on a dome-shaped screen with a head-slaved projector.

5.3.2 Specific results

5.3.2.1 Image deflection

As explained above, two-axis continuous tasks were used to control the head motion in a virtual environment. Image deflection was found to eliminate totally the degradation in head tracking performance with pure display lags up to 100 ms (Figure 5.4) and 380 ms (So and Griffin, 1992). Loss of field-of-view occurs with image deflection: as the display lag increases, the loss of field-of-view increases (Figure 5.7). With a 380 ms pure display lag, a loss of field-of-view of about 3·5 degrees r.m.s. was encountered. The required image deflection is proportional to both the lag and the head velocity. If these are large, the target may need to be deflected beyond the field-of-view; performance would then deteriorate rapidly. Also, deflection of graphics in three-dimensional perspective will generate parallax distortion. The effect of the loss of field-of-view was not apparent in the reported data (Figure 5.4). The study used a small circular target which, once captured within sight, was kept around the centre part of the total field-of-view. Any reduction of the field-of-view was not noticeable unless the target reached the edge of the displayed visual field. In addition, no parallax error was produced as the target was presented in two-dimensional graphics.

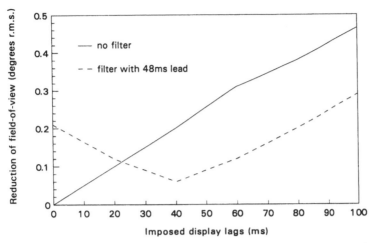

Figure 5.7 Reduction of field-of-view by image deflection with different display lags (with and without head position prediction, medians of six subjects).

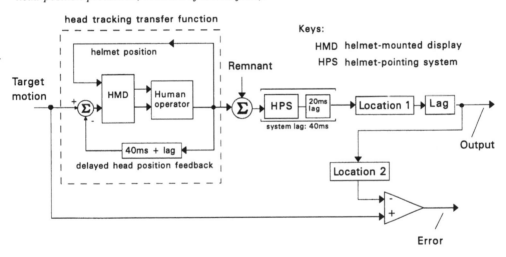

Figure 5.8 Model of a head tracking system illustrating the possible locations of head position predictors (adapted from So and Griffin (1994b)).

5.3.2.2 Head position prediction

The appropriate location to insert a head position predictor in a head-coupled VR system has been addressed. A quasi-linear model of a head tracking system is shown in Figure 5.8 and a predictor was placed at Location 1.

A simple predictor based on a first-order Taylor series expansion was shown to be unacceptably noisy. But its combined use with image deflection was beneficial (see section 5.3.2.3). Further studies with phase lead filters showed that when performing head tracking tasks, human operators compensated for any non-unity gains introduced by the predictors. The gain compensation was associated with additional lags in the tracking responses which, in turn, increased the tracking errors. It was concluded that an ideal predictor should have unity gain with phase lead within its operational frequency range. For head position prediction, this frequency range will be about 0–1 Hz (Wells and Griffin, 1987b). A phase lead filter with 48-ms lead, optimized to predict head position, reduced tracking error with pure display lags of 80 ms or more (Figure 5.4).

It is concluded that head position prediction can reduce tracking error with lags. Owing to the inability of a filter to predict the random part of the measured head movement, errors will always remain.

5.3.2.3 Combined image deflection and head position prediction

Studies with combined image deflection and head position prediction have shown that the two techniques complement each other and enhance the overall lag compensation capability. This technique utilizes head position prediction, to reduce large position errors due to the lag, and image deflection to remove remaining errors. The use of a phase lead filter has been shown to reduce the loss of field-of-view introduced by image deflection (Figure 5.7).

5.4 Summary

Lags are inevitable within head-coupled virtual reality systems, especially with head-slaved cameras. This impairs the ability to present space stationary images in a head-coupled virtual environment. The lags affect the realism of the simulations and discourage operators from making fast head movements. Human performance with visual search tasks is also degraded. Immersion in a head-coupled virtual environment with a large display lag may cause motion sickness.

With two-axis continuous head tracking tasks, display lags affect the tracking gain and phase responses and, therefore, performance measures such as mean radial tracking error and time-on-target.

Combined head position prediction and image deflection techniques have been shown to significantly improve tracking performance with lags.

References

Albrecht, R. E., 1989, An adaptive digital filter to predict pilot head look direction for helmet-mounted displays, unpublished MS thesis, University of Dayton, Dayton, OH.

Allen, J. H. and Hebb, R. C., 1983, Helmet-mounted display feasibility model, unpublished report TR-Navtraequipcen IH-338, Advanced Simulation Concepts Lab., Naval Training Equipment Center, Orlando, FL, February.

Bendat, J. S. and Piersol, A. G., 1986, *Random Data: Analysis and Measurement Procedures*, 2nd Edn, New York: John Wiley.

Benson, A. J. and Barnes, G. R., 1978, Vision during angular oscillation: the dynamic interaction of visual and vestibular mechanisms, *Aviation, Space and Environmental Medicine*, 49, 340-5.

Bryson, S. and Fisher, S. S., 1990, Defining, modelling and measuring system lag in virtual environments, in *Conference on Stereoscopic Display and Applications*, Santa Clara, USA, SPIE, Vol. 1256, pp. 98-109, San José, CA: SPIE.

Friedmann, M., Starner, T. and Pentland, A., 1992, Synchronization in virtual realities, *Presence*, 1 (1), 139-43.

Furness, T. A. and Lewis, C. H., 1978, Helmet-mounted display reading performance under whole-body vibration, paper presented at UK Informal Group Meeting on Human Response to Vibration, 18-20 September 1978, NIAE, Silsoe, Bedfordshire, England.

Gibson, C. P. and Furness, T. A., 1987, Effective control strategies for the virtual cockpit, Technical Report RAE TR-87037, Royal Aerospace Establishment, Farnborough, England.

Grunwald, A. J., Kohn, S. and Merhav, S. J., 1991, Visual field information in Nap-of-the-Earth flight by teleoperated helmet-mounted displays, in *Conference on Large-Screen-Projection, Avionic and Helmet-Mounted Displays*, San José, CA, 26-28 February 1991, SPIE, Vol. 1456, pp. 132-53, San José, CA: SPIE-The International Society for Optical Engineering.

Lewis, C. H., 1984, The development of an adaptive image stabilisation system for helmet-mounted displays, paper presented at UK Informal Group Meeting on Human Responses to Vibration, 21–22 September 1984, Heriot-Watt University, Edinburgh, Scotland.

Lewis, C. H., 1987, Unpublished data, Institute of Sound and Vibration Research, University of Southampton.

List, U. H., 1983, Non-linear prediction of head movements for helmet-mounted display, Operation Training Division, AFHRL Technical Paper 83-45, Williams AFB, AZ, USA.

Oman, C. M., 1990, Motion sickness: a synthesis and evaluation of the sensory conflict theory, *Canadian Journal of Physiology and Pharmacology*, **68**, 294–303.

Regan, E. C. and Price, K. R., 1993a, Some side-effects of immersion virtual reality, Report 93RO10, Army Personnel Research Establishment, Farnborough, England.

Regan, E. C. and Price, K. R., 1993b, Some side-effects of immersion virtual reality: the effects of increasing head movements, of rapid interaction, and of seating subjects, Report 93RO22, Army Personnel Research Establishment, Farnborough, England.

Smith, B. R. Jr, 1984, Digital head tracking and position prediction for helmet-mounted visual display systems, paper presented at AIAA 22nd Aerospace Sciences Meeting, 9–12 January, Reno, NV, 9–12 January 1984, New York: American Institute of Aeronautics and Astronautics.

So, R. H. Y. and Griffin, M. J., 1992, Compensating lags in head-coupled displays using head position prediction and image deflection, *Journal of Aircraft*, **29** (6), 1064–8.

So, R. H. Y. and Griffin, M. J., 1995, Effects of lags on human operator transfer functions with head-coupled systems, *Aviation, Space and Environmental Medicine*, in press.

So, R. H. Y. and Griffin, M. J., 1995, Use of phase lead filters to compensate lags in head-coupled visual displays, *IEEE Transactions on Systems, Man, and Cybernetics*, in press.

Wells, M. J. and Griffin, M. J., 1983, Vibration-induced eye motion, paper presented at the Annual Scientific Meeting of the Aerospace Medical Association, 23–26 May 1983, Houston, TX, pp. 73–4.

Wells, M. J. and Griffin, M. J., 1984, Benefits of helmet-mounted display image stabilisation under whole-body vibration, *Aviation, Space and Environmental Medicine*, **55**, 13–8.

Wells, M. J. and Griffin, M. J., 1987a, Flight trial of a helmet-mounted display image stabilisation system, *Aviation, Space and Environmental Medicine*, **58**, 319–22.

Wells, M. J. and Griffin, M. J., 1987b, A review and investigation of aiming and tracking performance with head-mounted sights, *IEEE Transactions on Systems, Man, and Cybernetics*, **SMC-17** (2), 210–21.

6

Perceptual cues and object recognition

John M. Findlay and Fiona N. Newell

6.1   Introduction

6.1.1 Simulation and recognition

Three disciplines with different aims have an interest in understanding the process of visual recognition. Two of these work in a way which is essentially analytic. Biological vision theorists are interested in how the brain (generally the human brain) carries out visual tasks. Solutions proposed must be subject to the constraints of neuro-anatomical and neurophysiological knowledge, for example the relatively slow speed of neurons when working as computing elements. Computational vision is the branch of artificial intelligence in which computational methods are proposed for analysing optical information to solve those problems which are solved by living visual systems. Workers in this area are not bound by constraints of those in the first area but nonetheless have traditionally derived fruitful insights from the study of vision in living systems. The third group is formed by designers constructing virtual reality systems. In contrast to the analytic approach of the two other cases, the interest here is in working synthetically to create a visual image or information structure which simulates the natural world, at least in some aspects.

Simulations of the visual world may take the ambitious approach of creating an exact copy of the optic array which reaches the eyes from a real world scene. The current achievements of the cinematographic and video industries have shown the potential of such a route. Hochberg and Brooks (1978) analysed these methods and accomplishments from the viewpoint of visual science. The restrictions of current techniques are familiar: limited field of view, absence of binocular disparity and above all no possibility of interactive action by the observer. To remove these restrictions is a problem of formidable complexity. For many purposes, however, it may be sufficient if information is presented which is adequate to fulfil a particular purpose. For example in the case of visual recognition, simulations might convey the presence of, say, a tree in the environment by an outline sketch (this could be considered a 'representation' in the term described in section 10.3).

Knowledge of the properties of human visual recognition can potentially facilitate this process. The remarkable fact that visual recognition can be successful when only a crude outline drawing is presented shows that much of the richness of normal visual information

113

is redundant in the recognition process. Redundancy is a ubiquitous feature of sensory processing and is often manifest in the presence of several alternative processes which achieve a particular goal. A familiar example in vision concerns the numerous 'cues' (disparity, shading, occlusion etc.) which support the perception of depth (see section 3.3). No one cue appears essential. Such redundancy is advantageous both in dealing with the diversity of visual environments which are encountered and also in providing back-up if damage occurs to the brain to render one process ineffective (see section 7.4.2).

Visual recognition is an extensive area and the scope of this discussion will be limited in several ways. Only recognition of rigid, static objects will be considered thus ignoring the important research area which has developed demonstrating the ability of biological visual systems to use visual movement to support recognition (Johansson, 1973; Cutting, 1986). The discussion will also concentrate on recognition of classes of objects, rather than within class recognition of exemplars (e.g. individual face recognition).

6.1.2 Visual recognition and other uses of vision

Visual recognition is shown most obviously in a situation where a name is assigned to an object in the visual world. The viewer may remark: 'I spy a wildebeeste'. For some workers it has seemed self-evident that the main aim of visual perception is such recognition. At the start of a classic paper, Bruner (1957) remarked, 'Perception consists of an act of categorization'. Categorization can convert visual experience into linguistic form and thus allow its communication. So it is not surprising to find the recognition aspect of vision emphasized in early discussions. It has become increasingly evident, however, that such categorical recognition is only one of many ways in which vision can be used. This is perhaps most clearly demonstrated in the clinical condition of visual agnosia, in which objects can no longer be recognized and named visually (Ellis and Young, 1988). Patients with this condition can often use vision perfectly adequately to guide actions such as reaching, grasping and locomoting.

Goodale and Milner (1992) show this particularly clearly in a remarkable case study with an agnosic patient. The patient lacked any conscious awareness of shape. Her abilities were only manifest when she was instructed to perform a visually guided action (such as aligning the shape with a slot in the experiment reported by Goodale *et al.*, 1991). This recognition/action distinction fits into a general framework which has emerged for neurobiological work on vision often described as the 'two visual systems' (see also section 7.3.2). In this framework, visual information is segregated in the posterior part of the cortex into two processing streams, described by their anatomical location as the dorsal and the ventral stream. The dorsal stream (occipito-parietal) is directed to the parietal part of the cortex and is concerned with action whereas the ventral stream (occipito-temporal) is directed to the temporal lobes of the cortex and subserves recognition (Ungerleider and Mishkin, 1982).

Such a separation makes sense of several problems in visual perception (see section 6.2.4) and also is an important point to bear in mind when considering theories of visual recognition. For example, although human vision is clearly three-dimensional and a variety of cues (stereo, shape from shading, etc.) give information about the relative depth of different parts of a viewed object, it cannot be assumed that this information is necessarily important for visual recognition. It might be the case, for example, that stereo information is used for visually guided action but not in general for visual recognition.

6.1.3   Constancies

The key problem of visual recognition is that a vast variety of different patterns of stimulation can elicit the same categorical classification. The light levels reflected from an object vary with the incident illumination sources, which change throughout the day and whenever nearby reflecting surfaces are moved. The orientation of the object's projection on the retina varies depending on both the object's position in the environment and the observer's viewpoint. As an observer walks around a 3-D object, different retinal projections of the image occur and the visual system needs to identify that these different images are from the same object. Likewise, retinal size is dependent on viewing distance. The general term 'object constancy' is used to denote the ability to recognize the invariance of objects in the face of this multitude of changes.

Recognition is possible on the basis of representations which are degraded in various ways, for example when part of the object is occluded. Except where specific camouflage is involved, recognition is remarkably unhindered by visual backgrounds or scene context although, as we discuss in section 6.3.3, context information can facilitate the recognition process. Finally, objects can be recognized from seemingly much impoverished representations of the real stimulus, such as line drawings.

A key question in understanding human object recognition is to know how peripheral visual mechanisms might operate to promote object constancy. It is generally accepted for example that retinal adaptation mechanisms ensure that the brain receives a spatial visual signal which is largely independent of the level of scene illumination. Marr's (1982) proposal that an early stage in vision is the formation of a primal sketch might be seen as a further mechanism for ignoring irrelevant illumination variations. Some workers have suggested that mental transformations equivalent to image rotation and zoom lens scaling may be achieved by basic visual mechanisms and may be important in visual recognition (as will be shown in sections 6.2.2 and 6.3.1).

6.1.4   Preliminaries to recognition

Under normal viewing situations, objects occur in a continuously structured visual scene. It is usually accepted that before an object can be recognized, it must be separated visually from the remainder of the scene, a process traditionally referred to as figure–ground segregation. Segregation is often regarded as a process which operates without any conscious attention or effort on the part of the viewer. It also operates across the whole visual field in parallel and may be thought of as a preliminary to the recognition process. Recognition, however, is always a function that is best performed in central vision. Thus another preliminary to recognition is the necessary orienting to the object that is to be recognized. Although recognition of objects in the visual periphery is possible (see section 6.3.2), in general objects that are to be identified are brought into central vision by means of orienting head and eye movements.

Preliminary analysis of a visual scene will thus consist of the segregation of figure–ground relationships in the scene, together with selection of targets for potential orienting responses. Recent work suggests that the same non-linear early visual processes are involved in both these functions (Findlay *et al.*, 1993). The remarkable efficiency of these preliminary processes justifies the simplification which is often made in theoretical and experimental work on object recognition to restrict consideration to isolated objects with well-defined contours.

6.2 Approaches to recognition

6.2.1 Visual processing and visual memory

A great diversity of approaches to visual recognition can be found and there is no fully satisfactory account of the process. This section follows several reviews (Pinker, 1984; Ullman,1989) in starting with a survey of the different types of theory-driven approaches. For visual recognition to occur, visual information must activate some form of visual memory. It is frequently assumed that visual processing solves the problem of object constancy, at least to a large extent, before the interaction with visual memory. Thus traditionally theories of visual recognition have been theories of visual information processing before the memorial stage is accessed. New perspectives concentrating on the way in which visual memory might be organized are discussed in section 6.2.6.

6.2.2 Alignment approaches

The idea that input data could be immediately matched to a memorial representation or template is frequently dismissed quickly in accounts of visual recognition. The myriad variations in input characteristics occurring in the natural world would seem to require an impossibly large number of stored representations. Some improvement can occur if the raw visual input is subject to a set of 'normalizing' transformations to bring it into a standardized or normalized form (section 7.4.3 also discusses the need for normalizing transformations in the broader context of sensory integration). In the case of a rigid two-dimensional image, just two transformations (size scaling, image plane rotation) would suffice for normalization, assuming some cue (such as a directed principal elongated axis) could be extracted from the image to guide the process. A further set of shear and bending transformations is needed to deal with depth plane rotations and non-rigid objects. The well-known work on mental rotation of Shepard and Metzler (1971) provides evidence of the mental capacity for rotational transformations of the required nature. But the rotation rates that have been demonstrated experimentally appear an order of magnitude too slow to assist the very rapid process of visual recognition.

The alignment idea has recently been revived in a more sophisticated form by Ullman (1989) who introduced an approach which he termed 'alignment of pictorial descriptions'. He points out that the transformations necessary to achieve rotation and size invariance can, in principle, be carried out for a rigid body using only a very limited amount of information (three non-collinear points or three corresponding features). He thus proposes that such alignment might occur with a simplified 'pictorial description' of the object. In this description, at least edge extraction has occurred and Ullman suggests other simplifying visual transformations (e.g. a 'hairy area' for the mane of a horse rather than a complete pictorial record). This bears some similarity to the idea of structural descriptions discussed below, but Ullman emphasizes the difference that pictorial descriptions are not symbolic.

6.2.3 Features

> That thing is round and nubbly in texture and orange in color . . . therefore it must be an orange.
> (Bruner, 1957)

In early thinking about visual recognition, it was often assumed that recognition occurred through analytic separation of aspects. Neisser (1967) gave further impetus to this approach by contrasting the power of the feature approach with the fragility of the template matching alternative. Thus in the 1960s and 1970s much effort was devoted to establishing components of an object that could be extracted in some way to provide a basis for recognition. Ideally such components would be invariant across the different transformations characterizing normal perception thus solving the problem of constancy. The search for invariant features characterizing complete nameable objects has been largely unsuccessful (Ullman, 1989) although an important recent idea (Biederman, 1987) is that invariant features may be present for components of objects rather than complete objects. This is given more attention in the following section.

The term 'feature' is not well defined and can be applied in a number of different ways. Straightforward segregation of spatial regions presents one possibility as instanced by the component lines of an alphabetic letter or the eyes, nose and mouth of a face. But it immediately becomes apparent in these examples that it is not just the presence of such features, but also the spatial arrangement of the features which is important in characterizing the object. This approach then fits more closely with theories involving structural descriptions.

Other ways of describing features are of course possible. One suggestion relates to the spatial distribution of light energy in the image. This can be given precise quantification in terms of spatial Fourier components. It is possible to analyse the distribution of light energy across spatial frequencies (e.g. in spatial frequency channels). The recognition that individual visual neurons are sensitive to patterns of distribution of light energy helped give rise to much work in the 1970s and 1980s in which the characteristics of the visual system were spelt out in these terms (Braddick *et al.*, 1978). Although one of the earliest papers in this area attempted to relate the distribution of energy across different channels to size constancy (Blakemore and Campbell, 1969) this approach to recognition has made only modest progress . An immediate obstacle is that objects are easily recognized in line drawing and silhouette forms when the light distribution Fourier components are entirely different. Although for many years Fourier approaches had great prominence, particularly in physiological work, more recent work (Tanaka, 1993) points to a more hybrid character for physiological units.

Finally, another way in which the term feature is used is shown by Bruner's quotation at the beginning of this section. Visual objects are characterized by non-localized properties such as surface colour, texture etc. It has recently become apparent that the visual cortex of the brain analyses some of these aspects in separate areas in a modular way (Zeki, 1993; see also section 7.3.2). Whilst this finding is clearly a breakthrough in understanding the biological substrate of vision, it is less clear how this information might be used in visual recognition. Treisman (1986), working on a paradigm of visual search, has contrasted the ease of search for single features (such as colour and orientation) with the difficulty of searching for conjunctions of features. She has also made the proposal that the conjunctions of features serve to specify objects, a suggestion which has attracted some physiological workers (Zeki, 1993). Experimental work provides very limited support. In the particular case of colour, some experiments by Biederman and Ju (1988) clearly refute the idea that the colour feature conjoins with other features to allow recognition. Biederman and Ju found that coloured pictures are no more rapidly recognized than monochrome, even when the colour is completely diagnostic (yellow bananas).

6.2.4 Affordances

James J. Gibson is a controversial figure who adopted a very idiosyncratic approach to visual perception. In a series of books (Gibson, 1950, 1966, 1979) he cast doubt on much traditional perceptual thinking and put forward his own 'ecological' approach. While he is recognized as having had many insights, his work has also been subject to a variety of criticisms (Cutting, 1986; Ramachandran, 1990; see sections 3.2.4 and 9.3.2 for a brief summary of the ecological approach to perception).

Much of Gibson's work relates to the perceptual control of action, but his ideas were formulated before the recognition/action distinction was widely recognized and he always believed his ideas applied equally to recognition. In his earlier work (Gibson and Gibson, 1955) stress was laid on perceptual invariants using an approach quite similar to that of the invariant features discussed above. But it did not prove possible to specify these invariants usefully (Cutting, 1986). In later work, he placed more emphasis on the concept of an 'affordance'. Roughly, an affordance specifies a possible action. Thus an object may be 'graspable', 'liftable' and 'eatable'. Gibson wished to stress that affordances are perceived directly, that is without intervention of memorial factors resulting from stimulus categorization. It has been pointed out that credibility is strained somewhat in applying this dictum to, for example, a postbox. It is interesting, however, to re-examine the notion of affordance in the light of more recent thinking.

Arguments have been advanced (in section 6.1.2) about the separation of perceptual routes for perception and action. Accepting this distinction, it is very clear that affordances relate by definition primarily to the way information is dealt with by the action route. Such information could also assist in recognition; indeed it could be said to entail a primitive form of recognition (objects that afford sitting on include rocks and benches, and so form a more extensive set than the set of 'chairs'). For some objects, such affordance information seems to form an important part of their categorization. Given the wide variety of visual forms which chairs can take, it is tempting to include the affordance of sitting as one of the key criteria by which chairs are recognized. Smets *et al.* in Chapter 9 show that they are trying to apply this approach to optimize a virtual environment for the 'action' of designing and creating.

6.2.5 Structural descriptions

Many visual objects can be segmented readily into component parts. A cat has a body, head, legs and a tail. We might achieve recognition of complex objects by some form of breakdown into simpler parts. But for this to be possible, we should need to segment the image into the components in an unambiguous way. Can such a segmentation occur before visual recognition? An important paper by Hoffman and Richards (1984) suggests a positive answer and shows that the topological relationship of *transversality*, based on visual properties such as points of maximum concavity, offers a route to a segmentation which accords with our intuitive expectations. Thus a part-based approach to recognition is not trapped into circularity and a number of workers have suggested that structural descriptions can offer a fruitful approach to recognition.

David Marr has been one of the most influential figures in recent work on vision, partly because of the sophistication of his analyses and partly because he took a wide perspective. He suggested the whole purpose of vision was to build three-dimensional axis-based representations of objects taking into account such depth cues as stereopsis (Marr, 1982). He considered such representations to be object-oriented, i.e. independent of any specific viewpoint and to form the memorial representations for recognition. His work with Nishihara on 'pipe-cleaner' figures (Marr and Nishihara, 1978) is widely cited as typifying the approach although it should be noted he also recognized the potential information contained in silhouettes (Marr, 1977).

The work of Biederman (1987) may be seen as following on from such a proposal. Biederman proposed a theory of object recognition, Recognition-by-Components, in which recognition of an object involved identifying a set component volumetric primitive shapes called 'geons'. Drawing on an analogy with phonemes and speech perception, he postulated that a small number (~ 36) of such primitives could be combined in such a way as to provide adequate characterization of all common objects. He has shown that many common classes of objects have unique geon-structural descriptions, although it is evident geon theory cannot be extended to discrimination between items within a class. Segmentation of the image into geon sub-components could occur using visual properties such as deep concavity as discussed above. Biederman suggested that the geon sub-components could themselves be recognized by non-accidental visual properties such as parallelism, symmetry and collinearity. Allowing for a certain tolerance in the measurements, such non-accidental properties do indeed constitute invariant features which reflect significant object information. The idea is attractive but experimental work explicitly designed to test the theory has not been supportive. Thus Bravo and Nakayama (1992) found no evidence that geons showed fast pop-out in a visual search task which might reasonably have been expected if they were detected in early stages of vision.

6.2.6 The activation of memory representations

Much early work paid rather little attention to the form in which object representations might be stored in memory, concentrating rather on the visual transformations that might achieve object constancy. Frequently an implicit assumption was of some sort of look-up table of feature lists with a separate entry for each category. This idea of discrete memorial representations had its physiological counterpart in the notorious 'grandmother cell', a hypothetical cell in the nervous system which was active if, and only if, a person's grandmother was present in the visual scene. Yet since many thousands of different objects can be recognized, locating the memory representation is not a trivial problem. Recent work on the properties of neural networks has shown alternative possibilities to those of discrete encoding and serial search.

One of the most powerful recent developments in neuroscience and in psychology has been the realization that networks of neuron-like elements possess important intrinsic capacities for computational and information processing tasks and in particular for information storage and retrieval. Complex pattern memory can be stored in neural networks so that retrieval based on partial information is possible—content-addressable memory. The recent text by Churchland and Sejnowski (1992) gives a lucid introduction to the excitement of the area.

It is possible for a number of different input patterns to a network to result in the same output pattern; exactly the property of generalization that is fundamental to the process of pattern recognition. Moreover, a network can be 'trained' to encode particular patterns. Rosenblatt (1958) emphasized the possibility of applying the network idea to visual perception in his classic work on perceptrons. Equally well known is the fact that a demonstration of the limitations of simple forms of perceptrons (Minsky and Papert, 1969) was interpreted too widely by researchers and funding bodies as a critique of the whole enterprise and led to a lack of interest in the area for a number of years (Papert, 1988). The subsequent revival of interest in neural networks initially focused attention on topics other than visual recognition (Rumelhart and McClelland, 1986). The reason for this might be attributed to the fact that it was not obvious how the network approach could deal with some of the fundamental problems of pattern recognition such as orientation constancy and size constancy.

Recently, several workers have explored the potential of network based models of the recognition process. Such models probably represent the most promising current approach to understanding this process. One of their major attractions is the potential for modelling both the recognition process and also the learning phase which gave rise to this process. For example, Edelman and Weinshall (1991) examined the effects of training self-organizing networks with a few specific views of wire frame objects. They found that such training produces a network which is capable of responding appropriately to intermediate views not experienced during the training. Moreover Edelman and Bulthoff (1992) found that their model showed similar generalization properties to those of human observers. Specifically, after training with two separated views of the same object, the system would recognize interpolated novel views but showed less successful extrapolation to novel views beyond the training views.

Poggio and Edelman (1990; Poggio *et al.*, 1992) have examined carefully the question of whether these network-based models can demonstrate the constancy properties required of object recognition. They use the powerful approach of 'generalized radial basis functions' as the first hidden layer stage of a network. Such functions are specifically suggested by the need to encode information optimally to generalize across transformations and also bear some resemblance to receptive fields found biologically. The network, by adjusting the weights of the summed basis functions during the training phase, is capable of demonstrating orientation-independent recognition when trained with a small number of specific views of an object.

6.3 *Experimental studies of visual recognition*

6.3.1 Object orientation

Only a small subset of the enormous research literature on human and animal visual recognition can be dealt with in this section. More extensive treatments are given by Neisser (1967), Seymour (1979), Shepp and Ballesteros (1989).

Mental chronometry is a method favoured by experimental psychologists. Careful measurement is made of the time required for human subjects to make some well-defined response and how this time varies with different conditions. An example of its use in the context of object recognition is reviewed by Jolicoeur (1992). Jolicoeur considers studies investigating the way in which the orientation at which an object, or object picture, is viewed affects the speed with which the object can be identified. In these studies, a visual pictorial representation of a familiar object is presented to a subject, whose task is to name the picture as rapidly as possible. Typically, a voice key is used to obtain an exact measure of the naming latency. The rationale is that any extra time taken for recognition of an object in an unusual view reflects in some way the extra processing involved and thus analysis of the time increment and the way it depends on object orientation should reveal something of the underlying processing. Although interpretation may not always be straightforward, the approach is productive.

Very many normally encountered objects possess a natural 'upright' view. How does the time increase when views other than upright one are presented? Several studies (Corballis *et al.*, 1978; Jolicoeur, 1988) have found naming time to show a consistent increase as the angle of presentation is increased so that the picture is tilted from the standard upright view. Clockwise or counterclockwise tilts have identical effects. This increase continues until, at 120-degree tilt, recognition is typically slower by about 200 ms than in the upright view. But the condition in which the picture is tilted through 180 degrees, i.e. inverted, is faster than continuation of the above process would predict, and indeed is faster than at a 120-degree tilt. The resultant plot of naming time against angle of tilt (measured from 0 to 360 degrees) thus shows an 'M-shaped' function (see Figure 9.1 of Jolicoeur (1992)).

By no means all studies have found such a dependence on orientation, however. In some cases, the orientation of presentation has neglible effects on naming time (Corballis and Nagourney, 1978; Koriat and Norman, 1985). Biederman and Gerhardstein (1993) used a 'priming paradigm' (in which the presentation of an object to be recognized is preceded by the presentation of a similar or different object) to determine whether the reduction in naming latencies which occurs with repeated viewing of an object depends on the objects being seen in the same view. No such view-dependence was found. A recent set of studies by Newell (1993; Newell and Findlay, 1993) has used a 3-D computer drafting program to generate realistic pictorial representations of common objects, which could be displayed in a variety of orientations, with rotations around the z axis (which left the principal axis orthogonal to the viewing axis) and rotations in depth around the x or y axis (introducing foreshortening, see Figure 6.1). She used a task of object-name matching which resulted in fast responses (300–500 ms) of match or mismatch. She found no effect of viewing orientation unless the representations were considerably foreshortened (viewing axis within 30 degrees of axis of elongation) in which case a slight slowing (up to 50 ms) occurred. In contrast to the massive view-dependent effects reported by previous researchers, the slight effect of foreshortening did not diminish with practice. Thus recognition generalized across a number of different views where the axis of object was maximally exposed or flattest. This suggests that these views are consistently represented in memory and other views are transformed to match the represented views.

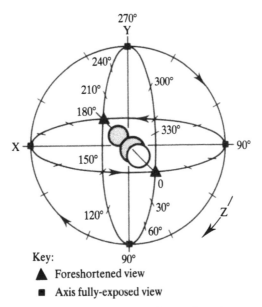

Key:
▲ Foreshortened view
■ Axis fully-exposed view

Figure 6.1   View-sphere illustrating the orientations and axes of rotations used in all the experiments from Newell (1993). The foreshortened, 0-degree, position shown above was the origin for all subsequent rotations.

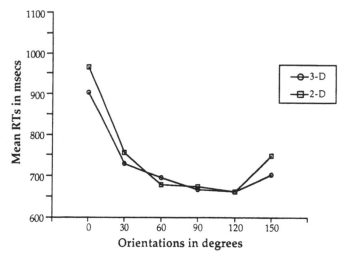

Figure 6.2   Data from Newell (1993) showing recognition times for objects presented at different orientations in depth in both drawing form (3-D) and silhouette form (2-D). Recognition is much slower at the 0-degree end-on axis view but is also significantly slowed when the objects are seen with a high degree of foreshortening (30-degree and 150-degree positions). Smaller amounts of foreshortening are responded to as quickly as in the axis orthogonal orientation (90 degrees).

The original aim of Newell's experiments was to test the idea that an object might be represented by a single, optimal, 'canonical' view (Palmer *et al.*, 1981). The finding that recognition occurred equally fast over a wide range of views does not support such a notion. Instead, it seems that, for elongated objects at least, recognition is equally fast over a wide range of views. These views may be collectively referred to as the *canonical aspect* of the object. Furthermore, Newell showed that silhouettes are recognized just as rapidly as the pictorial objects (see Figure 1 in Newell (1993)). So if features are used, these are not likely to be of a simple type, but must relate to form of contour. Other studies have similarly found that the recognition of objects depends on the 2-D projected view of the object's image, with some views more readily recognizable than others (Edelman and Bulthoff, 1992; Humphrey and Khan, 1992).

Thus object recognition latencies sometimes show dependence on orientation but sometimes do not. This suggests that the process may operate in two possible ways. Familiarity with the particular stimuli used seems to be one factor determining whether a dependence on orientation is found. In the Corballis *et al.* (1978) experiments, the M-shaped function was found at the start of the experiment but, after a few blocks, a much flatter function was produced. Such a result was also found by Jolicoeur and Milliken (1989). Jolicoeur (1992) suggested that a transition between two recognition routes occurs during early experience with visual objects. Initially a process is used in which mental rotation occurs. This operates rather slowly and Jolicoeur suggests that it is later superseded by some form of template matching, in which orientation-invariant features are learned. As discussed in an earlier section, however, it has proved difficult to specify such invariant features and possibly the view interpolation process proposed by Edelman's group (Edelman and Weinshall, 1991; Poggio *et al.*, 1992) could provide an alternative fast route to recognition which is view-independent.

The effects of size on recognition have received rather less study than those of orientation. When unfamiliar shapes are learned in a recognition memory experiment, increased latency is shown if the shape is presented at a different size (Jolicoeur, 1987; Milliken and Jolicoeur, 1992). Biederman and Cooper (1992), however, have shown an impressive size-invariance in picture-naming latencies, and also in the speeding of such latencies by an immediately prior presentation of the same stimulus, known as repetition priming. Repetition priming effects were found to be independent of whether or not the stimulus was previously seen at the same size, or at a different one. Biederman and Cooper argue that this supports the position that image shape, for purposes of recognition, is stored in a size-independent way.

Another research area, whose empirical findings shed some light on visual recognition, is cognitive neuropsychology; the study of the effects of brain damage on cognitive abilities. 'Visual agnosia' is the term used to describe the difficulties in visual recognition experienced in such cases. Warrington and Taylor (1973) reported that patients frequently experienced difficulties in recognizing unusual views of objects, i.e. views which are along the principal axis of an object, giving maximum foreshortening. This issue was further investigated by Humphreys and Riddoch (1987), who showed, by means of careful single case studies, that two separable patterns of loss occurred in visual agnosia. Patients with damage to the right hemisphere of the cortex were particularly affected by unusual views whereas patients with more diffuse bilateral damage were particularly affected by occlusion (when an object is partially hidden). Humphreys and Riddoch argue that the findings suggest two routes to recognition, one axis based and the other feature based. This conclusion is similar to that of Jolicoeur (1992) discussed above.

6.3.2 Recognition in peripheral vision

The work discussed so far has dealt almost entirely with recognition of high-resolution images in central vision and has ignored the distinction between central and peripheral vision. This has some justification because in general we look at an object in order to recognize it; in other words we direct our eyes to bring it into central vision. Recognition of objects away from the central visual field is worse than that at the centre not only because of the decline in visual acuity but also because of increased lateral interference between contours (Bouma, 1978).

A way in which peripheral recognition could be investigated in a realistic task was developed by Pollatsek *et al.* (1984). They presented a stimulus in the visual periphery to which an observer was required to orient and to name as rapidly as possible. The display was such that the subjects made an orienting saccadic eye movement to the target in the course of recognition. The findings showed 'preview advantage' whereby peripheral preview of an object speeded recognition time, showing that some useful information was extracted from the periphery. Pollatsek *et al.* (1984) then investigated the effect of 'saccade-contingent' display changes, in which the occurrence of the eye movement was monitored and the display modified as soon as the movement occurred. The modification took place before the saccade terminated and the subject was unaware of the change. The displayed object was altered in various ways. A preview advantage was found if one pictorial long thin object (e.g. a carrot) was replaced by another (a baseball bat). A preview advantage was also found if one picture (e.g. of a dog) was replaced a different, visually dissimilar picture of a similar object (e.g. another dog). Thus both visual and more abstract properties contributed to the preview advantage. In a follow-up study by Henderson *et al.* (1989) it was shown that the facilitation produced by peripheral preview only occurred if the preview was in the specific location to which the eyes were about to move (see also Rayner and Pollatsek, 1992). This finding could be interpreted as supporting the idea proposed by Treisman (1985, 1988) that without directing attention to a region of the visual periphery, certain operations (called by Treisman 'feature conjunction'; see also section 7.4.3) necessary for recognition cannot occur. The idea that visual attention moves to a peripheral location prior to an eye movement has received much recent support (Shepherd *et al.*, 1986; Henderson, 1993).

Treisman's suggestion is that only a small part of the visual field can receive the attentional operations necessary for full recognition. Support for this claim has come largely from studies of visual search and in particular the way in which the time needed to carry out a search is dependent upon the number of non-target items in the search display. When the distinction between target and non-target (distractor) items can be made in terms of simple features (e.g. a red target with green distractors), search times are unaffected by the number of distractors. When the difference between targets and non-targets depends on more complex feature combinations or object identity, the search time increases (often linearly) with the number of distractors, suggesting that items are searched serially.

The idea that, for anything other than simple feature searches, a serial deployment of attention is necessary, has been widely accepted although not universally (Enns and Rensink, 1990; Palmer *et al.*, 1993). But it is often not made explicit whether the attention search is covert (without moving the eyes) or involves overt eye movements. Findlay *et al.* (1995) have followed up earlier work which suggested that in visual search with free scanning, saccadic eye movements can only be directed to simple features (Williams, 1967; Gould and Dill, 1969). They found that in an object search task, the capacity to direct saccades to target objects is about as limited as Treisman's theory would predict and this ability is unaffected by the object's orientation provided it is not appreciably foreshortened. Recent work has shown that saccades can be carried out to feature conjunctions much better than Treisman's theory would predict (Findlay, 1995).

Retinal inhomogeneity is the way in which human vision meets the double requirements of a wide field of view and assimilation of fine visual detail. The question of what visual information is used to recognize objects in peripheral vision has both theoretical and practical significance. Faced with the problem of providing realistic visual simulations, designers have suggested that displays might be planned on a principle which reflects the structure of human vision (Haber, 1986) with high-resolution information presented only in the limited region at which the observer is looking. However, it is very difficult to predict on *a priori* considerations whereabouts an observer will choose to look in a display. As an alternative, the position of the observer's gaze can be monitored and used to control the position of the high-resolution region. Technical advances in eye-fixation recording (Green, 1992) render this approach feasible, if difficult and expensive (see also sections 2.3.2 and 2.4.3).

6.3.3 The effects of context on object recognition

Apart from changes in viewing position and lighting conditions, one of the major sources of variability in images of objects in the natural environment is the context. By context we can mean the visual context , for example when an occluding object modifies an object's contour, or semantic context, such as changes in the surrounding scene in which an object is placed. Despite such changes, the visual system shows a remarkable capacity to compensate, which is achieved in a number of ways. These may be grouped broadly into 'bottom-up' and 'top-down' processing. Bottom-up processing can be considered as processing which involves use of information in the stimulation at the eye without recourse to prior knowledge in memory. In other words, images are interpreted independently of what has been previously learned by the individual; interpretation relies on the sensory data alone. Top-down processing, on the other hand, relies on previous knowledge in the brain.

Consider the objects in Figure 6.3 below. The shapes in this demonstration are readily perceived as being whole shapes placed in such a way that they occlude each other. Occlusion can be demonstrated with nonsense shapes as well as with familiar shapes. Clearly, the visual system can handle occlusion without relying on top-down or higher cognitive information. The brain interprets the shapes as being superimposed on each other and can recover each full shape. Occluded shapes are rarely interpreted as the parts of shapes that are visible (see Figure 6.3C) but are instead seen as whole shapes. The term 'amodal completion' has been used to describe the fact that, perceptually, objects and their contours appear to continue behind the occluder. Kellman and Shipley (1991) have recently given an extensive theoretical treatment of this phenomenon.

In another recent demonstration of bottom-up processing, Nakayama *et al.* (1989), using a type of demonstration pioneered by Kanisza (1979), showed that if a face is occluded by stripes there is no effect on the recognition of that face. The visual system can cope with such occlusions. But if the parts of the face remaining visible are isolated and are superimposed on to a black background, then recognition of the face is more difficult even though exactly the same parts of the face are shown in both cases. This demonstrates that the solution of the occlusion problem depends on the ability to treat the occluder as a unitary visual object.

Other demonstrations of bottom-up processing show the effect of the three-dimensional scene context of an object. For example, Palmer *et al.* (1988) found that gravitationally defined squares and diamonds were more easily discriminated in diagonal arrays when pictorial depth information was added to the array. Without depth information, they found an interference effect between diagonal and either horizontal or vertical arrays of the shapes which was much reduced when depth information was added to the array. Similarly, Humphrey and Jolicoeur (1993) recently reported that the recognition of three-dimensional objects rotated in depth is improved if the objects are shown against a three-dimensional textured gradient. Humphrey and Jolicoeur argued that the recognition of objects rotated in depth is poor because of an improper interpretation of the three-dimensionality of the objects. Moreover, it is less difficult to resolve the principal axis of an object if its position

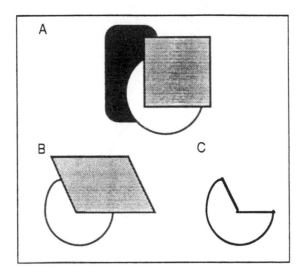

Figure 6.3   A demonstration of bottom-up processing in occlusion. In A, all the shapes are recoverable even though they are mostly occluded. The visual system rarely interprets the circle shown in B as C when occluded by another shape.

is specified in depth (Marr and Nishihara, 1978). These effects may be of little significance in the fast recognition of unambiguous objects, however. In a similar experiment using computer-generated three-dimensional shaded images of objects, Newell (1994) reported no improvement on recognition times for objects rotated in depth when the objects were shown against a three-dimensional textured background. The objects used as stimuli were three-dimensional paper-clip objects and three-dimensional familiar objects. That the recognition of the objects rotated in depth did not benefit from pictorial depth cueing suggested that the two-dimensional projected image of the object suffices for recognition purposes without the need for three-dimensional information.

Although bottom-up processing plays a large role in the interpretation of images, previously learned knowledge can also affect recognition by the top-down route. Consider the pattern in Figure 6.4. Before knowing the correct interpretation of the drawing an observer may report that the picture contains a pattern of circles and lines. Once told that the picture is of a Mexican on a bicycle seen from above then the patterns become objects.

Since the days of the Leeper pictures (shown in Neisser, 1967), textbooks of visual perception have routinely contained demonstrations of pictorial material which is meaningless without assistance of top-down processing form specific knowledge. One of the most powerful demonstrations is the well-known photograph by R. C. James (shown in Gregory, 1970) in which a Dalmatian dog can be resolved from an image of seemingly random blobs. It is uncommon that a naive observer will recognize the dog but, once the observer has been told that the image depicts a Dalmatian against a pebbled background, then it is very difficult to interpret the picture differently. There are many such demonstrations which show the effect of learning on recognition; however, it is important to empirically assess the role of knowledge or semantic processing on recognition.

In some early experiments, Biederman (1972) looked at the role of environmental context on the recognition of objects. He found that objects that were depicted in a coherent scene were recognized more easily than when the scene was segmented and jumbled. Although the target object was never segmented by the jumbling, subjects were faster at recognizing the object when shown in the coherent scene. In a later experiment, Biederman *et al.* (1982) reported that changes in physical and semantic context interfere with top-down processing and causes delays in recognition. An example of a physical context change would be placing a fire-hydrant on top of a post-box in a street scene. Subjects found it more difficult to

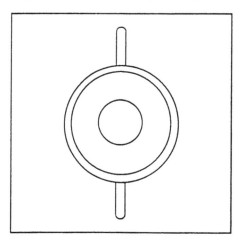

Figure 6.4 What is this object? The correct interpretation of this figure requires top-down processing. See text for solution to the problem.

perceive the target object when the correct physical context was violated. In the semantic condition, the fire-hydrant was placed in a kitchen scene thus violating the correct environmental context. Again subjects found it more difficult to recognize objects when shown in an inappropriate environmental context. An interpretation of Biederman's findings is that the physical and semantic changes are violating the rules previously learned by the visual system about where certain objects are likely to be placed in an image and also what are the objects' likely surroundings.

Conversely, recognition can be facilitated given the correct semantic context. Freidman (1979) showed that objects were faster recognized when presented in an appropriate scene context. This has also been demonstrated in priming experiments, especially by priming faces with associated faces. Bruce and Valentine (1986) reported that familiar faces are recognized more quickly (by about 100 ms) if the face was preceded by an associated face (such as Ernie Wise followed by Eric Morecambe), than if the face was preceded by unassociated or unfamiliar faces. In a similar study, Sperber *et al.* (1979) found that a picture was named more quickly if it was preceded by a semantically related picture compared with a condition when the object was preceded by an unrelated picture. Taken together, these studies suggest that information is somehow organized in the visual system so that items that are related in the real world are semantically related in memory. It could be argued that the visual system is organized in such a way that items in the real world that are highly likely to co-occur will be associated in memory. Once the system encounters an object, memory for other related objects is activated; thus the visual system is primed to expect the co-occurrence of such objects.

6.4 Conclusion

Like many effortless human activities that appear simple at first sight, visual recognition turns out to involve a number of extremely sophisticated processes which are still only partly understood. This chapter has attempted to survey our knowledge in this field without concealing our many areas of ignorance. Much of the information can be of use to designers of human–computer interfaces and virtual environments. As the world of computer games has demonstrated, it may often pay to take advantage of the remarkable human cognitive ability to make inferences from minimal information and from context. We would urge designers of virtual reality systems to plan intelligently, taking into account aspects of human visual psychology such as those we have discussed. In turn, since advances in an area are often dependent on improvements in technology, we may anticipate that as virtual reality systems of increasing sophistication are developed, these will provide new tools for scientists attempting to unravel the mysteries of human vision.

References

Biederman, I., 1972, Perceiving real world scenes, *Science*, **177**, 77–80.

Biederman, I., 1987, Recognition by components: a theory of human image understanding, *Psychological Review*, **94**, 115–45.

Biederman, I. and Cooper, E. E., 1992, Size invariance in visual object priming, *Journal of Experimental Psychology, Human Perception and Performance*, **18**, 121–33.

Biederman, I. and Gerhardstein, P. C., 1993, Recognizing depth-rotated objects: evidence for 3D viewpoint invariance, *Journal of Experimental Psychology, Human Perception and Performance*, **19**, 1162–82.

Simulated and virtual realities

Biederman, I. and Ju, G., 1988, Surface versus edge-based determinants of visual recognition, *Cognitive Psychology*, **20**, 38–64.

Biederman, I., Mezzanotte, R. L. and Rabinowitz, J. C., 1982, Scene perception: detecting and judging objects undergoing relational violations, *Cognitive Psychology*, **14**, 143–77.

Blakemore, C. and Campbell, F. W., 1969, On the existence of neurones in the human visual system selectively sensitive to the orientation and size of retinal images, *Journal of Physiology*, **203**, 237–60.

Bouma, H., 1978, Visual search and reading: eye movements and the functional visual field. A tutorial review, *Attention and Performance*, **11**, 115–47.

Braddick, O. J., Campbell, F. W. and Atkinson, J., 1978, Channels in vision: basic aspects, in Held, R., Leibowitz, H. and Teuber, H.-L. (Eds), *Handbook of Sensory Physiology*, Volume 8: *Perception*, pp. 3–38, Berlin: Springer-Verlag.

Bravo, M. J. and Nakayama, K., 1992, The role of attention in different visual search tasks, *Perception and Psychophysics*, **51**, 465–72.

Bruce, V. and Valentine, T., 1986, Semantic priming of familiar faces, *Quarterly Journal of Experimental Psychology*, **38A**, 125–50.

Bruner, J. S., 1957, On perceptual readiness, *Psychological Review*, **64**, 123–52.

Churchland, P. S. and Sejnowski, T. J., 1992, *The Computational Brain*, Cambridge, MA: MIT Press.

Corballis, M. C. and Nagourney, B. A., 1978, Latency to categorize disoriented alphanumeric characters as letters or digits, *Canadian Journal of Psychology*, **32**, 186–8.

Corballis, M. C., Zbrodoff, N. J., Shetzer, L. L. and Butler, P. B., 1978, Decisions about identity and orientation of rotated letters and digits, *Memory and Cognition*, **6**, 98–107.

Cutting, J. E., 1986, *Perception with an Eye for Motion*, Cambridge MA: MIT Press.

Edelman, S. and Bulthoff, H. H., 1992, Orientation dependence in the recognition of familiar and novel views of three-dimensional objects, *Vision Research*, **32** (12), 2385–400.

Edelman, S. and Poggio, T., 1992, Bringing the grandmother back into the picture: a memory based view of object recognition, *International Journal of Pattern Recognition and Artificial Intelligence*, **6**, 37–61.

Edelman, S. and Weinshall, D., 1991, A self-organizing multiple-view representation of 3D objects, *Biological Cybernetics*, **64**, 209–19.

Ellis, A. W. and Young, A. W., 1988, *Human Cognitive Neuropsychology*, Hove: Lawrence Erlbaum.

Enns, J. T. and Rensink, R. A., 1990, Influence of scene-based properties on visual search, *Science*, **247**, 721–3.

Findlay, J. M., 1995, Visual search: eye movements and peripheral vision, *Optometry and Vision Science*, in press.

Findlay, J. M., Brogan, D. and Wenban-Smith, M., 1993, The visual signal for saccadic eye movements emphasizes visual boundaries, *Perception and Psychophysics*, **53**, 633–41.

Findlay, J. M., Newell, F. N. and Scott, D. W., 1995, How much of the visual periphery is monitored during visual search?, in Gale, A., Astley, S., Carr, K. and Moorhead, I. (Eds), *Visual Search 3, Proceedings of the Third International Conference on Visual Search*, Nottingham: Taylor & Francis, in press.

Friedman, A., 1979, Framing pictures: the role of knowledge in automated encoding and memory from gist, *Journal of Experimental Psychology: General*, **3**, 316–55.

Gibson, J. J., 1950, *The Perception of the Visual World*, Boston, MA: Houghton Mifflin.

Gibson, J. J., 1966, *The Senses Considered as Perceptual Systems*, Boston, MA: Houghton Mifflin.

Gibson, J. J., 1979, *The Ecological Approach to Visual Perception*, Boston, MA: Houghton Mifflin.

Gibson, J. J. and Gibson, E. J., 1955, Perceptual learning: differentiation or enrichment, *Psychological Review*, **62**, 129–38.

Goodale, M. A. and Milner, A. D., 1992, Separate visual pathways for perception and action, *Trends in Neuroscience*, **15**, 20–5.

Goodale, M. A., Milner, A. D., Jakobson, L. S. and Carey, D. P., 1991, Object awareness, *Nature*, **352**, 154–6.

Gould, J. D. and Dill, A., 1969, Eye movement parameters and pattern discrimination, *Perception and Psychophysics*, **6**, 311–20.

Green, P., 1992, Review of eye fixation recording methods and equipment, University of Michigan, IVHS Technical Report No. 92-20.

Gregory, R. L., 1970, *The Intelligent Eye*, London: Weidenfeld & Nicolson.

Haber, R. N., 1986, Flight simulation, *Scientific American*, July 90–7.

Henderson, J. M., 1993, Visual attention and saccadic eye movements, in d'Ydewalle, G. and Van Rensbergen, J. (Eds), *Perception and Cognition. Advances in Eye Movement Research*, pp. 37–50, Amsterdam: Elsevier.

Henderson, J. M., Pollatsek, A. and Rayner, K., 1989, Covert visual attention and extrafoveal information use during object identification, *Perception and Psychophysics*, **45**, 196–208.

Hochberg, J. and Brooks, V., 1978, The perception of motion picture, in Carterette, E. and Freidman, M. (Eds), *Handbook of Perception*, Vol. 10: *Perceptual Ecology*, pp. 259–304, New York: Academic Press.

Hoffman, D. D. and Richards, W. A., 1984, Parts of recognition, *Cognition*, **18**, 65–96.

Humphrey, G. K. and Jolicoeur, P., 1993, An examination of the effects of axis foreshortening, monocular depth cues, and visual field on object identification, *Quarterly Journal of Experimental Psychology*, **46A** (1), 137–59.

Humphrey, G. K. and Khan, S. C., 1992, Recognising novel views of three-dimensional objects, *Canadian Journal of Psychology*, **46** (2), 170–90.

Humphreys, G. W. and Riddoch, M. J. (Eds), 1987, *Visual Object Processing*, London: Lawrence Erlbaum.

Johannson, G., 1973, Visual perception of biological motion and a model for its analysis, *Perception and Psychophysics*, **14**, 201–11.

Jolicoeur, P., 1987, A size congruency effect for visual shape, *Memory and Cognition*, **15**, 531–43.

Jolicoeur, P., 1988, Mental rotation and the identification of disoriented objects, *Memory and Cognition*, **13**, 289–303.

Jolicoeur, P., 1992, Identification of disoriented objects: a dual systems theory, in Humphreys, G. W. (Ed.), *Understanding Vision*, Oxford: Blackwell.

Jolicoeur, P., 1993, An examination of the effects of axis foreshortening, monocular depth cues, and visual-field on object identification, *Quarterly Journal of Experimental Psychology*, **46**, 137–59.

Jolicoeur, P. and Milliken, B., 1989, Identification of disoriented objects: effects of context of prior presentation, *Journal of Experimental Psychology: Learning, Memory and Cognition*, **15**, 200–10.

Kanisza, G., 1979, *Organization in Vision: Essays on Gestalt Perception*, New York: Praegar.

Kellman, P. J. and Shipley, T. F., 1991, A theory of visual interpolation in object perception, *Cognitive Psychology*, **23**, 141–221.

Koriat, A. and Norman, J., 1985, Mental rotation and visual familiarity, *Perception and Psychophysics*, **37**, 429–39.

Marr, D., 1977, Analysis of occluding contour, *Proceedings of the Royal Society*, **197B**, 441–75.

Marr, D., 1982, *Vision*, San Francisco: W. H. Freeman.

Marr, D. and Nishihara, H.K., 1978, Representation and recognition of the spatial organization of three-dimensional shapes, *Proceedings of the Royal Society*, **200B**, 269–94.

Milliken, B. and Jolicoeur, P., 1992, Size effects in visual recognition memory are determined by perceived size, *Memory and Cognition*, **20**, 83–95.

Minsky, M. and Papert, S., 1969, *Perceptrons. An Introduction to Computational Geometry*, Cambridge, MA: MIT Press.

Nakayama, K., Shimozo, S. and Silverman, G. H., 1989, Stereoscopic depth: its relation to image segmentation, grouping, and the recognition of occluded objects, *Perception*, **18**, 55.

Neisser, U., 1967, *Cognitive Psychology*, New York: Appleton-Century-Croft.

Newell, F., 1993, Perceptual recognition of objects in different orientations, PhD thesis, University of Durham.

Newell, F., 1994, The effect of background on the recognition of rotated objects, *Investigative Ophthalmology and Visual Science*, **35** (4), 1625.

Newell, F. N. and Findlay, J. M., 1993, Viewpoint invariance in object recognition, *Irish Journal of Psychology*, **13**, 494–507.

Palmer, J., Ames, C. T. and Lindsay, D. T., 1993, Measuring the effect of attention on simple visual search, *Journal of Experimental Psychology: Human Perception and Performance*, **19**, 108–30.

Palmer, S., Rosch, E. and Chase, P., 1981, Canonical perspective and the perception of objects, in Long, J. and Baddeley, A. (Eds), *Attention and Performance IX*, pp. 135–51, Hillsdale, NJ: Lawrence Erlbaum.

Palmer, S., Simone, E. and Kube, P., 1988, Reference frame effect on shape perception in two versus three dimensions, *Perception*, **17**, 147–63.

Papert, S., 1988, One AI or many?, in Graubard, S.R. (Ed.), *The Artificial Intelligence Debate*, Cambridge, MA: MIT Press.

Pinker, S., 1984, Visual cognition: an introduction, *Cognition*, **18**, 1–63.

Poggio, T. and Edelman, S., 1990, A network that learns to recognize three dimensional objects, *Nature*, **343**, 263–6.

Poggio,T., Edelman, S. and Fahle, M., 1992, Learning visual modules from examples: a framework for understanding adaptive visual performance, *Computer Vision, Graphics and Image Processing*, **56**, 22–30.

Pollatsek, A., Rayner, K. and Collins, W. E., 1984, Integrating pictorial information across saccadic eye movements, *Journal of Experimental Psychology, General*, **113**, 426–42.

Ramachandran, V. S., 1990, Visual perception in people and machines, in Blake, A. and Troscianko, T. (Eds), *AI and the Eye*, Chichester: John Wiley.

Rayner, K. and Pollatsek, A ., 1992, Eye movements and scene perception, *Canadian Journal of Psychology*, **46**, 342–76.

Rosenblatt, F., 1958, The perceptron: a probabilistic model for information storage and organization in the brain, *Psychological Review*, **65**, 386–408.

Rumelhart, D. E., McClelland, J. L. and the PDP Research Group, 1986, *Parallel Distributed Processing: Explorations in the Microstructure of Cognition*, Vol. I: *Foundations*, Cambridge, MA: MIT Press.

Seymour, P. K., 1979, *Human Visual Cognition*, London: Collier Macmillan.

Shepard, R. N. and Metzler, J., 1971, Mental rotation of three-dimensional objects, *Science*, **171**, 701–3.

Shepherd, M., Findlay, J. M. and Hockey, G. R. J., 1986, The relationship between eye movements and spatial attention, *Quarterly Journal of Experimental Psychology*, **38A**, 475–91.

Shepp, B. E. and Ballasteros, S. (Eds), 1989, *Object Perception: Structures and Processes*, Hillsdale, NJ: Lawrence Erlbaum.

Sperber, R. D., McCauley, C., Ragain, R. and Weil, C. M., 1979, Semantic priming effects in picture and word processing, *Memory and Cogntion*, **7**, 339–45.

Tanaka, K., 1993, Neuronal mechanisms of visual object recognition, *Science*, **262, 685**–8.

Treisman, A., 1985, Preattentive processing in vision, *Computer Vision, Graphics and Image Processing*, **31**, 156–77.

Treisman, A., 1986, Features and objects in visual processing, *Scientific American*, November, 106–13.

Treisman, A., 1988, Features and objects, *Quarterly Journal of Experimental Psychology*, **40A**, 201–37.

Ullman, S., 1989, Aligning pictorial descriptions: an approach to object recognition, *Cognition*, **32**, 193–254.

Ungerleider, L. G. and Mishkin, M., 1982, Two cortical visual systems, in Ingle, D. J., Goodale, M. A. and Mansfield, R. J. W. (Eds), *Analysis of Visual Behavior*, pp. 549–86, Cambridge, MA: MIT Press.

Warrington, E. K. and Taylor, A. M., 1973, The contribution of the right parietal lobe to object recognition, *Cortex*, **9**, 152–64.

Williams, L. G., 1967, The effects of target specification on objects fixated during visual search, in Sanders, A. F. (Ed.), *Attention and Performance I*, pp. 355–60, Amsterdam: North-Holland.

Zeki, S., 1993, *A Vision of the Brain*, Oxford: Blackwell.

7

Sensory-motor systems in virtual manipulation

Rupert England

7.1 Introduction

7.1.1 Understanding manipulation

There are many reasons why we would like to use our hands in a virtual environment. Smets *et al.*, in Chapter 9 of this book, discuss the value of 'direct manipulation' (section 9.3.2.2), and Ellis lists many applications for virtual environments which would benefit from the ability to directly manipulate virtual objects with our hands (sections 2.3 and 2.4.1). The problem is, though, that the technology does not exist which allows us to do this well, and we do not really know yet how to go about developing the most effective technology. Manipulation is a complex set of coordinated perceptions and responses, which is affected by many conditions in the real world (learning, environmental conditions, etc.). Manipulation in a virtual world has its own set of complications over and above those of the real world. This chapter presents an overview of the many issues which need to be understood if we are to develop the capability to directly manipulate a virtual object. These issues include the strategy for improving technology, the nature of human behaviour, human sensory systems, the processing of different senses into integrated perceptions, and the control of our responses. The physiological basis of perception and behaviour is described as this provides an important basis for analysing overt psychological phenomena (see section 6.1.1). Particular emphasis is given to the tactile and kinaesthetic modalities, as vision and audition are covered in more depth elsewhere in this book. The implications of these issues for the development of computer-generated virtual environments are discussed.

Stephen Ellis has already expressed his reservations in the terminological jungle surrounding 'virtual reality' (section 2.1.1), and he prefers the term 'virtual environments'. Since Gibson used the term 'cyberspace' in *Neuromancer* (Gibson, 1984) it has often been used synonymously with virtual reality, mostly in connection with computer-generated synthetic worlds or virtual environments,[1] and this term will be employed in this chapter to refer only to immersive, computer-generated virtual reality. Virtual reality is used as a broader term to include any virtual environment, such as remote immersive viewing for telepresence, as well as cyberspace.

7.1.2   Human factors strategy for the optimization of virtual reality

Although many of the concepts behind virtual reality are not new (e.g. stereo and immersive displays, both fixed and mobile;[2] computer-generated pictorial displays[3]), VR is still an underdeveloped technology. Chapters 2, 3, 4 and 5 discuss visually related technological limitations and developments, and Chapter 8 discusses the capabilities of auditory display technology. Although the current problems of manual interaction with virtual reality will ultimately be solved by technological progress, some of the most difficult technological problems will continue to need improvement for some time to come.

The developers of virtual reality can attempt to overcome the temporary technological limitations by using human factors research to enhance existing technology. As Ellis (sections 2.3.7 and 2.4.1) and Findlay and Newell (section 6.1.1) have emphasized, different tasks place different requirements upon the capabilities of any virtual environment. For some simple tasks, a minimal cyberspace will suffice; many applications, however, will require good information to more than one sense. These more demanding applications can benefit most from virtual reality if human factors research is undertaken to analyse task components, and identify precisely what an operator needs in order to complete a task. These 'user requirements' can then be evaluated with respect to what the virtual reality platform can provide. In this manner, any mismatch or unfulfilled requirement may be identified and corrected. Where technological limitations prevent such rectification, informed enhancement or compromise can be adopted to overcome the shortfalls. For example, a cyberspace can be enhanced with information from artificial cues (e.g. a sound or colour to represent thermal properties of a virtual object), or technological trade-offs can be balanced to find the compromise which will allow the best performance possible.

Matching technology to user requirements makes the user's task easier and improves task performance. Naturally some means of assessing task performance is required before such improvements can be verified. While performance on simple task components may be easily assessed, how is one to evaluate a complex activity such as manipulative interaction? This problem confronted teleoperation experts evaluating the performance of operators with a broad range of remotely controlled manipulators, from master–slave actuators to semi-autonomous robots (see Vertut and Coiffet, 1985). The solution involved operational definitions of a system's sensitivities and potential, together with what the operator could achieve. The problem also exists in the field of medicine, where dexterity tests are devised to assess trainee surgeons conducting endoscopic operations (e.g. how long does it take to tie four surgical knots, what is the quality of the knots, how many errors did they make[4]). Developers and researchers can employ similar techniques in virtual reality.

The identification of user requirements also provides a route which can direct the development and evolution of technology. By identifying and quantifying problem areas, human factors and associated research can provide the focus for driving engineering research. In this way, not only can we make best use of what we currently have, but we are also more likely to achieve the improvements needed within a shorter term.

7.1.3   Real problems for virtual solutions

Until technology can provide the required perceptual information in virtual reality, we should perhaps in the meantime be making better use of the knowledge acquired to enhance performance in the real world when conditions are sub-optimal. Such conditions can exist not only on the surface of the planet (e.g. in poor weather conditions) but below it (i.e. mining underground or diving underwater) and above it (i.e. in the atmosphere or in space).

One example of disrupted natural feedback is experienced by astronauts in free-fall orbit around the Earth, who experience weightlessness. Research has been conducted to assess astronauts' performance on estimating the mass of an object in the absence of gravity cues (Ross and Reschke, 1982). These missing cues can be artificially introduced through the interpretation of inertial cues when the object is shaken from side to side. Visual interpretation and judgement of distance are also affected in space where there is high luminance contrast and shadow cues are degraded, and where there are relatively few, if any, interpositional objects or structures for assisting in making relative depth judgements (for a review of visual cues to depth, see Sekuler and Blake (1990, p. 209); also section 3.3).

On the surface of the Earth, similar problems confront us with heavy rain, fog, blizzards, etc., all of which can alter our perceptions either by degrading visibility, making objects and surface features hard to see, or by altering their attributes in such a way as to affect how they are interpreted (e.g. making them seem further away (Ross, 1975), or changing their temperature and apparent weight (see section 7.3.5.5)). Other senses can also be affected (e.g. increased air pressure may alter auditory and olfactory perception, as well as skin pressure displacement thresholds). Cold can affect how we manipulate objects as it numbs the sensitivity of the tactile cutaneous receptors, as well as slowing motor activity.

One way of countering poor visibility is to make things more conspicuous or to improve depth perception with the aid of 'visibility' panels (e.g. warning panels on the back of long or wide-load lorries (Lee, 1976)). Protective clothing shields us from weather extremes but alters how we can perform certain tasks (e.g. wearing thick gloves reduces the capacity for fine manipulation). Ergonomic databases such as MIL-STD 1472C (1981) are used by many engineers to optimize control panel configuration for operators wearing different clothing. A logical extension of these databases would be into the virtual reality domain. A 'virtual reality database' could prescribe designs for virtual objects and device configuration for different virtual reality technologies and for different tasks. This database for sub-optimal virtual reality would stand alongside the database for optimum technological specifications suggested by Ellis (Figure 2.21).

7.1.4 Real behaviour for virtual interaction

Many scientific and cultural advances are founded upon simple concepts and observations. A simple experimental paradigm (e.g. assessing motion after-effects) can reveal intricacies of nature and behaviour (e.g. motion perception: see Wade *et al.* (1993)). Basic questions being investigated in the real world must be revisited in the particular conditions of virtual reality. Thus experimental research must identify: how, when, and where sensory inputs are attended to, integrated and perceived; what effect displayed (and imperfect) information has on perception; how sensory dominance is affected by the quality of sensory inputs; where and how responsive behaviours are controlled. An understanding of these issues through basic research will allow us to enhance or find optimum compromises for technology.

7.2 Perceptuo-motor behaviour

7.2.1 Stimulus and response

Behavioural patterns range from simple and reflexive to complex and voluntarily controlled. All types of behaviour are used in our interaction with environments, whether real or virtual.

In order to create an effective virtual environment it is important to understand how these behaviours are controlled. Although complex behaviour may be more interesting and may be the target of virtual reality designers, simple behaviour can have a large effect on the sense of realism, the usability and the 'side-effects' of virtual reality. This is because reflexive behaviour cannot easily be adapted to the particular conditions of a virtual environment. We should therefore ensure that the virtual environment is adapted to natural reflexive behaviour.

Simple reflexive behaviour is common to all animals. Unicellular organisms, such as amoebae or paramecia, respond to sensory stimuli (or sensory 'feedback'). The response may be positive (i.e. towards the stimulus) or negative (i.e. away from it). Such 'stimulus–response' (S–R) behaviour appears quite simple. Some feature of the stimulus (e.g. the presence of a highly directional bright light) causes a molecular change to the exposed surface of the organism, which induces a specific motor response. Other stimuli may produce other specific responses. In higher animals S–R behaviour is also evident: shine a bright light into someone's eye and it will cause the pupil to contract and the eyelid to blink. Perhaps the light will also result in a hand being raised to shield the eye, or a head movement away from the light. In this case we can observe a number of possible responses to a single stimulus. Some are clearly reflexive, while others are voluntary. It seems that, during the course of evolution, an intermediary process has evolved, lying between the stimulus (input) and response (output). The study of this intermediary information-processing system, together with the sensory input and motor output systems, is the concern of perceptuo-motor research.

Perceptuo-motor behaviour is a multidisciplinary research field, which includes the study of sensation, perception, information processing and attention, motor control, physiology, engineering, and learning. Major theories in each of these areas have been combined in the interests of explaining the achievement of coordinated behaviour. Many of these individual areas have experienced clashes between contrasting theories (e.g. the issues of 'direct' versus 'indirect' perception: see section 3.2) and the field of perceptuo-motor behaviour itself has been similarly affected. The last twenty years have witnessed the rise and fall of many motor paradigms and, as the current rivalries between the opposing camps of 'movement system' theorists and 'action system' theorists attest (see Reed, 1982; Meijer and Roth, 1988), these conflicts continue.[5] According to Abernethy and Sparrow (1992) motor behaviour is set for an inevitable paradigm-crisis. The challenges confronting perceptuo-motor research are exciting, especially with the emergence of virtual reality, where many of the real-world perceptual cues we take for granted are absent.

7.2.2 Sensory feedback

During the course of evolution we have developed senses which allow us to interpret our environment. Senses are groups of modalities which respond to specific forms of environmental energy. For example, there are modalities for different portions of the electromagnetic spectrum (heat, visible light) and for vibratory stimuli (sound, physical contact). There are also modalities for our 'internal' environment, such as baroreceptors which respond to blood pressure. The sense of touch includes modalities responsive to vibration, displacement, temperature and pain. Stimulation is transduced by the different receptors into electrochemical impulses and are eventually interpreted as meaningful constructs.

Sherrington (1906) described three categories of sensations:

- those coming from the external environment (transduced by 'exteroceptors');
- those coming from within the body ('interoceptors');
- and those describing the body's posture and action ('proprioceptors').

Exteroception uses the primary senses of vision, touch, audition, olfaction, and gustation. Interoception depends on internal receptors (e.g. chemoreceptors that monitor blood sugar level). Proprioception comprises two subdivisions: kinaesthesia, a sense defining body and limb posture and limb movements, which is based on feedback derived from joints and muscles; and the vestibular sense, which defines balance and head position, and is derived from saccular, utriclar, and semicircular canal feedback. (These sensory systems are all described in section 7.3, below.)

The terminology has become somewhat confusing, however, as some authors consider Sherrington's terms to overlap. Gibson (1966, 1979) considered vision to be both extero-ceptive and proprioceptive. Lee and Aronson (1974) suggested a further category of 'exproprioception' to describe the feedback about position or body movement relative to the environment. 'Visual kinaesthesia' gives visual information about body position (Gibson, 1950), and auditory sensation can provide feedback about movements (e.g. contact sounds between limbs or objects and surfaces such as footsteps: see Henderson (1977); or reflected sound giving elevation cues: see section 8.2.3). It seems appropriate then to suggest an 'auditory kinaesthesia' as well as visual kinaesthesia, which contributes to exproprioception.

Major types of 'sensory-motor' feedback for manipulation include cutaneous and kinaesthetic receptors, vestibular apparatus, vision and audition (e.g. Winstein and Schmidt, 1989). The information from many of these sources overlaps and offers redundancy, thus decreasing the uncertainty of perceptual cues (i.e. improving the signal-to-noise ratio). The loss of such redundancy, as may happen in virtual reality, could lead to changes in the performance of manipulation tasks.

Engineering terminology has also been used to describe feedback in biological systems. 'Open-loop' behaviour occurs in the absence of useful feedback, and is ballistic (once programmed and initiated it cannot be modified, as is the case with a rocket which can be aimed at the moon, but cannot be controlled after launch). 'Closed-loop' behaviour occurs when useful feedback is employed during the behaviour; it incorporates a comparative process between a desired status (e.g. a moon-trajectory) and an actual status (e.g. an off-course trajectory) and attempts to correct movement through a reduction in the error signal (returning the rocket to a moon-trajectory). This reduction in deviation from a desired state is an example of closed-loop control employing 'negative feedback'; a biological example of this form of feedback is used when reaching for an object with your hand. Sometimes in closed-loop systems a deviation from a status is amplified, rather than reduced, as is exemplified by an explosion, or by the errors often made when trying to trace by hand a shape seen only via a mirror. This type of closed-loop control is termed 'positive feedback'. Woodworth (1899), describing the performance on a task requiring subjects to move a stylus rapidly between two points, distinguished two forms of motor behaviour: 'initial impulse' and 'current control' movements. Initial impulse behaviour can be considered as open-loop, current control behaviour as closed-loop with negative feedback.

Another associated term is 'feedforward', which can occur in open-loop situations (i.e. where no feedback is available) and where system performance is relatively stable (Rosenbaum, 1991). Stability allows accurate prediction and hence feedforward (e.g. a constant target velocity and trajectory allows one to anticipate where the target will be). The vestibulo-ocular reflex (see section 7.3.4) is controlled by feedforward (Cotman and

McGaugh, 1980, p. 478). The nervous system employs feedback and feedforward in synaptic inhibitory mechanisms (Carpenter, 1987).

7.2.3 Skilled behaviour

A novice confronted by a complex new task is faced with a number of initially separate task elements. Somehow these have to come together in a coordinated manner. How is this done? There are stages in skill acquisition which result in the grouping of associated task elements into sub-tasks (e.g. in driving, the sub-task of changing gear may group task elements like depressing the clutch, locating the gear stick, selecting the appropriate gear, releasing the clutch, etc.). Some groupings may not be appropriate; the novice learns which ones to concentrate upon. Legge and Barber (1976) refer to this organizational aspect of skill acquisition in terms of 'functional units' of skill. They suggest that as a skill improves, changes occur within these functional units (e.g. there is an increase in the number of movement sequences controlled by a single functional unit (see p. 118)).

A key feature of this learning process concerns feedback. During the early stages of skill acquisition the dependence upon feedback is high. For example, in locating the gear stick the learner must look for it and coordinate both reaching and grasping activities in order to manipulate it. Over time, the learner acquires a more accurate spatial representation ('mental model'), and selecting the gear no longer requires visual feedback. Tactual feedback (see section 7.3.5.1) becomes perfectly adequate for this sub-task. This 'fixing' of the gear stick's location in the driver's behaviour pattern is equivalent to the 'hardwiring' of an invariant subset of the driving skill. On demand, the stick-locating subset can be activated, triggering a behavioural 'fixed action pattern' (see Lorenz and Tinbergen, 1938): it becomes automated.

7.2.4 Perceptuo-motor behaviour in virtual reality

System designers and developers have examined the nature of human sensory feedback loops and have attempted to incorporate interfacing components to retain these loops in virtual reality. A primary impetus came from Sutherland's graphical computer interface research (Sutherland, 1965, 1968) which retained head-movement and visual feedback loops. Indeed the main approach to cyberspace, and virtual reality in general, has been from a visual perspective, as seen through the development of, among others: head-slaved stereoscopic visual systems in teleoperation activities (e.g. Comeau and Bryan, 1961; Goertz *et al.*, 1965), visually oriented interaction methods in computer research (e.g. Krueger, 1991), and pictorial information displays in aviation projects (e.g. Furness and Kocian, 1986).

Interaction with virtual environments slowly received more attention with the development of peripherals such as the wand-like 'Sorcerer's Apprentice' (Vickers, 1973), and the inclusion of audio feedback for auditory target localization (e.g. in the 'Super Cockpit': Furness and Kocian (1986)). It was, however, with the development of devices like the 'Sayre Glove' (see Kreuger (1991) for a 1976 photograph from Sandin, Defanti and Sayre) and Zimmerman's (1980) design, which was translated into VPL's commercially available 'DataGlove', that NASA's highly interactive virtual environment research programme captured the scientific headlines (e.g. Fisher *et al.*, 1986, 1988).

The feedback loops currently available for perceptuo-motor behaviour in virtual reality are very different from those in the real world. Even when sensory feedback is closed-loop (e.g. vision), the loop is indirect and unreliable (see, for example, Chapter 5). In other cases, the loop is wide open (e.g. touch, smell, taste, and even hearing). This means that it is more difficult to learn to control perceptuo-motor behaviour in virtual reality, especially those behaviours which are carried out in the real world under different feedback loops.

As mentioned above, with the acquisition of skill, certain behaviours require less perceptual feedback. This might mean that the perceptual requirements in virtual reality will be less for skilled behaviour which was learned in the real world. Even experienced drivers, however, glance occasionally at the gear stick, especially if in a new car. This could reinforce or re-calibrate the mental model defining its location. In this way, the retention of skilled behaviour is subject to continual refinement and reinforcement. This might have interesting effects in virtual reality. Skills learned in the real world (e.g. fine manipulation) could become incorrectly 'updated' through inadequate virtual feedback in a simulation of a real world task. Using real world skills in virtual reality with poor feedback would in fact be de-skilling with respect to the ability to use this skill in the real world. This, of course, could have serious implications when using virtual reality for training purposes.

7.3 Sensory neurophysiology

7.3.1 Peripheral and central nervous systems

It is with sensory physiology that technology must interact in order to provide virtual reality. Issues such as where in the nervous system particular behaviours are controlled, or how a sense of 'wetness' can be simulated, require physiological explanation. It is important to know which aspects of a virtual environment people can learn to deal with, and which aspects will affect natural behaviour patterns. The study of the physiological basis of sensation and behaviour will allow us to optimally design the physical interface between virtual reality technology and the human 'machine'. The following overview of sensory neurophysiology will also assist in understanding subsequent discussions (in sections 7.4 to 7.8) of sensory integration and the control of responses. It should be pointed out that much physiological knowledge has been acquired from the study of other animals. Eckert and Randall (1978) describe the fundamental processes that are collectively termed 'life' as processes that are shared in common by all animals. In this sense the molecular and electrical activities that produce a nerve impulse in the human brain are fundamentally the same as those producing neural impulses in squids or rats. Much of our knowledge regarding human physiology has originated from what we have learned from both vertebrates and invertebrates. Thus, the discussion that follows cites a number of animal physiological studies because of their relevance to human physiology.[6]

The variety of sensory feedback mentioned in the previous sections depends upon specialized receptor cells. Eyes contain specialized visual receptors, the rods and cones; ears contain specialized auditory receptors, the inner and outer hair cells; the nose contains olfactory receptor cells; and the skin contains a range of cutaneous receptors (e.g. mechanoreceptors, thermoreceptors and nociceptors (pain receptors); see section 7.3.5).

Physiological studies reveal much about the structure and functioning of the sensory and motor systems. As a simple metaphor, the neural structure of these systems can be imagined as either ends of a tree: at one end is the (sensory) input system with a mass of root-hairs which converge into major roots that become concentrated into the trunk (the central nervous system); at the other end is the (motor) output system with branches

138   *Simulated and virtual realities*

dividing into an ever-increasing number of twigs and leaf stems. The sophistication of neural morphology and function is highest in the central nervous system (CNS).

The neuron is frequently described as the basic functional unit of the nervous system. It comes in many shapes and sizes (e.g. multipolar or bipolar, according to the configuration of the axon) but typically comprises impulse-receiving processes called dendrites, a cell body, and an axon (i.e. a fibre) along which an electrochemical neural impulse is transmitted (see Figure 7.1). Axon length may vary from a few microns up to ~ 1 m and may be encapsulated in 'Schwann cells' containing myelin (a fatty insulating substance). The presence or absence of myelin determines the white or grey colouring of neural tissue respectively (e.g. white matter or grey matter evident in a cross-section of the spinal cord). Axons terminate in varying numbers of presynaptic terminal branches, the synapses of which adjoin other neurons via 'synaptic junctions' on the dendrites or cell bodies. A synaptic 'cleft' physically separates the neurons at these junctions but impulses cross by chemical transmission (transmitter substance).

Nerves permeating the body comprise bundles of neurons. According to location and function they are divided between different nervous systems. While the neurons of the brain and spinal cord form the CNS, the neurons conducting impulses between the CNS and the rest of the body form the peripheral nervous system (PNS). The neurons in the PNS which convey impulses towards the CNS are called sensory or afferent fibres, and neurons conveying impulses away from the CNS are called motor or efferent fibres. There are 31 major pairs of PNS nerves, both efferent and afferent, connected to the spinal cord and a further 12 pairs (cranial nerves) connected to the brain. Collections of nerve cell bodies are termed nuclei (if they occur within the CNS) and ganglia (if they occur outside the CNS). Nerve fibres are classified according to the size of their diameter (for example, Aβ, Aδ and C fibres, or groups I, II, III and IV fibres).

The CNS is responsible for receiving and integrating sensory information and for producing appropriate responses in the light of past experience (see Romanes, 1976). A brief description of CNS morphology will assist in understanding the sensory-motor pathways described later on.

The CNS comprises four areas: the forebrain, midbrain, hindbrain, and spinal cord. The forebrain includes the cerebrum (which has two hemispheres connected by the corpus callosum), and groups of nuclei called the thalamus (which has one lobe per cerebral hemisphere), the hypothalamus and the mis-named basal ganglia. The bark-like surface of each cerebral hemisphere earns it the name of cortex. The midbrain includes the crura cerebri and the colliculi. The hindbrain includes the pons, the cerebellum (which also has two hemispheres) and the medulla (see Figure 7.2).

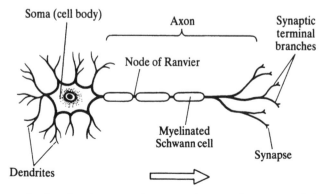

Figure 7.1 Generalized multipolar (motor) neuron. Arrow indicates signal direction.

The two cerebral hemispheres (forebrain) are connected to the pons (hindbrain) by the crura cerebri which are two broad neural bundles constituting the anterior portion of the midbrain. The oculomotor nerves emerge from this region of the midbrain. On the posterior surface of the midbrain (the tectum) are the four small swellings known as the colliculi. These are buried between the cerebrum and the cerebellum. The cerebellum plays a vital role in motor control.

The midbrain, pons, and medulla oblongata (the conical protrusion from the pons to the medulla) are also referred to as the 'brainstem'. The ventral surface of the medulla oblongata is divided into two longitudinal ridges astride a deep median fissure. These ridges are known as the 'pyramids' because they taper to a point. They are bundles of nerve fibres originating in the cerebral cortex and descending through the cerebrum, crura cerebri, and pons. Some of the descending pyramidal fibres cross over the fissure at the 'pyramidal decussation', to form the 'lateral corticospinal tract'; others continue without crossing to form the 'anterior corticospinal tracts'. Both sets of tracts, because of their location, are also referred to as the pyramidal tracts and are concerned with manipulation and other fine skilled voluntary movements (Carpenter, 1987, p. 229). A number of pathways, both ascending and descending the CNS, cross over at various stages, so that for some neurological processing the right side of the CNS is concerned with the left side of the body and vice versa (this is known as contralateral processing).

Impulses from the bodily senses, collectively termed somatosensory impulses, ascend via the spinal cord and brainstem (i.e. reticular formation) to the thalami and cortex and are classified as either 'lemniscal' or 'reticular'. While the former are rapid and reliable, the half-dozen reticular pathways convey more diffuse impulses (Cotman and McGaugh, 1980). Associated with general states of activity, especially in the cortex, the reticular activating system (RAS) also receives inputs from across the brain and projects to the cortex via the thalami, modifying levels of arousal (Carpenter, 1987).

A cross-section through the spinal cord reveals a medially pinched oval structure (see Figure 7.3). Embedded within the white interior (myelinated fibres) is an H-shape of grey matter (unmyelinated). The upper tips of the H lie dorsally (posterior) and are referred to as the dorsal horns; ventral horns lie anteriorly. Afferent fibres have cell bodies in the dorsal root ganglia (outside the cord) with most fibres entering the dorsal horns (on both sides). Efferents exit primarily via the ventral horns and have their cell bodies within the

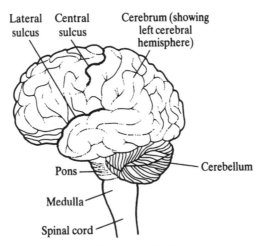

Figure 7.2   Stylized lateral view (sagittal plane) of the brain and spinal cord.

cord. Fibres have collateral connections and also travel up and down the cord in tracts, which are arranged in dorsal, lateral, and ventral columns. Their destinations determine their name (e.g. 'spinocerebellar': cord to cerebellum; 'corticospinal': cortex to cord).

The convolutions in the cortex arise from twisting ridges, or 'gyri', and trenches, or 'sulci'. Some of these (e.g. the central and lateral sulci) have a relatively constant location in different individuals and therefore assist in identifying cortical lobe divisions (i.e. frontal lobe, parietal lobe, etc.; see Walsch, 1978). Both cerebral cortices are divided into similar lobes, and neurons with specific sensory-motor functions have been associated with these (e.g. motor neurons in the frontal lobe; see Figure 7.4).

Sensory impulses from the receptors are carried along the afferent and ascending pathways towards the sensory areas. The primary visual sensory pathways converge on the visual cortex (also known as striate cortex) located in the occipital lobes; the auditory pathways project to the auditory cortex in the temporal lobes; and the somatosensory (bodily sensation, e.g. touch) pathways project towards the primary somatosensory cortex (SI) in the parietal lobes (i.e. postcentral gyrus).

A portion of the cortex in the frontal lobe (i.e. precentral gyrus) is the primary motor cortex (MI), and other cortex in the frontal lobe is dedicated to secondary sensory-motor regions (SII, MII). Much of the remaining unassigned cortex in and surrounding these

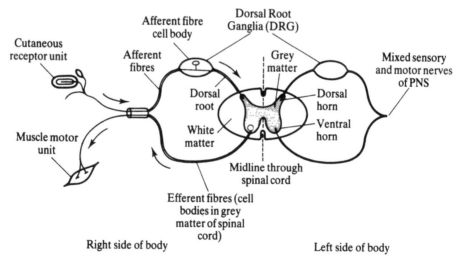

Figure 7.3   Schematic transverse section through the spinal cord. Arrows indicate direction of impulses.

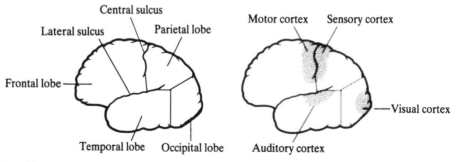

Figure 7.4   Lobic divisions of the brain and sensory-motor cortex.

regions is known as 'association cortex' (Cotman and McGaugh, 1980), responsible for inter-sensory associations, memory and speech.

The distinction between sensory and motor structures is somewhat arbitrary as many neurons (e.g. in the superior colliculus) share sensory-motor properties. Evidence opposing classical sensory and motor differentiation in the cortex shows some areas may serve both roles (e.g. Brooks, 1969; Evarts and Fromm, 1977). Laszlo (1992, p. 48) reports that these findings are largely ignored by psychologists working on motor control. One example of shared responsibility is the neurons responsible for controlling saccadic eye movements, which also respond to sensory stimuli (Mohler and Wurtz, 1976). Despite common morphology, however, neural activity is temporally differentiated by utilizing different impulse frequencies, timing patterns, durations, etc. (Stein and Meredith, 1993). Section 7.5.4 discusses the association between sensory and motor neural structures. Luria (1973) proposed a tri-zonal division of the postcentral cortex into zones: a primary zone (projections for modally specific sensations), a secondary zone (intra-sensory integration), and a tertiary zone (for inter-sensory integration).

An important point (which will be revisited in section 7.3.6) is that not all incoming signals reach the upper cortical levels of the CNS, at least not directly. Some incoming stimuli effectively short-circuit the higher sensory levels and route instead through intermediary neurons in the spinal cord, directly to motor output. These stimuli are the reflex-response initiators. Higher level reflexive behaviours also exist and the superior colliculi are known to play an important role in these.

Within the nervous system, a subsystem of nervous control known as the autonomic nervous system (ANS) controls involuntary muscles, endocrine glands and the heart. Unlike voluntary muscles, involuntary muscles (e.g. around the blood vessels and in the eyes) are usually innervated by complementary dual efferents, one excitatory and one inhibitory, which correspond respectively to the parasympathetic and sympathetic components of the ANS (see Green, 1977, p. 204). Thus the parasympathetic system controls constriction of the pupil of the eye and the sympathetic system inhibits this constriction.

7.3.2 Vision

Visual perception is discussed in detail in other chapters (Chapters 3, 4, 5, 6), so this section will be limited to the underlying neurophysiology.

Light falling onto the eyes is focused onto the retina (see section 4.3.1 for a description of the optics of the eye). The light sensitive retinal surface of each eye comprises about 125–127 million specialized photoreceptors: rods (sensitive to dim light) and cones (sensitive to bright light and colours) which communicate in turn with horizontal, bipolar, and amacrine cells before converging on the 1 million or so retinal ganglion cells which transfer the neural impulses along the optic nerve (Hubel, 1988). There are approximately 120 million rods and 7 million cones for each eye (Bruce and Green, 1990, p. 23), and they are distributed differently over the retina. Cone distribution is concentrated in the central visual field, in a ~ 1 mm wide retinal indentation referred to as the fovea. It is the region of the eye with highest acuity and sensitivity to colour. Rods are mainly distributed in the surrounding periphery, which can respond to lower light levels than can the fovea. A receptor-free region known as the blind spot marks the exit-point of the optic nerve.

Two retinal pathways can be described: a highly specific three-stage direct path (receptor-to-bipolar-to-ganglion cell) and a diffuse four- or five-stage indirect path involving horizontal and amacrine cells. At the fovea, a single cone feeds a single ganglion cell (in the optic nerve), while further in the periphery many receptors converge on a single bipolar and many bipolars converge on single ganglion cells. Overall, this results in a 125:1 retinal receptor to optic nerve convergence (Hubel, 1988).

The optic nerves from both eyes meet at the optic chiasma. Chiasma derives from the Greek word 'chi' for cross (Sekuler and Blake, 1990) and it is at this location that a cross-over of the optic nerves occurs. Each eye has a left and right visual hemi-field, with the fovea being represented in both. While the left eye's left (temporal) hemi-field exits the chiasma ipsilaterally along the left optic tract, the right (nasal) hemi-field crosses over to exit the chiasma contralaterally along the right optic tract. Fibres from the right eye's hemi-fields behave conversely (see Figure 7.5). Thus both hemispheres receive input from both eyes, but only from one half of the visual scene (Hubel, 1988).

Most of the optic tract fibres project to the lateral geniculate nucleus (LGN) one of many nuclear bodies that collectively form the thalamus (of which there are two: left and right).

The cortical response to retinal stimuli reveals a topographical mapping of the retina onto the visual cortex, although the map is highly distorted. For example, ~ 15 mm of cortex corresponds to 1° at the fovea, compared with the 1 mm of cortex corresponding to 1° at peripheral eccentricities of the retina (Lennie *et al.*, 1990). Such mapping is referred to as retinotopy (or visuotopy: Stein and Meredith, 1993), and is also evident in the three-layered superficial layer of the superior colliculus (Stein and Meredith, 1993).

The pattern of sensitivity of a single neuron, or the collective sensitivity of a group of neurons converging on a higher level neuron, is known as a 'receptive field'. At low levels, visual receptive fields respond to simple, specific stimulation (such as whether the centre of the field is stimulated when the periphery is not: e.g. Kuffler (1953), Enroth-Cugell and Robson (1966); for a review see Lennie *et al.* (1990)). Neuronal convergence results in receptive fields of increasing size or complexity with each successive level up the processing hierarchy (e.g. hypercomplex cortical cells: Hubel, 1988).

There is evidence that there are different visual pathways and different visual systems. Bruce and Green (1990) describe two prominent visual pathways. One projects from the retina to the superior colliculi (the retinotectal pathway), and appears to have a central role in the head–eye turning reflex (i.e. turning towards the source of visual stimulation: Romanes (1976)). The other pathway projects from the retina to the visual cortex, via the LGN (the geniculostriate pathway). Damage to this pathway can result in apparent blindness, but with the ability still to point to visually identified objects ('blindsight':

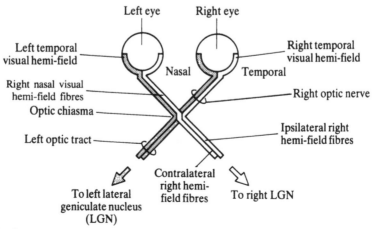

Figure 7.5 Cross-over of visual pathways at optic chiasma.

see Weiskrantz *et al.* (1974)). The topographical mapping is evident throughout both pathways.

The primary visual area of the striate cortex (V1) is not the end of the visual pathway(s). Some twenty or more areas in the extrastriate cortex beyond V1 have been identified including: V2, V3, V4 and the middle temporal (MT) areas (e.g. see Lennie *et al.*, 1990). Many maintain retinal topography and have receptive fields responsive to very specific complex stimuli (e.g. faces: Perrett *et al.* (1982)). Reciprocal projections extend between these extrastriate areas and V1. Two major subdivisions of the geniculostriate pathway have been suggested: the ventral and dorsal streams (Ungerleider and Mishkin, 1982; see also section 6.1.2). The dorsal stream interconnects the spatiotemporal (motion) sensitivies of specific extrastriate visual areas. This forms a potential motion analysis pathway. The ventral stream interconnects visual areas best suited for form and colour analysis (e.g. Zeki, 1973). The origins of these separate pathways can be traced back to the parvocellular layer (dorsal stream) and the magnocellular layer (ventral stream) in the LGN (Lennie *et al.*, 1990). Goodale and Milner (1992) cite evidence, however, that these streams may not be as independent as first thought. A third pathway to the primary visual cortex has also been described by Casagrande (1994).

The concept of different visual systems is not new. Research in the 1960s examining the role of the superior colliculus also resulted in an indication of separate visual systems. The primary role of the superior colliculus in the head–eye turning reflex was observed and investigated by Sprague and Meikle (1965), and Schneider (1967, 1969). Schneider showed that while damage to the superior colliculus showed contralateral neglect of this orientation response (i.e. damage to the left half of the colliculus resulted in neglect of the right visual field), lesions in the visual cortex resulted in discriminatory problems between visual patterns. These early experiments thus led to the proposal of two separate visual systems, one locatory in nature (vision for 'where?') and one inspectory (vision for 'what?').

The division of the visual system into parallel 'what' and 'where' subsystems was also advocated by Ingle (1967), Trevarthen (1968), and Held (1968). This differentiation provided:

- focal vision (foveal exploratory vision for identification and discrimination, which also enables fine manipulation: the 'what?' system);
- ambient vision (peripheral detection of motion, responsible for orienting in space and guiding gross movements: the 'where?' system).

Conscious (focal) vision appears to be a function of the cortex and ambient vision appears a function of the superior colliculi (and possibly other structures). Current thinking still reflects this visual dualism, although the proposed physiological correlates have shifted from a tectal–cortical divide to a ventral–dorsal divide of the extrastriate (post-VI) pathways (e.g. Ungerleider and Mishkin, 1982).

7.3.3 Audition

The normal human auditory system is binaural. Each ear is a tri-functional structure comprising the outer, middle and inner ear. The structure of the ear is described in Chapter 8 (section 8.2.1).

The ciliated hair cells attached to the basilar membrane in the cochlea can be of two types. Inner hair cells (IHC) form a single row that runs the length of the basilar membrane, near where the tectorial membrane attaches to the cochlear duct. They number approximately 3500. Outer hair cells (OHC) are aligned in 3–5 rows and number approximately 12 000. The majority (90–95%) of the auditory nerve connections are with the IHCs; only 5–10% are with OHCs (Sekuler and Blake, 1990). Each auditory nerve comprises some 50 000 nerve axons. As with retinal ganglia cells, impulses pour continuously along the axons; hearing a sound thus denotes a change in the background activity and represents an increased firing pattern. Each axon has an associated minimal intensity sensitivity and this varies for different frequencies.

The left auditory nerve projects to the left cochlear nucleus. The right side mirrors this configuration. Primary aural processing is monaural, but processing beyond this level (e.g. in the superior olivary nucleus, inferior colliculus, medial geniculate nucleus (MGN), and auditory cortex) receives input binaurally (Schiffman, 1990). Binaural cues include interaural time differences and interaural intensity differences (both important for localization). Neurons sensitive to small binaural intensity differences have been identified (e.g. Masterton and Imig, 1984).

With progression towards the auditory cortex, the analysis and interpretation of auditory sensations change. Initial processing is of frequency analysis but the neural sensitivities to the temporal and intensity characteristics of the impulses steadily attenuate along the cortical pathways in favour of much more specific neural sensitivities (e.g. to complex sounds such as vocalizations). Auditory signals passing through the MGN, which also receives input from the reticular system, may be influenced by arousal (e.g. fatigue may inhibit perception of a sound).

See Chapter 8 for a review of audition and auditory virtual environments.

7.3.4 Vestibular system and orientation

Attached to the cochlea within the inner ear lie fluid-filled organs for maintaining balance, posture, and head orientation. These organs belong to the vestibular system, and comprise: the saccule (or sacculus), utricle (or utriculus), and three orthogonal semicircular canals. The primary role of the saccule and utricle is to provide feedback concerning linear motion and body position with respect to gravity, while that of the semicircular canals is to provide feedback about head orientation based upon rotary acceleration (Schiffman, 1990). Gravity-sensing receptors employ free-moving elements within fluid-filled cavities. These cavities are called otocysts, and they enclose free-moving otoliths, which react to gravity and linear motion. Embedded within the inner lining of the saccule and utricle lie sensory hair cells which make up the 'macula', numbering some 19 000 in the saccule; 33 000 in the utricle (Howard, 1986). Here the high-density otoliths are suspended above the hair cells by a gelatinous mass; their inertia causes hair cell deformation during movements, triggering neural firing. The semicircular canals are connected to the utricle, widening at one end into the 'ampulla'. Within the ampulla lies the free-swinging 'cupula', a gelatinous encapsulated tuft of sensory hairs attached to vestibular nerve fibres at the cupula's base (the crista). Thus a rotary head movement displaces the cupula, deforming the hair cells and initiating a signal to the medulla and cerebellum in the brain, from where motor efferents may conduct impulses to body muscles (neck, limbs, etc.) or to ocular muscles (for eye movements). When head rotation stops or when the rate of rotation is constant differential hair cell deformation terminates. The vestibular organs control the adaptive vestibulo-ocular reflex which is responsible for maintaining visual stability of a tracked object during head movements (see section 5.1.2.1).

The vestibulo-ocular reflex (VOR) is a reflexive movement of the eyes which allows us to maintain a steady fixation on a visual stimulus even if we are moving around. Signals from the vestibular system trigger compensatory movements of the eyes. A metric for assessing the VOR is its gain (velocity of eye movements divided by velocity of head movements). A gain of $1 \cdot 0$ results in optimal compensation. As the VOR is adaptive, the gain can change. For example, if visual images are viewed via a lens, doubling the lens power doubles the size of the image on the retina, as well as the velocity of any image movements across the retina. This requires an increase in the gain (doubling from $1 \cdot 0$ to $2 \cdot 0$) in order to maintain optimal performance. In order for this adaptation to occur, a simultaneous exposure to both head and eye movements is required. The nervous system attempts to correlate the two movements (see Lisberger (1988), cited by Rosenbaum (1991, p. 27)).

Earth's gravitational field is a stable environmental feature which organisms have evolved to utilize. Organisms in unnatural environments, such as zero gravity, experience problems (space or motion sickness). Thus these vestibular cues appear to have a strong influence on behaviour and perception, but there is evidence that visual cues can be stronger. In a static flight simulator, visual cues for motion can soon override vestibular cues for non-motion, although the conflict between the two senses can result in simulator sickness (see Pausch *et al.*, 1992). It is possible to provide a vestibular sensation of movement with movement platforms which travel only small distances but can control acceleration and deceleration to stimulate the vestibular system (see section 2.3.2). Thermal stimulation, through irrigation of the auditory canal with water, is known to cause sensations of movement. Temperature changes cause expansion or contraction of the endolymph within the semicircular canals and thus cupula displacement.

In immersive virtual reality, head motion is tracked and reported to the image generator which then creates the new viewpoint. As the head moves the normal VOR will operate, but there may be image generation processing delays. Slow head movements with consistent image update lags may allow compensatory gain adjustments, but the more 'natural' high accelerative head movements (see McDonnell *et al.*, 1990)[7] cannot currently be matched by existing virtual reality technology. Even slow head movements may result in variable rather than constant lag which makes it difficult for VOR adaptation. (See Chapter 5 for a review of human factors research on display lags.)

7.3.5 Taction

7.3.5.1 Tactile and kinaesthetic perception

The term taction is used here in the same sense as Boff and Lincoln (1988) used the term 'tactual perception', that is, to include the tactile and the kinaesthetic modalities. It is important to be aware of the difference between tactile and kinaesthetic perception, but at the same time to realize that these two senses are often used together, as, for example, when touching objects. From taction we can discern an object's solidity, shape, surface texture and even what it is made of. Although we can perceive some object attributes passively (i.e. by being touched by an object) we gain most tactual information by active touching. Gibson (1962) suggested that 'active touch' is the tactual equivalent of ocular scanning for vision. It is interesting to note that both Katz (1925) and Revesz (1950) considered the hand itself to be a sensory organ, and that tactual explorations provided a 'haptic' experience. Etymologically, 'haptic' originates from Greek and means 'able to lay hold of' (Gibson, 1966, p. 97). Haptic perception is therefore a combination of tactile and kinaesthetic feedback, but is also a subset of taction, as not all tactile and kinaesthetic perception is concerned with touching objects.

Tactile information comes in part from a variety of cutaneous receptors found at varying depths throughout the skin. Their distribution varies tremendously (e.g. the lips and fingertips are densely packed with receptors; the skin covering the back is sparsely supplied). Johansson and Vallbo (1983) have reported the glabrous (hairless) surface of a hand to contain as many as 17 000 receptors. Two types of neuro-tactile properties have been identified by Vallbo and Hagbarth (1968): slowly adapting (SA) fibres and rapidly adapting (RA) fibres. It is known that some fibres (punctate fibres) support small well-defined receptive fields, while others (diffuse fibres) have larger fields with poorer boundary definition. Mapping of the skin has revealed a distribution of the punctate fields (called spots) varying in their sensitivities to physical stimulation, such as warm thermal spots, cold thermal spots (Postman and Egan, 1949).

Sensory fibres (afferents) innervating the skin are categorized by size: unmyelinated fibres are ~0·24-1·3 μm in diameter; myelinated ones are ~1·3-14 μm in diameter. The latter are divided into: Group I fibres with a mean diameter of 8 μm; and Group II fibres with a mean diameter of 2-2·5 μm (Iggo, 1982). The peripheral termination of all these afferents can be either free-nerve endings or encapsulations. Free endings, abundant in hairy skin, are usually innervated by small, unmyelinated afferents (e.g. Group C and A) while encapsulated receptor afferents are mainly Group Aβ (Group II) (see Carpenter, 1987).

Tactual sensations are detected by three main groups of receptors: mechanoreceptors, thermoreceptors and nociceptors; with at least 15 morphological and functional distinctions (Iggo, 1982). With further reference to Iggo, these may be described as follows:

1. Mechanoreceptors
   - RA mechanoreceptors: Pacinian corpuscles (PCs)
                                             hair follicles
                                             Meissner corpuscles (MCs)
                                             Krause end bulbs
   - SA mechanoreceptors: SA I: Merkel cells/disks
                                             SA II: Ruffini endings
   - C mechanoreceptors
   - additional sources of mechanoreception: Golgi tendon organs
                                             muscle spindles
                                             joint receptors
2. Thermoreceptors
3. Nociceptors
   - Mechanical nociceptors
   - Thermal and mechanothermal nociceptors

These receptors are described in more detail below.

7.3.5.2 *Mechanoreceptors*

The mechanoreceptors (innervated by larger myelinated afferents) are sensitive to skin deformation, pressure and hair movement, and fall into two categories depending on response characteristics of their afferent fibres: RA mechanoreceptors (with rapidly adapting fibres) and SA mechanoreceptors (with slowly adapting fibres).

RA mechanoreceptors
RA mechanoreceotors are present in both hairy and glabrous skin (hairless fine-ridged skin of the hands and feet) and respond maximally to movement. Displacement velocity is frequency-coded. A constant velocity gives a proportional discharge of impulses, following a power function relationship:

$$R = aS^b$$

where R is the response, S is the stimulus, a is a constant, and b is the exponent (Iggo, 1982). RA mechanoreceptors include: Pacinian corpuscles; hair follicles; Meissner corpuscles; Krause end bulbs.

Pacinian corpuscles (PCs). Pacinian corpuscles are small, grey, encapsulated and pearl-shaped ($\sim 0 \cdot 5$–2 mm long) found deep in the skin (Iggo, 1982) and also in the mobile joints of the skeletal system (Schiffman, 1990). The lamellar structure surrounding their elongated nerve terminal allows successive layers to slide over one another, so sustained deformation causes only transient discharge. PCs can detect mechanical vibration over a frequency range of 70–1000 Hz but remain insensitive to large stimuli outside this range. The threshold varies with frequency but is < 1 μm for frequencies of 200–400 Hz. Blowing softly on the skin is sufficient to excite PCs. Winstein and Schmidt (1989) describe PCs as having an important role in adaptive responses during precision grip (citing Johansson and Westling, 1987). PCs are insensitive to changing skin temperature.

Hair follicles. Nerve endings wrap around the root of the hair follicle with rod-like terminations. They attach to the basement membrane of the root sheath and corium, and are enclosed in Schwann cells. These units discharge only when the hair moves. Several kinds have been identified according to differences in sensitivity, rate of firing, and size of receptive fields. Hair follicle units are equivalent to the (glabrous) Meissner corpuscles.

Meissner corpuscles (MCs). Meissner corpuscles are encapsulated, comprising a helical array of sheet-like nerve endings interspersed with insulative Schwann cells, arranged orthogonally to the MC's longitudinal axis. MCs are found in glabrous skin, packed into dermal papillae filling the epidermal grooves. The upper surface of the MCs transmits movement of the overlying epidermis to the nerve endings, via collagen fibres. They are velocity-sensitive, discharging only during skin movement, with a vibration sensitivity ranging between 10 and 200 Hz. Sekuler and Blake (1990) report that while pre-teen children have a density of 40–50 MCs mm^{-2}, 50-year-old adults have only ~ 10 mm^{-2}. This drop corresponds with age-related loss of tactile sensitivity (Thornbury and Mistretta, 1981).

Krause end bulbs. Krause end bulbs are found in glabrous skin and come in two types: cylindrical and globular. They have a simple lamellar capsule structure, with axons terminating in either a rod (within cylindrical capsules) or spiral (within globular capsules). They were once thought to be cold thermoreceptors (e.g. Von Frey, 1895: see Kenshalo, 1971), but on further investigation none were found beneath cold-sensitive dermal regions (Kenshalo, 1971, p. 121) and they have now been shown to be velocity-sensitive like MCs (Iggo, 1982). They have a sensitivity range between 10–100 Hz and do not respond to sustained skin deformation.

SA mechanoreceptors
These slowly adapting mechanoreceptors are sensitive to both skin deformation and sustained displacement. They provide longer-term information about skin changes. As skin displacement velocity levels to a sustained rate of deformation, the rate of firing decreases to one that is proportional to the degree of deformation. This rate is determined by the type of SA unit involved. Iggo (1982) describes two kinds based on adaptation time differences: SA I and SA II.

SA I. These units are usually inactive when no stimulus is present. They fire impulses in irregular trains, and can respond at very high frequencies to rapid stroking. Cooling is known to briefly enhance their response to sustained deformation. This effect of skin temperature on SA units explains the temperature–weight illusion (see section 7.3.5.5). Merkel cells are an example of SA I receptors.

Merkel cells/disks. Merkel cells are found at the base of the epidermis. The SA I afferent axon that supplies them terminates in a disk-shaped expansion, known as a Merkel disk, which is located at the base of each Merkel cell.

SA II. These units have a resting-level discharge rate (~ 2–20 s^{-1}), have low response thresholds to skin stretching, and fire impulses in regular trains. Ruffini endings are an example of SA II receptors.

Ruffini endings. Ruffini endings are encapsulated units found deeper in the dermis than MCs and Krause end bulbs. They provide a continuous intensity measure of sustained pressure and dermal tension. Similar receptors are found in and near certain joints. Ruffini endings were once mistaken for warm-thermoreceptors (Von Frey, 1895: see Kenshalo, 1971).

C-mechanoreceptors
C-mechanoreceptors are found in the dermal and epidermal layers of hairy skin, are small and numerous, have no visible endings, have small receptive fields, and respond slowly to deformation or hair movement. Their sensitivity decreases with repeated stimulation until after 20–30 s they become unresponsive, unlike the continuous responses from SA I and SA II units. They are thought to have a role in the sensation of itching (Iggo, 1982).

Additional sources of mechanoreception
In addition to the above are the mechanoreceptors found in the viscera (e.g. stretch receptors, flow receptors and also some Pacinian corpuscles) and skeletal muscles. The latter provide kinaesthetic data from Golgi tendon organs, muscle spindles, and joint receptors.

Golgi tendon organs. The Golgi tendon organs (GTOs) are mechanoreceptors composed of small tendon fascicles and are found in tendons at musculo-tendon interfaces. These receptors are found in series with muscle fibres, with ~ 10 muscle fibres per receptor. They are extremely sensitive to muscle contraction (tension) and communicate with the CNS via single Ib afferent nerves which ultimately inhibit the alpha motor neurons of the contracting muscle. The response appears to counter the muscle spindle activity.

Muscle spindles. Laszlo (1992) describes muscle spindles as the main kinaesthetic receptor (citing: Burgess and Clark, 1969; McCloskey, 1978; Matthews, 1964; Praske *et al.*, 1988). These receptors comprise fluid-filled capsules, of ~ 2–4 mm length and ~ 2–300 μm diameter, which terminate on exterior sheaths of neighbouring muscle fibres. Inside each is a small number of modified muscle fibres (intrafusals). These have contractile ends and non-contractile centres with nuclei arranged either in rows or bunches. This nucleic distribution distinguishes between two types of intrafusal fibres: nuclear chains and nuclear bags. A typical muscle spindle has ~4 chains and ~3 bags. Muscle spindles are connected to two kinds of afferents: large primary fibres (group Ia) with annulospiral terminations on nuclear bags and chains; and small secondary fibres (group II) terminating in annulospirals and flower-sprays, primarily on nuclear chains. Primary afferents are RA, firing maximally at a frequency proportional to the rate of stretch, while secondary afferents are non-adapting, discharging in proportion to degree of muscle stretch. Muscle spindles also receive efferent innervation from fibres which are ~6 μm in diameter and come in two forms: static (linked to chains) and dynamic (bags). These fusimotor fibres, when stimulated, elicit a stretch-reflex and also increase stretch sensitivity. Thus the CNS uses them to control muscle spindle sensitivity and the adaptation properties of their afferents (Carpenter, 1987).

Joint receptors. Joint receptors relay limb position from specialized endings in the ligaments and capsules of joints. Their multiple morphologies are innervated by group II afferents (RA for rate of change, and SA for limb position). Thus limb position is partly frequency-coded and partly determined by spatial location of units discharging.

7.3.5.3 Thermoreceptors

Thermoreceptors discharge continuously at given skin temperatures and variably with a change in temperature. They do not respond to mechanical or algogenic (i.e. painful chemical) stimuli. Receptive fields of individual receptors are very small and spot-like (e.g. ~1 mm^2), corresponding to recognized warm and cold spots (Iggo, 1982). They are non-overlapping, and separated by large areas of skin. On the hand cold spots are 5–10 mm apart, while warm spots are some 15 mm apart (Carpenter, 1987). Cold receptors respond maximally between 25 and 30°C. Their firing frequency increases as the temperature falls. They discharge bursts of grouped impulses that are temperature-dependent. Warm receptors respond maximally between 40 and 42°C. Their firing frequency increases with rising temperature and they discharge regular impulse patterns. The dynamic sensitivity of thermoreceptors enables them to detect slow thermal changes (e.g. <1°C in 30 s). Thermoreceptive afferents are small. In non-human primates cold units are myelinated with 1–2 μm diameter fibres while warm units are unmyelinated (Iggo, 1982). Thermoreceptive C fibres exhibit similar properties to A fibres but more of them respond to cooling.

Thermo-perception
The perceived temperature is not an object's true temperature. It is based upon the temperature differential between the object and the area of the skin in contact with it. The thermal receptors respond to the object's thermal conductivity (i.e. how much heat is drawn from the skin). In 1960 Gibson referred to this thermal sensation as 'touch temperature' (Gibson, 1966). A consequence of this is that an object's temperature can be perceived differently depending on whether the skin is hot or cold.

If the skin is at a stable temperature it can be considered as being at physiological zero. Thermal stimulation represents a change from this. Temperatures above are perceived as being warmer, those below as being colder. The physiological zero is not, however, a point but a range extending across a neutral zone in which no thermal variations are perceived. The range extends 0·1°C, from +0·05°C to −0·05°C around the stable temperature (Postman and Egan, 1949). Schiffman (1990) reports the size of the neutral zone can be as wide as 8°C.

7.3.5.4 Nociceptors

While some C fibres respond to temperature and light touch (along with A fibres), the remaining Aδ and C fibres serve nociception providing the sensations of pricking and burning pain (Carpenter, 1987). Although mechano- and thermo-receptors respond to intense stimulation, they respond maximally to less intense (innocuous) stimuli. Thus while these units may contribute to the pain experience it is from receptor units discharging under intense levels of stimulation (called nociceptors) that pain is sensed. The fact that morphine desensitizes nociceptors but not other cutaneous receptors reveals the modal independence of nociceptors. Schiffman (1990) states there are more pain spots than pressure and thermal ones. Two main classes of nociceptors have been identified: mechanical and thermal (or mechanothermal) nociceptors, both of which terminate in free-nerve endings.

Mechanical nociceptors
Mechanical nociceptors respond to epidermal puncture by sharp objects (e.g. pin-pricks), to firm pressure, (e.g. squeezing) and to crushing of the skin, but are insensitive to light touch (e.g. stroking) and skin temperatures (unless very high, e.g. $>50°C$, and even then there may be a 30 s firing delay). In primates mechanical nociceptors can have myelinated or unmyelinated afferents. They can provide rapid warning of injury with axon conduction at 50 m s^{-1} for Group II afferents (Iggo, 1982).

Thermal and mechanothermal nociceptors
Whilst thermal and mechanothermal nociceptors respond primarily to severe thermal stimuli ($2-3°C$ and $43-45°C$) they also respond to mechanical nociceptive stimuli. Their predominantly unmyelinated afferents provide maximum conduction velocities of $\sim 10 \text{ m s}^{-1}$ (Iggo, 1982). Polymodal mechanothermal nociceptors can respond to mechanical, thermal or algogenic (e.g. acid or histamine) stimulation (Iggo, 1982).

7.3.5.5 Receptor combinations

Combinations of different cutaneous receptors provide our experiences of many tactile qualities (e.g. oiliness, wetness, etc.). Sullivan's study (Sullivan, 1923) on the perception of liquidity (i.e. vaporous, wet, oily substances), semi-liquidity (i.e. gelatinous, slimy, greasy), semi-solidity (i.e. muddy, mushy, soggy), and solidity (i.e. doughy, soggy, dry) concludes that all perceptions of liquidity are based on the fusion of pressure and temperature. The role of this fusion is weaker in the perception of solidity, in which the primary input 'pattern' is from pressure. Sullivan describes these symbolically. Thus, where (p) represents weak pressure cues and (P) strong pressure cues, and where (T) represents strong temperature cues and (t) weak temperature cues (Sullivan, 1923):

$$\text{Liquidity} \propto \text{Solidity}$$

$$pT \quad \propto \quad Pt$$

The sensation of burning hot comes from the simultaneous excitation of both warm and cold receptors. The cold spots have two thresholds responding to low and high (i.e. not warm) temperature (Carpenter, 1987).

Winstein and Schmidt (1989) describe evidence that cutaneous receptors on the fingers and thumbs serve a supportive role for kinaesthetic sensations elicited from other nearby sources. They report a study by Clark *et al.* (1983) showing that if fingertips are anaesthetized, this impairs people's sensitivity to slow joint rotations (but anaesthetizing the skin surrounding the interphalangeal joints does not have this effect). Also, anaesthetizing the skin of the thumb diminishes subjects' sensitivity to slow displacements of the index finger. In contrast, anaesthetic injected into the joint results in no observed impairment.

Stevens (1979) describes a weight–temperature illusion which arises from the interaction between thermo- and mechano-receptors: a cold coin feels heavier than the same coin when it is warm.

7.3.5.6 Tactual pathways and projections

The complex interaction between the tactual receptors is evident from the organization of the tactual pathways and projections into the cortex. Each of the 31 pairs of dorsal roots in the somatosensory pathways is responsible for innervating a specific region of skin (a dermatome). For example, for each half of the body, there are 12 thoracic nerves which correspond with 12 dermatomes. Each point on the body is mapped by at least two dorsal roots (some by three, e.g. the nipple). Touch dermatomes overlap more frequently than thermal or nociceptive ones (Carpenter, 1987).

Four main classes of large tactual neurons have been identified in the dorsal horn: Class 1 neurons respond only to cutaneous mechanoreceptors; Class 2 respond to mechano-receptors and nociceptors (and may continue firing for several seconds after termination of noxious stimuli); Class 3 respond only to nociceptors; and Class 4 respond to thermoreceptors. The receptive fields of individual dorsal horn neurons are topographically organized (Iggo, 1982).

Columns and tracts

The spatial arrangement of fibres in the cord varies according to their size (larger axons are located medially and smaller axons laterally). This segregates the larger mechano-receptor afferents from the smaller nociceptive and thermoreceptive ones.

Large ascending fibres (A fibres) send branches upwards in the dorsal columns that terminate in the dorsal column nuclei (the cuneate and gracile nuclei) of the medulla oblongata. Here, second-order fibres cross sides and continue upwards as the medial lemniscus to the ventral posterolateral (VPL) thalamus and then on to the cortex. These A fibres and second-order fibres form the lemniscal system which is concerned with touch (see section 7.3.1). A second somatosensory pathway, the anterolateral system, involves smaller afferents which synapse with second- and third-order neurons and has axons which also cross sides to ascend as part of the spinothalamic projection. This pathway subdivides into the neo- and palaeo-spinothalamic pathways, which are concerned with temperature, pain and light touch (Carpenter, 1987).

Iggo (1982) describes the following ascending sensory tracts in the spinal cord: spinocervical tract; dorsal column; spinoreticular tract; spinothalamic tract. The spinoreticular and spinothalamic tracts are both crossed. The former terminates in the lateral cervical and dorsal column nuclei prior to cross-over and enters the VPL thalamic nucleus. The spinothalamic tract projects to three thalamic centres: VPL; Po (posterior group); and IL (intralaminar nuclei). The pathways using this tract continue to the somatosensory cortex.

Dorsal columns. Formed primarily by large myelinated afferents for cutaneous mechanoreceptors, these columns have an important tactile role. The axons terminate in a somatotopic map configuration within the dorsal column nuclei. While the trunk and legs are represented in the gracile nucleus, the arms are represented in the cuneate nucleus. The nuclei differentiate into anterior reticular regions and posterior clusters. Primary afferents make up the clusters whilst the reticular zone comprises second-order afferents originating from Class 1 and 2 dorsal neurons. Ascending fibres from the gracile and cuneate nuclei project via the medial lemniscus to the ventrobasal nuclei of the thalamus (or VPL), maintaining their somatotopy. From here they project to the somatosensory cortex.

Spinothalamic tract. This tract comprises axons ascending from neurons in the contralateral dorsal horn (classes 1–4), and thus carries all cutaneous impulses. Nociceptive and thermoreceptive inputs to the cord cross over within two vertebral segments. Damage to this tract leads to a contralateral loss of both pain and thermoreception. It projects up through the brainstem to the thalamus, terminating in three somatosensory regions: the VPL; the Po; and the IL nuclei. Interactions between different cutaneous modalities occur in each of these areas, along with input from other sensory tracts.

Spinocervical tract. This is an ipsilateral spinal pathway, comprising mainly classes 1 and 2 dorsal neurons (conducting mechanoreceptive and nociceptive impulses). Hair follicle afferents are abundant in this tract, which projects via the dorsolateral funiculus to the lateral cervical nucleus. It is here that ascending axons decussate, traversing the medial lemniscus to terminate in the contralateral thalamic VPL.

Spinoreticular tract. Neurons located more deeply within the spinal cord project to the contralateral reticular formation of the brainstem. Owing to convergence of superficial dorsal neurons, the receptive fields of afferents in this tract can be very complex (e.g. large, bilateral and multimodal: they may include mechanoreceptors and nociceptors). The tract is thought to influence sensory awareness (e.g. via arousal and alertness).

7.3.5.7 Somatosensory cortex

Somatosensory area SI
The final destination of cutaneous inputs is in the cortex. On histological grounds, Brodmann classified the postcentral gyrus into areas (numbered: 3a, 3b, 2 and 1) which constitute the primary somatosensory area (SI). Most afferents projecting to SI originate in the thalamic VPL and IL nuclei. The SI cortex responds to contralateral body stimulation, and to stimuli to both sides of the scalp and face. Representation in the somatosensory cortex is distorted with larger areas dedicated to lips and fingers.

There is a columnar organization of the cortex, with all cells in a column responding to stimuli from the same modality. Each column receives input from specific afferents, from 'commissural' fibres which bring output from corresponding columns in the opposite hemisphere, and from other ipsihemispheric columns.

The primary projection of the specific cutaneous receptors is to Area 3b, with each associated cortical column being driven by a particular modality or by a specific receptor from a specific skin location. Other areas are associated as follows:

- Area 3a receives primary projections from muscle and joint receptors;
- Area 1 receives convergent projections from both cutaneous afferents and from muscle and joint receptors;
- Area 2 receives highly convergent projections and cells have complex properties (e.g. feature interaction);
- Areas 3 to 1 all project to parietal association areas (Area 5 and then to Area 7);
- Area 7 receives input from other cortical sensory areas, and responds to cutaneous, visual, and auditory stimuli.

Somatosensory area SII

A secondary somatosensory projection area (SII) is found on the superior wall of the lateral (or Sylvian) sulcus. It is smaller than SI and shows extensive convergence of afferent input and receptive fields. SII represents the body bilaterally and is differentiated into anterior and posterior divisions: the anterior SII is driven tactilely, receiving the same thalamic VPL input as that projected to SI. The posterior SII responds to nociceptive stimuli, receiving input from the Po thalamic nuclei. SI also receives input from the contralateral SI. SII receives input from both SI and SII.

7.3.5.8 Efferent control of sensory input

The afferent somatosensory pathways are subject to continuous supraspinal control. Efferents originating in the somatosensory cortex control lower levels; pyramidal tract efferents can directly control sensory nuclei in the medulla and cord; and brainstem efferents can exert more indirect control. Corticothalamic efferents can control which afferent impulses reach the cortex and which are ignored. Such switching mechanisms, recognizable in many relay stations along the sensory pathways, improve 'sensory contrast'. Iggo (1982) describes how the skin of the glans of the penis and clitoris shows an absence of Pacinian corpuscles and Merkel disks. This makes it relatively insensitive to light touch but the sensations experienced testify to the ability of the CNS to modify sensory impulses from these (and other) regions. This modifying role of the CNS is also prominent in the perception of pain (hence the variance in people's pain thresholds).

7.3.5.9 How many tactual systems?

It has been shown that there are at least two visual systems (see section 7.3.2). Taction comprises both kinaesthesia and tactile perception. Are there possible sub-categories of these senses? How useful is it to consider taction as a single sense? There is some reason to suggest that there are several possible tactual senses, and this has been discussed for many years. Thompson (1967) describes Head's (1920) functional differentiation of somatosensory sensations into two divisions: epicritic (specific light touch and pressure) and protopathic (primitive, diffuse sensations of touch, pain and temperature).

Defining the senses according to the physical stimulation, the sensation evoked, or the physiological processing seldom results in clear sensory divisions (Carpenter, 1987; see also Lee and Lishman, 1975, p. 87). Thus, although we may define sensations as kinaesthetic or tactile, specific receptors contribute to both (e.g. Ruffini endings). Also, kinaesthetic perception not only gains sensory input from its own dedicated receptors but also from other senses (e.g. visual kinaesthesia, auditory kinaesthesia: see section 7.2.2). (If vision and audition can supply corresponding feedback for kinaesthesia, then the remaining senses might do likewise, giving 'olfactory' and even 'gustatory' kinaesthesia.)

On this basis kinaesthesia could be differentiated as two processes:

- *Primary kinaesthesia* — from dedicated local receptive elements (e.g. muscle spindles, articular receptors, GTOs, vestibular system, etc.).
- *Secondary kinaesthesia* — combination of primary kinaesthesia with other senses (e.g. visual kinaesthesia, auditory kinaesthesia).

7.3.6 Multiple senses

There may be more senses than are generally recognized; also the distinction between senses may not always be clear-cut. Vision, for example, combines the visual cues of retinal disparity with the oculomotor cues of accommodation and convergence when sensing depth or distance. In this sense, visual processing uses sensory information from both vision and the kinaesthesia. Similarly, manipulating objects combines tactile and kinaesthetic sensations (and often vision). Vision and audition are bilateral, and processing from the two sides is combined. It was suggested earlier that senses might have primary and secondary inputs (section 7.3.5.9), and ambient and focal systems (section 7.3.2; see also section 6.1.2).

In view of these complexities, it would seem more appropriate to investigate the 'pluralistic' nature of sensory perception, rather than one sense in isolation.

The dualism described by ambient and focal processing can be generalized to conscious and unconscious activity. We often feel our actions are conscious and voluntary, but many behaviours (as in a reflex) are not under our direct, conscious control. Spinal preparations (severing of the spinal cord from the brain) show that complex coordinated activity can still occur without cortical control (e.g. a wiping reflex in frogs, and spinal 'walking' mechanisms: see Shik *et al.* (1966)). This implies that some activities that we assume to be under conscious control may actually be reflexive and higher-level processing of sensory input is by-passed. If sensory feedback becomes degraded (as in virtual reality) then certain reflexive components of such 'complex' motor behaviours may be inappropriately activated or absent, resulting in uncoordinated behaviour which is not easily modified by conscious (or high level) control. This theme will be considered further in sections 7.5 and 7.8.1 below.

7.4 The nature of perceptual integration

7.4.1 Multi-sensory cues

Interaction with the environment must involve the combined use of cues from different senses. Each sense has its own range of perceptual cues. The visual system responds to a multitude of visual cues pertaining to the perception of shape, form, size, colour, depth, motion, etc.; the tactual system responds to a variety of tactile and kinaesthetic cues (e.g. contact, thermal conductivity, solidity, shape, texture, weight). Perception of an environment is, however, not restricted to the intra-sensory interpretation of cues. Cues from different senses are somehow interpreted together (as with the example given previously of stereoscopic distance perception: section 7.3.6).

7.4.2 Cue redundancies, hierarchies and dominance

In a natural environment, there is a rich source of cues providing more information than necessary for desired behaviours ('redundancy'). These cues can interact to reduce any ambiguities. Studies with visual illusions have shown that when cues are reduced to a minimum and there is some ambiguity, stimuli can be perceived in different ways. Thus six triangles forming a hexagram can be perceived as separate triangles (local percept) or as a cube without hidden-line removal (global percept) (see also Navon, 1977). Multiple cues offer high levels of redundancy and can improve signal-to-noise ratios (e.g. lip reading was shown to modify the auditory cortex, enhancing auditory perception: Sams *et al.* (1991), described in Stein and Meredith (1993)).

One interesting issue is how these different sources of information are all combined to form what we might call holistic percepts. Cinema sound systems are currently undergoing a revolution as special effects become more demanding. The central sound source for speech, however, does not follow the screen speaker's position and yet is perceived to emanate from the character's mouth. In this case, the spatially separated sources of sound and vision are perceptually integrated. This particular form of sensory integration is called the 'ventriloquist effect' (Howard and Templeton, 1966). It is fairly tolerant, as the sound and vision sources can also be temporally separated or even unrelated, as in a dubbed foreign film. Even after we become aware of the odd lip movements that seemingly have no connection with the temporally asynchronous soundtrack, we can still accept the speech as belonging to the character concerned. It is probably fair to assume that a high level of top-down processing is overriding such irregularities and maintaining one's suspension of disbelief. With eyes closed, the sound source is more accurately located (e.g. coming from the foot of the screen). The ventriloquist effect is thus an example of visual 'dominance', or the visual 'capture' of attention (Tastevin, 1937; Hay *et al.*, 1965; Posner *et al.*, 1976). Diderot, in 1749, offered early support for the existence of sensory dominance (Morgan, 1977). Gibson (1933), Rock and Victor (1964), Rock and Harris (1967) and Power (1981) have demonstrated a dominance of vision over taction when one object is seen but another is felt.

If the quality of information in a sense is poor, as is usually the case with virtual environments, it might be expected that the dominance of that sense would be reduced (see Ellis's Figure 2.5 in Chapter 2). Indeed, Heller (1983, 1992) reported tactual capture when vision is blurred, with normal taction dominating impaired vision. This raises concerns for virtual environments. If sensory information is impoverished compared to the real world, this may affect the ability to integrate information from different senses by upsetting the natural dominance hierarchy (or hierarchies). Poor feedback may undermine the level of dominance of a sense or cue (but it may be possible to compensate for the poverty of one sense by the enhancement of another, for example, objects changing colour when touched).

Lee and Lishman (1975) describe a common assumption that vision plays a minor role in balance (i.e. that balance is mainly controlled by the vestibular and kinaesthetic systems). They report on an extension of the Lee and Aronson (1974) study on infant balance and provide evidence that vision plays an integral role in human stance control. This 'visual proprioceptive control' is shown to dominate over non-visual information (toddlers were caused to fall over in response to the visual stimulus of a 'swinging room': Lee and Aronson (1974, p. 530)). Lee and Lishman (1975) also describe the fine-tuning role visual proprioception plays in learning a new stance (i.e. ankle–foot proprioception).

Katz (see Kreuger, 1970, p. 341) observed that different materials (e.g. leather, cloth, rubber, paper) can be easily confused by blindfolded subjects. This suggests that vision has a major role in making things feel the way they do.

In immersive virtual reality, if there are spatial or temporal mismatches in feedback to different senses (e.g. the virtual hand is not quite in the same position as the real hand), adaptation of the subordinate sense to the dominant sense can occur. Adaptation is more difficult in semi-immersive virtual reality, where conflicts between cues within a sense can occur (e.g. seeing a real hand control a virtual object: Schmandt (1983)).

7.4.3 Integration processes

One way in which we can control how we process sensory information is through the use of attention. If we direct our attention to certain events or objects we can enhance the perception of some things and reduce or avoid perception of other things. In fact, some qualities of our environment can only be perceived if we use our attention to integrate various characteristics which define them ('feature integration' as proposed by Treisman and Gelade (1980): see section 6.3.2). For example, if an object is identified by a particular combination of shape and colour (a small round copper penny among other coins), we need to combine our perception of shape and colour in order to recognize it. Not all perception requires attention, however, and many studies have shown that there is a difference between perceptual processing which can be carried out in parallel and that which can only be carried out in series (e.g. Broadbent, 1958; Neisser, 1967). It is the latter which is usually associated with the use of attention, because we can apparently only attend to a limited amount of information at one time. Attention can be a focus of processing on a particular spatial location, on a particular characteristic (e.g. all green things), or on a particular object (e.g. a heavy, rattling, oval-shaped wrapped birthday present). Research has shown that it might be possible to divide or switch attention rapidly between different locations or senses, so that we can perform concurrent tasks (e.g. Shaffer, 1975). It has been suggested that attention may be modified like a searchlight to be spread either thinly and widely or concentrated in a smaller area (Posner *et al.*,1980). Attention may be split between different spatial locations (Muller and Findlay, 1987).

Other studies have shown that more attention is required at the early stages of learning a behaviour, and as skill is acquired, behaviour becomes 'automated' and can be carried out as a parallel activity (Schneider and Shiffrin, 1977). This does not mean that we can carry out an unlimited number of automated activities. The concept of 'resources' has been used to explain why some activities can be carried out in parallel while others cannot. Some models of resources suggest that there is a central 'pool' of effort, which can be distributed to different activities according to the difficulty or demand of each (e.g. Kahneman, 1973). When this pool is used up, effort must be removed from some activities in order to carry out new activities. Other models of resources (e.g. Navon and Gopher, 1979) suggest that there are separate pools of effort for different sets of activities, based on which senses are being used (e.g. visual tasks or auditory tasks). If one pool is fully utilized by one activity, this does not necessarily affect the availability of effort from another pool for another activity. Thus two activities using the same pool may interfere with each other, while a third activity using a different pool may be unaffected.

Thus attention might allow the integration of sensory information by focusing processing, and by relating parallel activities.

Given the widely variant characteristics of the senses, how do we combine them to perceive an integrated environment and to achieve coordinated behaviour? It is possible that the unmatched senses may be transformed by some processing function that effectively 'normalizes' them. The normalized senses would thus be superimposable or integratable. This notion gains support from Stein and Meredith (1993) who propose that the combination of different modal sources must require some form of commonality for them to integrate efficiently. The nature of such a hypothetical integration function would be crucial to the understanding of modal interaction in manipulative tasks (which require visuo-motor coordination). Previous studies have investigated cross-modal matching but it is still uncertain as to where and when any integration function is applied or indeed what the nature of the integration function is (e.g. Connolly and Jones, 1970; Newell *et al.*, 1979; Greening, 1993).

Stein and Meredith (1993) have suggested that if stimulus intensity is represented in the CNS by frequency coding of action potentials then perhaps multi-modal stimuli can be judged on the same scale because of commonalities in neural activities (e.g. frequency coding). This may well represent the means by which such proposed scaling functions operate.

It may be that integration is carried out at different processing levels, according to the particular behaviour it serves. This proposition would seem quite plausible in light of the different visual systems proposed by Schneider (1967), Trevarthen (1968), Ungerleider and Mishkin (1982), Casagrande (1994) and others, and the finding of physiological correlates for such divisions. Specific sets of cues (both inter- and intra-sensory), may integrate more efficiently through such functions. This would represent a cross-sensory elaboration of Treisman and Gelade's feature integration theory (Treisman and Gelade, 1980).

This would imply that multi-sensory perception may not always be cortical. Some of the physiology associated with sensation has been described (section 7.3), tracing afferent pathways to cortical regions and to their 'perceptual centres'. Stein and Meredith (1993) suggest that some sensory integration occurs at the superior colliculus, which is a lower centre of processing than the cortex. In addition, as we have seen in the descriptions of the sensory pathways, a great deal of integration (e.g. neural convergence) occurs en route to, and at the cortex (see sections 7.3.2, 7.3.3, 7.3.5.6). Given that conscious action appears to be primarily cortical, this suggests some sensory integration is subconscious.

There is evidence that different senses are not always appropriately integrated, as some people experience 'synaesthesia'. For example, on hearing certain musical tones these people can experience seeing colours, or feeling cutaneous sensations. The latter is an example of 'sonogenic synaesthesia' (Stein and Meredith, 1993). Many types of synaesthesia have been reported. The experiences are non-volitional and appear to be permanent. Synaesthesia is important because it represents a case where, as its meaning implies, a sensation is experienced in a part of the body other than the one that has been stimulated. How is this possible? Does this represent sensory confusion? Given the convergence of sensory input pathways, evoked phantom (synaesthetic) sensations appear quite plausible phenomena (i.e. through neuronal cross-talk).

The superior colliculus is one of many sites in the CNS where visual, auditory and somatosensory stimuli converge (Stein, 1984), sharing neurons and mapped representations. Shared neurons can be involved in multiple circuits performing multiple roles. This is made possible by state-dependent impulses that alter a neuron's membrane characteristics so that it can be used by different circuits at different times (Hooper and Moulins, 1989). Multi-sensory neurons can convey both inter- and intra-modal data simultaneously. Despite this sharing, the uniqueness of each input can be spatially retained even in the presence of others (Stein and Meredith, 1993). Thus an auditory cue and a visual cue will not affect the representation of each other's receptive fields, although interactions do occur and these can enhance or depress the overall level of activity within the neuron. In an examination of cat sensory neurons in the superior colliculus, Stein and Meredith (1993) report finding a neuron that responded to very low intensity auditory stimuli but whose response was suppressed when the eyes were covered or the room lights were turned off. The neuron apparently only responded bimodally. The authors refer to some amusing conclusions regarding this late-night discovery: cats are deaf at night; blind cats are deaf!

Further experimentation revealed that responses to bimodal stimuli were far more vigorous (in terms of number of impulses, peak frequency, and duration) and reliable (Meredith and Stein, 1986) than responses to unimodal stimuli. Thus while a visual response occurred on only 6 out of 16 trials with few impulses, and auditory responses were even less reliable than visual ones, combined visual–auditory responses were always prevalent on every trial (mean number of impulses up by 1207%: Stein and Meredith (1993, p. 124). Such response enhancements were evident in every multi-sensory grouping (i.e. visual–auditory; somatosensory–auditory; somatosensory-visual; and trimodal). Stein and Meredith

(1993) also describe cases in which a stimulus did not produce a response at all when presented alone, but did when combined with stimulation in other modalities. In contrast, multi-sensory inhibition was also observed, with neurons which are excited by one modality remaining inactive in the presence of stimuli from other modalities (p. 128).

This research has important implications for multi-sensory interactions and stimulus effectiveness in virtual reality. Inter-modal enhancement occurs most when one modality has minimally effective stimuli (Stein and Meredith, 1993, p. 143) and so in circumstances where many such stimuli exist, strong inter-modal responses may occur. In virtual reality, as in the real world, these will combine with other cues in other modalities to produce a strong response. If the stimuli in a virtual environment are not accurate or appropriately matched, however, minimally effective stimuli may be strengthened by inappropriate multi-modal interactions and thus be detrimental to the perception of virtual reality.

Animal evolution has maintained a high proportion of topographic representation throughout the pathways and brain. Even ancient structures such as the midbrain do this, suggesting it is a fundamental quality of processing. Stein and Meredith (1993) conclude that such sensory maps and their subsequent multi-sensory alignment provide a basis for enhancing stimuli in spatial and temporal registration while inhibiting those that are not spatially or temporally related. This facilitates the detection of meaningful but weak stimuli, whilst screening distracting ones. Thus in virtual reality the reaction not only to weak virtual stimuli but also to misregistered ones may also prove to be of concern.

7.5   The nature of response

7.5.1   Coordinated and skilled responses

The usual connotation of the term response is of reaction, and of body movement (motor behaviour) such as walking, smiling, grasping, or manipulating some control device such as a steering wheel. These motor outputs are not random, but are coordinated and repeatable. The achievement of coordination suggests that some sort association between functions has been 'programmed'. But the form of such programming is unknown. Are there specific programmes for specific responses? How are such programmes stored and then selected? According to Paillard (1960, cited by Laszlo, 1992) no two response-programmes are ever totally identical. Body posture may be slightly different from when the response was used before, so too may be the level of dynamic activity, the degree of muscle fatigue, the level of arousal and motivation, etc. If each response is different, then an enormous repertoire of one-off response-programmes would be necessary. The notion of a specific hardwired response for every eventuality also seems impracticable. If the programme is not specific, a flexible form of storing coordinated behaviour is necessary.

How can coordinated motor behaviour arise so rapidly in light of the complexity of motor control (e.g. motor equivalence: the ability to perform a given task in a number of different ways)? What process decides which of the numerous possible alternative movements (Bernstein, 1967) to select and activate? The number of muscle fibres that would require activating in order to implement a simple finger movement would be immense. It seems implausible that every individual muscle response should require cortical involvement.

7.5.2   Types of motor response

What types of motor behaviour are there? Painting is usually a conscious, voluntary process but snatching your hand away from a hot kettle requires little if any thought. Usually you become aware of a reflex (involuntary action) during or after it has been initiated. These two types of motor response delimit two extremes. Does anything lie in between? Can further distinctions be made within voluntary and involuntary actions? In what category are well-learned or skilled behaviours (e.g. playing the guitar or piano)? It would appear that some motor responses can be performed without conscious attention and yet clearly, they are not pure reflexes. Innate reflexes may be considered physiologically hardwired while well-learned motor behaviours are 'programmed'.

The 'searchlight' metaphor employed in the field of visual attention (see section 7.4.3), could also be applied to motor control. According to this metaphor, processes being actively attended to are 'illuminated', while other processes take place 'in the dark'. Thus some motor responses may be conscious, while others are carried out without awareness. Similarly, Trevarthen's (1968) dichotomy of focal and ambient vision can be extended to motor control. In this way, focal motor responses are described as illuminated and conscious behaviours, whilst ambient motor responses can be considered reflexive and/or non-reflexive, automatic and/or semi-conscious.

The nature of motor control for focal or ambient responses is very important as it dictates the requirements for interacting in virtual reality. If a response is focal and feedback is poor, then the response may not be correctly activated, but it will be still open to conscious error correction. Alternatively, if a motor response is ambient then, once triggered, it may function correctly irrespective of feedback quality because it is a fixed-action pattern. It may however be incorrectly triggered or omitted because it is not under conscious control.

7.5.3 Types of motor control

Abernethy and Sparrow (1992) describe four main themes of motor control theories:

1. closed-loop theory (e.g. Adams, 1971);
2. motor-programming/schema theories (e.g. Henry and Rogers, 1960; Keele, 1968, Schmidt, 1975, 1980);
3. the impulse-variability theory of Schmidt *et al.*, (1979);
4. dynamical oscillatory theories of motor control (e.g. Kelso *et al.*, 1981; Kugler *et al.*, 1982; Kugler and Turvey, 1986; Schmidt *et al.*, 1993).

Adams (1971) proposed that motor learning and control was dependent on refining the reference (perceptual trace) through negative feedback. One problem with this closed-loop theory is in explaining how animals and humans produce movements in the absence of feedback. There have been a number of reports describing the performance of motor activities in the absence of feedback, including deafferented (relevant afferent nerves severed) monkeys which still could reach, climb, and walk (e.g. Taub, 1976). Studies have also been made of deafferented people, including a gunshot injured patient (e.g. Lashley, 1917) and people with degenerative neural diseases who, even with no limb (kinaesthetic) feedback, could draw in the air and make rapid arm movements (e.g. Rothwell *et al.*, 1982). These rapid arm movements, however, were clumsy and, after 30 s, of poor accuracy (Rothwell *et al.*, 1982, described by Summers, 1989). Thus feedback would seem to be necessary for fine control.

Open-loop theories (such as 2 and 3 above) stress the sequencing and timing aspects of behaviour which they consider to be controlled by motor programmes. It has been proposed that the CNS contains neural circuits called central pattern generators (CPG) which can produce different patterns of coordinated interlimb motion (Grillner, 1985). There is evidence for CPGs controlling locomotion (e.g. the brainstem stimulation of a spinally prepared cat can produce coordinated walking (Shik *et al.*, 1966)). Summers (1989) suggests therefore that a complete walking programme exists in the spinal cord. Even CPG behaviour, however, still requires feedback for fine-tuning (Grillner, 1985).

Keele's original theory of motor control (Keele, 1968) was that the motor programme was centrally stored and pre-structured, comprising a sequence of efferent commands which on execution allowed specific movement patterns to occur without the need for sensory input. This concurred with the studies showing corrective movements before the feedback could be processed (e.g. Pew, 1966).

One problem with Keele's theory was that each action would require a programme to run it. Given the enormous range and number of action possibilities, this would require a fair degree of neural representation and would pose problems for new actions. These anomalies are described by Abernethy and Sparrow (1992; citing MacNeilage, 1970, and Schmidt, 1982), and they suggest a less specific and more flexible programming refinement, with only essential (variable) features requiring storage.

Schmidt's (1975) 'schema' theory of motor learning was a hybrid of open and closed theories (see also Keele and Summers, 1976). The schema concept centres on the storage of a generalized rather than a specific motor programme, which could be enhanced via available feedback. The evidence in favour of this theory is at best equivocal (e.g. van Rossum (1989) in Abernethy and Sparrow (1992)).

Reflex-stepping behaviour has been observed in young infants (Thelen and Fisher, 1982). Thelen (1987) reports that one-month-old babies reveal a walking pattern when their weight is supported. This evidence suggests the existence of an innate pattern generating mechanism that undergoes differentiation and integration with other developing systems (Thelen *et al.*, 1987). Carpenter (1987) suggests that one of the first actions of the developing brain is to suppress this primitive 'walking' drive to allow the brain to develop its own more sophisticated walking patterns that draw upon integrated feedback.

Studies have shown an inability to perform certain types of movements following some apraxias (i.e. damage to the cerebral cortex). Carpenter (1987) employs an analogy of hierarchical control with an army, which reflects the classic symptoms of motor cortical damage. Drawing upon wide sources of information, the general makes global plans while the subordinate officers detail the logistics and tactics, and their subordinates elaborate on them further with their detailed local knowledge. The consequences of the death of the general are different from the death of a soldier. The general's death will have a delayed effect as the army still functions as though nothing is wrong. Only after time do subtle effects become evident through a lack of long-term planning. The death of a soldier or platoon results in an immediate specific loss of function. Further, the loss of the general results in new behaviours as subordinates initiate unrestrained activities.

The open-loop impulse-variability theory of Schmidt *et al.* (1979) focuses on the initial impulse of a given movement (e.g. reaching). In an extension of Fitts' aiming task (Fitts, 1954), Schmidt *et al.* required subjects to respond within ~200 ms (thus isolating activity from most feedback) and to be as accurate as possible. The characteristic flinging and coasting motion of the arm to the target area comprises an initial force component, the magnitude and duration of which may vary, and which are under independent control. The subject's task is therefore to minimize the variability in time and force. This model fails to explain observed corrective sub-movements, dependent on feedback (Rosenbaum, 1991). Abernethy and Sparrow (1992) report revisions of this theory proposing an open/closed-loop hybrid (e.g. Meyer *et al.*, 1988).

Dynamic oscillatory theories of motor control (e.g. Kelso *et al.*, 1981; Kugler *et al.*, 1982; Kugler and Turvey, 1986) represent a fundamental shift away from centrally stored movement kinematics (i.e. motor schema and programmes) in favour of emergent properties of the dynamics of the underlying motor systems (i.e. the collective physical properties of functional muscle groups involved in a given action). An early example is the mass–spring model posited by Bizzi *et al.* (1976), which likens muscles to loaded springs and describes critical length–tension points for agonist–antagonist muscle pairs as the basis for accurate movement localization. This model has problems with multi-directional movements and has been superseded by a more robust variant centred on oscillatory processes (e.g. Kelso *et al.*, 1981). This approach attempts to use physics to describe the selection of action categories. Its operating principles derive from physics and biology; its conceptual basis is rooted in the hybridization between Bernstein's views on coordination (Bernstein, 1967) and Gibson's (1979) views on perception (Abernethy and Sparrow, 1992). This ecological approach, reliant on dynamic system analysis, emphasizes a perception–action coupling of biological motion (e.g. a lion directly perceiving an abnormal gait pattern in a fleeing herd of prey, homes in on it as it will be easiest to catch: Rosenbaum (1991). This is also known as an 'action systems' theory, and is in opposition to the inferential, contructionist, information-processing approach to perception and action (see also section 9.3.2.1).

7.5.4   Feedback processing times

Studies of human performance have addressed aspects of both complex (molar) and simple (molecular) behaviours (Laszlo, 1992). The latter include investigations of response times to stimuli which may be of simple or multiple-choice design. These were first investigated by Helmholtz (1850) and Donders (1868) (see Legge and Barber, 1976). These reaction time (RT) studies reveal approximate durations for the processing of sensory feedback information. Suggested visual processing durations (see Carlton (1992) and Glencross and Barrett (1992) for reviews) from such RT studies include: 450 ms by Woodworth (1899); 400 ms by Vince (1948); 190–260 ms by Keele and Posner (1968); 100–150 ms by Whiting *et al.* (1970); 290 ms (165 ms discarded result) by Beggs and Howarth (1970); 164 ms by Smith and Bowen (1980); 110–170 ms by Cordo (1987); 190–210 ms by Carlton (1992, p. 4).

Slower movements without visual feedback may benefit from kinaesthetic feedback. Chernikoff and Taylor (1952) and Gibbs (1965) provide evidence for kinaesthetic RTs of 110–130 ms. Time to react to kinaesthetic feedback suggested by Glencross (1977) was 100 ms or greater. Movements faster than the visual processing time (e.g. 190 ms) may therefore be under kinaesthetic control.

7.5.5   Motor neurophysiology

Motor neurophysiology shows how associations between different sensory and motor areas have been established. The number of different motor areas at different levels of processing indicate the various control mechanisms for motor behaviour.

Fritsch and Hitzig (1870) identified cortical motor areas after stimulating the frontal lobes of dogs and eliciting contralateral muscle activity. Sherrington (1906) described the functional relationship between the cerebellum and cortical motor regions, and according to Thompson (1967, p. 314) was responsible for differentiating the motor-sensory cortex. Thus Brodmann's Areas 4 and 6 (along the precentral gyrus) were ascribed to motor cortex as they were known to control movement, and Areas 1, 2, and 3 (along the postcentral gyrus), known to receive cutaneous projections from lemniscal pathways, were defined as somatosensory cortex. Despite this distinction, the motor and sensory cortices are highly interconnected, and Woolsey (1958) preferred to consider both as somatic cortex, with an anterior 'somatic motor-sensory' and a posterior 'somatic sensory-motor' dichotomy. In keeping with the earlier description of SI for primary sensory cortex, primary motor cortex is referred to as MI.

The primary motor cortex was initially thought to comprise only the large Betz cells of Area 4; Area 6 was considered premotor. Stimulation of Area 6 produces more complex movements than elicited from MI (Fulton, 1949, in Thompson, 1967). Woolsey *et al.* (1950a, 1950b) observed however that this region, but not Area 4, controls the back musculature, and is really part of MI. While MI projects to the distal musculature like the fingers (Rosenbaum, 1991), premotor cortex projects to proximal musculature (e.g. trunk and shoulders). The indications are that premotor cortex assists postural commands for specific motor activities and in the selection of movement trajectories. Wiesendanger (1981) reports on premotor cortical ablations in cats which fail to recognize alternative indirect trajectories for obstructed paws.

Another motor region (above and anterior to MI) was examined by Woolsey *et al.* (1950a, 1950b), and Penfield and Welch (1951). Referred to as the supplementary motor area (Thompson, 1967) or MII, like MI it organizes body musculature topographically (e.g. Penfield and Rasmussen, 1950); it also appears to be involved in planning and production of complex movements (Wiesendanger, 1987). Rosenbaum (1991) reports evidence of this involvement from positronic emission tomography (PET) scans. MII is observed to be extremely active during finger movements. Even imagined movements cause high blood flow to MII whilst that flowing to MI drops to normal (Roland *et al.*, 1980). Thompson (1967) describes MII being found to be independent of MI (e.g. stimulating MII can elicit MI type movements even when MI is destroyed) and more bilaterally organized (Carpenter, 1987).

The topographic cortical representation of sensory input is evident in motor cortex as neurons which correspond to synergistic muscles are systematically organized in cortical columns (Carpenter, 1987). The proportionately large area of motor cortex attests to its involvement in manipulation (e.g. grasping). Motor mapping, like somatosensory mapping, is largely dedicated to the hands (and mouth). Lesions to Area 6 frequently result in uninhibited grasping (which is also commonly observed in young children).

Motor cortex conducts a coordinative function between skin and other spinal afferents and the gain of stretch reflexes (Carpenter, 1987). The segmental nature of the spinal cord[8] is unsuited to such a coordinative role, unlike the high levels of cortical convergence and divergence (e.g. a cortical monkey pyramidal cell can receive some 60 000 synapses). The cortex is also more suited for modifying behaviour through experience and frequent association. Cutaneous sensations reach SI next to MI. The analysis of cutaneous feedback requires information about the movements being made. Tactual perception of texture (e.g. Lederman, 1981) and common sensations such as softness and stickiness, rely on knowing how much finger force is applied. These data are conveyed by the pyramidal tract and interneurons between the cortical areas.

Motor cortex comprises six histologically different layers that communicate in varying degrees. Pyramidal cells in the fifth layer down, layer V, form the primary projection efferents from the cortex to sub-cortical structures (e.g. the ventrolateral and ventro-anterior thalamic nuclei). Motor cortex and thalamus function as one unit. Ascending input to thalamic motor nuclei are not sensory but originate from basal ganglia, the cerebellum, and the reticular formation. The role of the basal ganglia appears to be in initiating movements, modulating their global scale, and regulating perceptuo-motor interactions. Evidence for this comes from diseases affecting them (e.g. Huntington's chorea: uncontrollable movements; Parkinson's disease: resting tremor, slow movements, rigidity). The cerebellum regulates muscle tone, coordination, timing, and learning. Its loss results in impaired muscle tone and coordination deficits (e.g. poor balance, slurred speech, over-reaching, and intention tremor), in impaired eye–hand coordination and learning, and in timing difficulties (Rosenbaum, 1991).

Motor cortical cells fire prior to movement onset (e.g. Deecke *et al.*, 1969), with impulses related to force and direction of movement. This prior response suggests a trigger rather than controlling role (Rosenbaum, 1991). Some motor cortex cells synapse with only one motor-unit (e.g. Asanuma, 1981) and this contrasts with the view that they controlled entire movements (e.g. Evarts, 1967).

7.6 The nature of manipulation – classification

As indicated in the introduction (7.1), if manipulation is to be possible in virtual reality, we need not only to be able to understand how manipulation is controlled, but also to be able to analyse and classify manipulation itself.

Physical manipulation may be indirect (i.e. manipulating with tools and robots) or direct (i.e. with bare hands). Although the use of interactive glove devices (IGDs), such as VPL's DataGlove or Exos's Dexterous HandMaster, is, strictly speaking, tool use and therefore indirect, the nature of the tool is in this case designed to be hand-like so that in this discussion we may consider it synonymous with direct physical manipulation.

Methods for analysing manipulation may be drawn from a number of areas including examination of the literature on manipulation-related research, medical case histories, conducting observational studies on the various forms of manipulative activity, or running experiments assessing the effects of impaired feedback on manipulation.

Much of the literature on manipulation is medical with anatomical descriptions and assessment techniques for patients suffering from rheumatoid and arthritic bone disorders (e.g. Dickson and Nicolle, 1972) and neural damage. Of greater interest for virtual manipulation are the papers providing functional descriptions (e.g. McBride, 1942; Griffiths, 1943; Slocum and Pratt, 1946; Napier, 1956; Landsmeer, 1962; Long *et al.*, 1970; Rosenbloom and Horton, 1971). There are very few reports on the classification of manipulation itself (although see the intrinsic manipulation descriptions of Elliott (1979) and Elliott and Connolly (1984)). Ergonomics journals offer the occasional report on optimal design characteristics for hand tools (e.g. Fransson-Hall and Kilbom, 1993) but these tend to be of limited value for a broadly applicable analysis of manipulative behaviour. What is needed is a manipulation choreography, or task analysis method. While video and film can assist in observation and recording they do not provide any methodical format for assessment or analysis.

In order to help devise a method for analysing manipulation which could be used to specify requirements for particular activities in virtual reality, literature reviews and video analysis of numerous manipulations were carried out (England, 1992). This study resulted in the development of a taxonomy for analysing manipulation referred to as MHMAT: the Manipulative Hand Movement Analysis Tool. The tool simplifies analysis by first breaking manipulative behaviour into the four primary phases of reaching, grasping, manipulating and releasing. Manipulative behaviour varies enormously but it is possible to classify all these activities according to the four primary phases and to further primitive components of each phase, such as:

- the number of hands and the amount of upper body movement (degrees of freedom) involved;
- the nature of the grasp (e.g. whether it employed a pincer (thumb–index finger) or palmar (whole hand) grip);
- the type of manipulation (e.g. did it involve an intrinsic roll such as rolling an object between thumb and finger; or an extrinsic action such as wielding a hammer?).

Thus different tasks can be classified and differentiated according to their respective key phases and sub-phase components using MHMAT, and any common features identified.

The primary purpose of MHMAT is to provide a framework upon which to map the varying perceptual requirements of specific manipulations, with the objective of providing a source of reference for designers of virtual systems. Thus when a requirement for a particular virtual task arises, the designers can analyse the manipulation task and identify the specific perceptual requirements which must be fulfilled if the task is to be successfully performed. If the given virtual reality system cannot support the appropriate feedback then recommendations for alternatives (e.g. enhancements) may be available. This perceptual requirements guide is currently under development.[9]

7.7 *Manipulating virtual objects*

7.7.1 Taction for virtual manipulation

Touch was once considered the primary, or dominant, sense (Berkeley, 1709). The variety and number of specialized tactual receptors are far greater than visual receptors. Although vision is the more developed sense, tactual sensitivities are great (consider the perception of fine textures, walking into a cobweb, twiddling a hair or silk thread, feeling a gentle breeze). For virtual reality, stimulating this highly complex sense is even more of a challenge than stimulating vision.

7.7.2 Tactual feedback technology

Four formats of tactile stimulators currently exist: miniature solenoids which drive a small pin against the skin simulating contact (Foley, 1987), shape memory alloys which contact the skin when energized (e.g. TiNi Alloy Company's Tactor device), piezoelectric buzzers/voice coils (vibrotactile), and pneumatic air-bladders which aim to provide contact feedback on inflation (e.g. the 'Teletact II' device). These formats rely primarily on the 'blanket' stimulation of Meissner and Pacinian corpuscles, Merkel disks, and Ruffini endings, although other receptors may be activated. Pneumatics tend to provide the poorest level of sensitivity (resolution) but the other formats are also far from matching the acuity offered by the 17 000 receptors reported by Johansson and Vallbo (1983) as populating the glabrous skin of the hand. For instance, Foley (1987, p. 87) reported the solenoid actuator as being a third of an inch thick.

Kinaesthetic feedback devices are even less developed, but recent devices include the Safire system from Exos Inc.

7.7.3 Tactile-kinaesthetic associations and trade-offs

Katz (1936) describes a Russian physician, Professor Hausmann of Minsk, who observed that touch (i.e. tactile) information was not critical in palpatory examinations. Palpating the surface of a patient's body (e.g. abdomen) enables the doctor to learn of intestinal pathological changes. Katz reports that Hausmann performed palpatory examinations with anaesthetized fingers and 'still got entirely satisfactory results' (1936, p. 148). He concludes such examinations are at least partially controlled by kinaesthetic feedback. This is very interesting for virtual reality. If true, it suggests that tactile cues are less critical for virtual shape perception and that we should be developing kinaesthetic feedback devices for such application requirements.

In contrast to Katz's observations, if digital cutaneous receptors do interact with kinaesthetic feedback as the anaesthetic experiments conducted by Clark *et al.* (1983: see section 7.3.5.5) suggest, then conflict between sensory cues may emerge when poorly calibrated or insensitive virtual reality systems are used. For example, natural movements will involve learned cutaneous–kinaesthetic relationships. Thus a slow joint rotation will have an associated relationship between tactile (cutaneous) feedback and kinaesthetic joint information. If blanket vibrotactile stimulation overstimulates the cutaneous receptors of a fingertip then any learned relationship as the finger flexes may temporarily falter. This would remove the supportive feedback to the kinaesthetic sense and potentially impair manipulative accuracy and performance on certain fine manipulation tasks. To use our hands as manipulators would be difficult if they became uncontrollable as a consequence of tactile stimulation!

The situation in virtual reality is additionally complex, as the novel tactile stimulation may have to be associated with uncertain kinaesthetic information as a consequence of system lag (where visual kinaesthesia will conflict with primary kinaesthesia).

7.7.4 Side-effects of vibrotactile stimulation

In addition to causing a loss of kinaesthetic interaction, tactile over-stimulation may produce a state of temporary and localized anaesthesia (the effects of which have been considered above). The sensitivity of certain receptors (RA-type) is known to fall after stimulation (e.g. after a contact induced displacement). There are two and possibly more effects of over-stimulating such receptors:

1. the cell will excite and then dampen (rapidly adapt) and cease responding despite further stimulation;
2. if vibrotactile stimulators do maintain the activation level of these receptors, instead of achieving a steady sensation of contact (as when an object is grasped), the continued activation or re-activation may create the impression that the object is continually being re-contacted and that the object must have previously fallen out of contact. The perception of grasping must therefore be undermined in that the grasp is never fully achieved and stable.

Goodwin *et al.* (1972) demonstrated that vibration stimuli applied to muscle tendons produced an illusion of movement in a restrained limb. Although this research was conducted at the macro-level (whole limb) it would seem likely that the effects could occur at a more elementary level (see also DiZio *et al.*, 1993). It is possible that some forms of tactile stimulation being considered for virtual reality may actually create unintentional illusory finger movements. These would result in conflicts between visual and kinaesthetic sensory cues and would distort the perception of finger position and action, affecting manipulation accuracy. These effects may currently go unnoticed because of the gross nature of current virtual manipulation, but they may surface as the quality of feedback begins to improve with the arrival of new technologies.

Winstein and Schmidt (1989) describe Pacinian corpuscle sensitivity to vibration as important for coordinated adaptive responses for precision gripping (e.g. the 'slip-induced grasping reflex'). This reflexive response was observed by Westling and Johansson (1984) and Johansson and Westling (1987). Subjects lifting a small cube between thumb and index finger were observed to make corrective grips when the object began to slip. Cutaneous finger tip receptors detected the slip and triggered a rapid compensatory response (e.g. 60–80 ms after slip onset). Subjects were, reportedly, frequently unaware of the event. Vibrotactile stimulation is likely to trigger such adaptive responses inappropriately, impairing manipulative performance and even causing finger fatigue!

Some illusions created by vibration could, however, be exploited for virtual reality. Lackner (1988) showed that by vibrating muscles, proprioceptive information can be distorted to cause illusions of body shape and orientation (e.g. if a person is grasping his or her nose, vibration applied to the arm can result in the nose feeling very long: a 'Pinocchio' illusion).

Judging by the range of sensitivities to velocity and acceleration of skin displacement receptors, the human response characteristics to variable frequency vibrotactile stimulators will be varied, with some frequencies eliciting stronger effects than others. Sensitivity is found to be greatest around 200 Hz, according to Sekuler and Blake (1990, p. 361). They also suggest that the cutaneous receptive fields most sensitive to this frequency lie not in the fingertips but in the palms. Should virtual reality developers interested in manipulation concentrate on developing interactive glove devices with palm-rich transducers for enhanced haptic perception?

7.7.5 The nociceptive experience

While the use of pain as a means of sensory feedback is undesirable, it is possible that it could supply critical components for tactual activity. It may be that pain is primarily a survivability feature (as indicated by the case of the young woman who had no sensation of pain and experienced severe orthopaedic difficulties as a consequence; see Baxter and Olszewski (1960)), but a secondary function might be in developmental and learning-orientated phenomena (e.g. grasping too hard is unnecessary and painful).

The survivability function of pain suggests that information from nociceptors would have priority. An intense sensation of pain may cause distraction. In the simulation of an absolutely critical 'one-chance-only' task, nociceptive feedback may provide a more rapid awareness and thus response than other modalities. Thus, in a simulated teleoperation task involving the location-critical insertion or removal of a safety mechanism and where error might result in catastrophic consequences, narrow-band thermal nociception may provide one means of ensuring caution. Its use would obviously have to be correctly and 'safely' applied. Excessive thermal feedback to a hand-controller, causing it to be dropped because it was too hot could itself cause a disaster! If, however, the operator manipulatively strayed into a dangerous task area and was too preoccupied to notice, then thermal nociception might be beneficial. In such circumstances the nociceptive sensations might alert the operator to the danger, allowing a controlled withdrawal to safer regions; or a sudden non-harmful thermal shock (i.e. rapid change from a tolerable hot extreme to a tolerable cold extreme) might cause the operator to release the hand-controller, preventing further progression into the danger zone.

While the use of such feedback may provide an important and as yet untapped source of sensory stimulation, the ethical concerns are both obvious and indeed serious. One important capability of virtual reality is to enable simulated exposure to danger for training purposes. Thus pilots learn how to respond to engine failure on a flight simulator, and firemen might visualize the best route through a burning building using a virtual environment. To introduce real danger would counter these advantages (although Sutherland's 'Ultimate Display' did discuss deadly virtual bullets: Sutherland (1965)). The suggestion being made here, though, is not to introduce severe pain but perhaps mildly unpleasant sensations as an alerting medium. The danger will ultimately lie in the hands of the purveyors and administrators of such systems, and it may be wise to consider a cautionary axiom: most good intentions are ultimately abused.

7.7.6 The thermal experience

Other than the possible link with nociception discussed above, thermal perception has some obvious implications for virtual reality: tactile devices (e.g. mini Peltier devices) could gradually change skin temperature prior to contact with a virtual object and then on simulated contact, radically change again to simulate greater thermal differentials and thus the perception of temperature, and other properties of the stimulus, such as weight and hardness (see the weight– temperature illusion in section 7.3.5.5).

7.7.7 Integration and coordination

The nature of cues, their interdependence and dominance all determine what we perceive. These factors contribute to how the information conveyed by each cue is integrated into global percepts. Strategies for the integration of cues were discussed in section 7.4.

Why might such strategies be important for virtual reality? Consider the following example. Kinaesthetic processing must include a representation of the size and scale of the body, hand, etc., as these do the sensing. If the sensor scale of a hand is disrupted by a device which affects the units of measurement (e.g. some interactive 'glove' devices effectively make fingers fatter or stiffer), the subsequent perceptual judgements will depend on whether the integration process can adjust to compensate, or whether it is a long-term (slowly learned) function. Similarly, if the distance between the two eyes is not matched by the stereo visual display, visual judgements will be affected. If the integration process is not rapidly flexible, this will affect people's ability to interact with virtual reality when the information provided to each sense is not consistent with the real world (because of distortion by devices).

The quality of cue presentation in virtual reality is poor compared to the real world, and may therefore demand more serial focused attention, resulting in more time-costly interaction. If attention is distracted briefly, this could result in mismatched cues (illusory conjunctions) that cannot be top-down corrected because either the stimulus is novel or there is insufficient information. Any task performance dependent on the affected cues could be detrimentally affected by such illusions.

7.8 Conclusions and recommendations

7.8.1 Conscious and subconscious processes

Over previous sections the consequences of disrupting the natural data exchange between incoming sensations and outgoing motor commands have been considered. This disruption affects not just how sensations are detected, but also how they are integrated with other sensations and how they are ultimately perceived and responded to. In most instances, the perception of events dictates the nature of the response. Perception is not always a conscious phenomenon and many responses that are made occur with little conscious involvement. These are less open to top-down modification. Manipulative interaction with virtual environments will thus require some long-term adaptation for low-level processing to efficiently process virtual information, as well as requiring top-down conscious attention to assist in perception and response. The discussion of motor control has highlighted differences in motor activities (e.g. volitional-focal, ambient, and reflexive behaviours). The 'triggers' for these activities are very often sub-cortical. Inappropriate feedback may activate neurons controlling specific sub-behaviours which are inappropriate for the ongoing global complex activity.

The more that people have to adapt and modify their behaviour for interaction in virtual reality, the further they move away from what we might call naturalistic interaction (i.e. interaction in the real world). This means that in such scenarios, people have to acquire new interactive skills. If the virtual reality simulates a stressful environment, or is in itself stressful, then people may unconsciously revert to earlier learned interactive skills; skills which are inappropriate or inadequate for virtual reality interaction.

7.8.2 Multiple senses

The visual sense is considered by many to be the most important. It is evident, however, that senses do not just provide information but also serve to confirm the 'perceptions' of the other senses. Such confirmation comes from redundancy and improves the signal-to-noise ratios of sensory inputs (and motor outputs). The degree of redundancy is thus an important factor for virtual reality designers and developers to consider.

The suggestion of sensory pluralism questions the usefulness of considering senses independently and shifts the emphasis towards multi-sensory sets of cues for specific objects or information. Identification of these object-centred sets of cues therefore becomes an important objective if we are to ensure their inclusion for enhancement of people's performance in virtual reality.

Physiological evidence supporting the notion of weak cross-modal signals being more readily integrated with others than with strong signals has been described. The implications suggest that when feedback is impoverished, as in virtual reality, the weak signals may be inappropriately integrated providing an unreliable and inconsistent perceptual experience.

Inappropriate or mismatched cue presentation in different senses has been discussed with respect to feature integration theory, and the illusory conjunctions which may be result when top-down processing is unable to compensate because of unfamiliarity with the virtual environment.

7.8.3 Tactual feedback

For a fine manipulative virtual reality task, current feedback to all senses needs to be significantly improved, in terms of both quantity (stimulate more kinds of receptors) and quality. For example, tactual feedback for the hands should be based on finger tracking which can match the digital degrees of freedom. This would ensure that many real-world manipulative behaviours could be replicated in virtual reality. In addition to finger tracking the levels of tactile resolution need to be raised. If an actuator is $0 \cdot 85$ cm in diameter (Foley, 1987) then any smaller virtual object will, if detected at all, still be considered $0 \cdot 85$ cm across. The packing density of such devices will be very important. Inaccurate feedback, especially if processed at low levels, may corrupt learned skills.

Tactile feedback at present employs low-resolution pneumatic bladders or vibrotactile stimulators. Although these devices create an impression of contact, its nature is not a sensation we are usually confronted with. This may introduce manipulative behavioural artefacts which impede and hinder (e.g. corrective grasping reflexes) rather than improve our interactive experience.

The reports of vibration-induced illusions of limb movement may be transferable to illusions of finger movements, illusions which in a manipulative task may initiate corrective measures that are themselves responsible for performance disruption.

Current tactile feedback devices concentrate on exciting mechanoreceptive displacement detectors. These are not the only types of cutaneous receptors from which our perceptual systems may gain valuable information. Thermoreceptors would also appear to have a highly functional role in the perception of a physical object's attributes. These attributes do not only concern an object's relative temperature, but also, when combined with other cutaneous inputs (e.g. pressure), suggest other characteristics such as viscosity, solidity, mass, etc.

Free-nerve endings and nociceptors may also have a key functional role to play in virtual interaction, much as they do in our everyday experience and may even enhance the sense of realism. Even Gibson (1966) considered cutaneous pain to convey useful information, but ethical issues are raised concerning the abuse of nociception in virtual reality.

7.8.4 Implementation

It must be remembered that the degree of feedback required will obviously depend on the task requirements (see section 7.1.2). A simple task may not require much operator involvement or interaction at all. If, however, the operator is expected to be immersed in virtual reality for extended periods, for example whilst working on a prototype design, or training in a simulation, then the need for multi-sensory, good-quality feedback will be critical. Some tasks may be enhanced by the presence of smell and taste, two senses which have not been discussed at all. Olfaction does in fact have a very powerful effect upon our perceptions, through association and through pheromones. Heilig (1962) introduced olfaction into his *Sensorama*; currently the *Scentax* facility of the *ZyberFantasy Ride* is using olfaction in an erotic application of virtual reality (Howells, 1994).

The cost of ignoring the need for multi-sensory and high-quality feedback is not only that the task may be difficult or uncomfortable to perform, but that the perceptuo-motor behaviour may be disrupted for periods extending outside the time spent in virtual reality. Fine direct manipulation in virtual reality represents the greatest challenge to developers; it will bring the greatest rewards in terms of the effective and widespread use of virtual reality.

Notes

1. Cyberspace may thus be thought of as a subset of virtual reality with the latter including all forms of representation from imagination to film, from graphics to holography.
2. Wheatstone and Brewster stereoscopes (see Wade, 1988); Heilig, 1962; Goertz *et al.*, 1965; Comeau and Bryan, 1961; Sutherland, 1968; Janow and Malone, 1973; Fisher *et al.*, 1988; Bolas and Fisher, 1990.
3. Sutherland, 1965, 1968; weapons targeting systems (Price, 1975); Wright–Patterson's proposed Super Cockpit project initiated in 1977 (see Furness and Kocian, 1986); NASA's early virtual reality projects (e.g. Fisher *et al.*, 1986); university research projects (e.g. Chung *et al.*, 1989).
4. Discussion with Professor Cuschieri describing endoscopic surgery training at Nine Wells Hospital, Dundee, Scotland (1994).
5. *Movement systems* are representative of an information processing approach to motor control, while the *action systems* approach concerns an ecological approach to perception and action.

6. Physiology: Greek, meaning 'inquiry into nature'; study of living organisms and changes occurring during activity (Green, 1977). It encompasses the mechanisms that operate in living organisms at all levels (i.e. from sub-cellular to the integrated whole animal) (Eckert and Randall, 1978).
7. Head tracking velocities recorded exceeding $360°$ s^{-1} (elevation) and $601°$ s^{-1} (azimuth); data for acceleration reportedly at $2452°$ s^{-2} (elevation) and $4753°$ s^{-2} (azimuth).
8. The spinal cord contains reflexes for locomotion, limb movements and visceral functions. Spinal reflexes are under descending cortical control.
9. This work is ongoing at the Sowerby Research Centre, British Aerospace (Ops.) Ltd, Bristol.

References

Abernethy, B. and Sparrow, W. A., 1992, The rise and fall of dominant paradigms in motor behaviour research, in Summers, J. J. (Ed.), *Approaches to the Study of Motor Control and Learning*, Advances in Psychology 84, Stelmach, G. E. and Vroon, P. A., series editors, pp. 3–45, Amsterdam: North-Holland, Elsevier Science Publishers.

Adams, J.A., 1971, A closed-loop theory of motor learning, *Journal of Motor Behavior*, **3**, 111–50.

Asanuma, H., 1981, The pyramidal tract, in Brooks, V. B. (Ed.), *Handbook of Physiology*, Vol. 2(1), pp. 703–33, Bethesda, MD: American Physiological Society.

Baxter, D. W. and Olszewski, J., 1960, Congenital universal insensitivity to pain, *Brain*, **83**, 381–93.

Beggs, W. D. A. and Howarth, C. I., 1970, Movement control in man in a repetitive motor task, *Nature*, **221**, 752–3.

Berkeley, G., 1709, An essay towards a new theory of vision, in Ayres, M. R. (Ed.), 1980, *Berkeley: Philosophical Works*, 4th Edn, London: Dent and Sons.

Bernstein, N. A., 1967, *The Co-ordination and Regulation of Movements*, Oxford: Pergamon Press.

Bizzi, E., Polit, A. and Morasso, P., 1976, Mechanisms underlying achievement of final head position, *Journal of Neurophysiology*, **39**, 435–44.

Boff, K. R. and Lincoln, J. E., 1988, *Engineering Data Compendium: Human Perception and Performance*, Wright-Patterson AFB, OH: AAMRL.

Bolas, M. T. and Fisher, S. S., 1990, Head-coupled remote stereoscopic camera system for telepresence applications, *SPIE*, Vol. 1256, *Stereoscopic Displays and Applications*, pp. 113–23, San José, CA: SPIE..

Broadbent, D., 1958, *Perception and Communication*, Oxford: Pergamon.

Brooks, V. B., 1969, Information processing in the motorsensory cortex, in Leibovic, K. N. (Ed.), *Information Processing in the Nervous System*, pp. 231–43, New York: Springer-Verlag.

Bruce, V. and Green, P. R., 1990, *Visual Perception. Physiology, Psychology and Ecology*, 2nd Edn, Hove: Lawrence Erlbaum.

Burgess, P. R. and Clark, J. F., 1969, Characteristics of knee joint receptors in the cat, *Journal of Physiology*, **203**, 317–35.

Carlton, L. G., 1992, Visual Processing Time and the Control of Movement, in Proteau, L. and Elliot, D. (Eds), *Vision and Motor Control*, Advances in Psychology 85, Stelmach, G. E. and Vroon, P. A., series editors, pp. 3–31, Amsterdam: North-Holland, Elsevier Science Publishers.

Carpenter, R. H. S., 1987, *Neurophysiology (Physiological Principles in Medicine)*, London: Edward Arnold.

Casagrande, V. A., 1994, A 3rd parallel pathway to primate area – V1, *Trends in Neurosciences*, **17** (7), 305–10.

Chernikoff, R. and Taylor, F. V., 1952, Reaction time to kinaesthetic stimulation resulting from sudden arm displacement, *Journal of Experimental Psychology*, **43**, 1–8.

Chung, J. C., Harris, M. R., Brooks, F. P., Fuchs, H., Kelley, M. T., Hughes, J., Ouh-young, M., Cheung, C., Holloway, R. L. and Pique, M., 1989, Exploring virtual worlds with head-mounted displays, *SPIE*, Vol.1083, *Three-Dimensional Visualization and Display Technologies*, pp. 42–52.

Clark, F. J., Burgess, R. C. and Chapin, J.W., 1983, Human's lack of sense of static-position of the fingers, *Society for Neuroscience Abstracts*, 1033.

Comeau, C. P. and Bryan, J. S., 1961, Headsight television system provides remote surveillance, *Electronics Magazine*, 10 November, 86–90.

Connolly, K. and Jones, B., 1970, A developmental study of afferent–reafferent integration, *British Journal of Psychology*, **61** (2), 259–66.

Cordo, P. J., 1987, Mechanisms controlling accurate changes in elbow torque in humans, *Journal of Neuroscience*, **7**, 432–42.

Cotman, C. W. and McGaugh, J. L., 1980, *Behavioral Neuroscience. An Introduction*, London: Academic Press.

Deecke, L., Scheid, P. and Kornhuber, H. H., 1969, Distribution of readiness potential, premotion positivity and motor potential of the human cerebral cortex preceding voluntary finger movements, *Experimental Brain Research*, **7**, 158–68.

DeFanti, T. and Sandin, D., 1977, USNEA R60-34-163 Final Project Report (in Krueger, 1991, p. 82).

Dickson, R. A. and Nicolle, F. V., 1972, The assessment of hand function, Part 1 – Measurement of individual digits, *The Hand*, **4** (3), 207–14.

DiZio, P., Lathan, C. E. and Lackner, J., 1993, The role of the brachial muscle spindle signals in assignment of visual direction, *Journal of Neurophysiology*, **70** (4), 1578–84.

Donders, F. C., 1868, Die Schnelligkeit psychischer Processe, *Archiv für Anatomie und Physiologie*, 657–81.

Eckert, R. and Randall, D., 1978, *Animal Physiology*, San Francisco: W. H. Freeman.

Elliott, J. M., 1979, Motor skills in theory and practice, in Oborne, D. J., Gruneberg, M. M. and Eiser, J. R. (Eds), *Research in Psychology and Medicine*, Vol. II, London: Academic Press.

Elliott, J. M. and Connolly, K. J., 1984, A classification of manipulative hand movements, *Developmental Medicine and Child Neurology*, **26**, 283–96.

England, R. D., 1992, Improving interaction in virtual environments 1: Analysing manipulative hand movements with the hand movement assessment tool (MHMAT), Internal Report No. JS 12232, Sowerby Research Centre, British Aerospace PLC, Bristol, UK, September.

Enroth-Cugell, C. and Robson, J. G., 1966, The contrast sensitivity of retinal ganglion cells of the cat, *Journal of Physiology*, **187**, 517–52.

Evarts, E. V., 1967, Representation of movement and muscles by pyramidal tract neurons of the precentral motor cortex, in Yahr, M. D. and Purpura, D. P. (Eds), *Neurophysiological Basis of Normal and Abnormal Motor Activities*, New York: Raven Press.

Evarts, E. V. and Fromm, C., 1977, Sensory responses in motor cortex neurons during precise motor control, *Neuroscience Letters*, **5**, 267–72.

Fisher, S. S., McGreevy, M., Humphries, J. and Robinett, W., 1986, Virtual environment display system, in *ACM Workshop on Interactive 3D Graphics*, Chapel Hill, NC, 23–24 October 1986, pp. 1–11, New York:ACM.

Fisher, S. S., Wenzel, E. M., Coler, C. and McGreevy, M. W., 1988, Virtual interface environment workstations, *Proceedings of the Human Factors Society – 32nd Annual Meeting*, pp. 91–5, Santa Monica, CA: Human Factors Society.

Fitts, P. M., 1954, The information capacity of the human motor system in controlling the amplitude of movement, *Journal of Experimental Psychology*, **47**, 381–91.

Foley, J. D., 1987,Interfaces for advanced computing, *Scientific American*, **257** (4), 82–90.

Fransson-Hall, C. and Kilbom, Å., 1993, Sensitivity of the hand to surface pressure, *Applied Ergonomics*, **24** (3), 181–9.

Fritsch, G. and Hitzig, E., 1870, Ueber die elektrische Erregbarkeit des Grosshirns, *Archiv. Anat. Physiol. wiss. Med.*, **37**, 300–32.

Fulton, J. F., 1949, *Functional Localization in the Frontal Lobes and Cerebellum*, Oxford: Clarendon Press.

Furness, T. A. and Kocian, D. F., 1986, Putting humans into virtual space, *Proceedings of the Society for Computer Simulation Aerospace Conference*, San Diego, CA, pp. 214–30.

Gibbs, C. B., 1965, Probability learning in step-input tracking, *British Journal of Psychology*, **56**, 233–42.

Gibson, J. J., 1933, Adaptation, aftereffect and contrast in the perception of curved lines, *Journal of Experimental Psychology*, **16**, 1–31.

Gibson, J. J., 1950, *The Perception of the Visual World*, Boston, MA: Houghton Mifflin.

Gibson, J. J., 1962, Observations on active touch, *Psychological Review*, **69** (6), 477–91.

Gibson, J. J., 1966, *The Senses Considered as Perceptual Systems*, Boston, MA: Houghton Mifflin.

Gibson, J. J., 1979, *The Ecological Approach to Visual Perception*, Boston, MA: Houghton Mifflin.

Gibson, W., 1984, *Neuromancer*, London: Grafton Books.

Glencross, D. J., 1977, Control of skilled movement, *Psychological Bulletin*, **84**, 14–29.

Glencross, D. and Barrett, N., 1992, Processing of visual feedback in rapid movement, in Summers,

J. J. (Ed.), *Approaches to the Study of Motor Control and Learning*, Advances in Psychology 84, Stelmach, G. E. and Vroon, P. A., series editors, pp. 289–311, Amsterdam: North-Holland, Elsevier Science Publishers.

Goertz, R., Potts, C., Mingesz, D. and Lindberg, J. F., 1965, An experimental head-controlled TV system to provide viewing for a manipulator operator, *Proceedings of the 13th Conference on Remote Systems Technology*.

Goodale, M. A. and Milner, A. D., 1992, Separate visual pathways for perception and action, *Trends in Neurosciences*, **15** (1), 20–5.

Goodwin, G. M., McCloskey, D. I. and Matthews, P. B. C., 1972, The contribution of muscle afferents to kinaesthesia shown by vibration induced illusions of movement and by the effects of paralysing joint afferents, *Brain*, **95**, 705–48.

Green, J. H., 1977, *An Introduction to Human Physiology*, 4th Edn, Oxford: Oxford University Press.

Greening, S., 1993, The translation of modality specific information, *Proceedings of the British Psychological Society*, **1**, 3.

Griffiths, H. E., 1943, Treatment of the injured workman, *Lancet*, **i**, 729.

Grillner, S., 1985, Neurobiological bases of rhythmic motor acts in vertebrates, *Science*, **228**, 143–9.

Hay, J. C., Pick, H. L. and Ikeda, K., 1965, Visual capture produced by prism spectacles, *Psychonomic Science*, **2**, 215–16.

Head, H., 1920, *Studies in Neurology*, London: Oxford University Press.

Heilig, M., 1962, Sensorama simulator, US Patent No. 3,050,870, 28 August.

Held, J. R., 1968, Dissociation of visual functions by deprivation and rearrangement, *Pyschologische Forschung*, **31**, 338–48.

Heller, M. A., 1983, Haptic dominance in form perception with blurred vision, *Perception*, **12**, 607–13.

Heller, M. A., 1992, Haptic dominance in form perception, vision versus proprioception, *Perception*, **21**, 655–60.

Helmholtz, H. von, 1850, On the methods of measuring very small portions of time and their applications to physiological processes, *Philosophical Magazine*, **6**, 313–25.

Henderson, S. E., 1977, Role of feedback in the development and maintenance of a complex skill, *Journal of Experimental Psychology: Human Perception and Performance*, **3**, 224–33.

Henry, F. M. and Rogers, D. E., 1960, Increased response latency for complicated movements and a 'Memory Drum Theory' of neuromotor reaction, *Research Quarterly*, **31**, 448-58.

Hooper, S. L. and Moulins, M., 1989, Switching of a neuron from one network to another by sensory-induced changes in membrane properties, *Science*, **244**, 1587–9.

Howard, I. P., 1986, The perception of posture, self motion and the visual vertical, in Boff, K. R., Kaufman, L. and Thomas, J. P. (Eds), *Handbook of Perception and Human Performance*, Vol. I: *Sensory Processes and Perception*, New York: John Wiley.

Howard, I. P. and Templeton, W. B., 1966, *Human Spatial Orientation*, London: John Wiley.

Howells, J., 1994, Show you a good time, dear?, *The Guardian*, 10 November.

Hubel, D. H., 1988, *Eye, Brain and Vision*, Scientific American Library series (22), New York: Scientific American Library.

Iggo, A., 1982, Cutaneous sensory mechanisms, in Barlow, H. B. (Ed.), *The Senses*, pp. 369–408, Cambridge: Cambridge University Press.

Ingle, D., 1967, Two visual mechanisms underlying the behavior of fish, *Pyschologische Forschung*, **31**, 44–51.

Janow, C. and Malone, T. B., 1973, Human factor roles in design of teleoperator systems, in Ranc, M. P. and Malone, T. B., (Eds), *Proceedings of the 17th Annual Meeting of the Human Factors Society*, pp. 305–14, Santa Monica, CA: Human Factors Society.

Johansson, R. S. and Vallbo, A. B., 1983, Tactile sensory coding in the glabrous skin of the human hand, *Trends in Neurosciences*, **6**, 27–32.

Johansson, R. S. and Westling, G., 1987, Signals in tactile afferents from the fingers eliciting adaptive motor responses during precision grip, *Experimental Brain Research*, **66**, 141–54.

Kahneman, D., 1973, *Attention and Effort*, Englewood Cliffs, NJ: Prentice Hall.

Katz, D., 1925, Der Aufbau der Tastwelt (see Krueger, 1970).

Katz, D., 1936, A sense of touch, *British Journal of Physical Medicine*, Dec, 146–8.

Keele, S. W., 1968, Movement control in skilled motor performance, *Psychological Bulletin*, **70** (1), 387–403.

Keele, S. W. and Posner, 1968, Processing of visual feedback in rapid movements, *Journal of Experimental Psychology*, **77**, 155–8.

Keele, S. W. and Summers, J. J., 1976, The structure of motor programs, in Stelmach, G.E. (Ed.), *Motor Control: Issues and Trends*, pp. 109–42, New York: Academic Press.

Kelso, J. A. S., Holt, K. G., Rubin, P. and Kugler, P. N., 1981, Patterns of human interlimb coordination emerge from the properties of nonlinear, limit-cycle oscillatory processes: Theory and data, *Journal of Motor Behavior*, **13**, 226–61.

Kenshalo, D. R., 1971, The cutaneous senses, in Kling, J. W. and Riggs, L. A. (Eds), *Woodworth and Schlosberg's Experimental Psychology*, 3rd Edn, New York: Holt, Rinehart & Winston.

Krueger, L. E., 1970, David Katz's Der Aufbau der Tastwelt (The world of touch): A synopsis, *Perception and Psychophysics*, **7** (6), 337–41.

Krueger, M. W., 1991, *Artificial Reality II*, 2nd Edn, Reading, MA: Addison-Wesley.

Kuffler, S. W., 1953, Discharge patterns and functional organization of mammalian retina, *Journal of Neurophysiology*, **16**, 37–68.

Kugler, P. N. and Turvey, M. T., 1986, *Information, Natural Law and the Self-Assembly of Rhythmic Movement*, Hillsdale, NJ: Lawrence Erlbaum.

Kugler, P. N., Kelso, J. A. S. and Turvey, M. T., 1982, On coordination and control in naturally developing systems, in Kelso, J. A. S. and Clark, J. E. (Eds), *The Development of Movement Control and Coordination*, pp. 5-78, New York: John Wiley.

Lackner, J., 1988, Some proprioceptive influences on the perceptual representation of body shape and orientation, *Brain*, **111**, 281–97.

Landsmeer, J. M. F., 1962, Power grip and precision handling, *Annals of the Rheumatic Diseases*, **21**, 164–70.

Lashley, K. S., 1917, The accuracy of movement in the absence of excitation from the moving organ, *American Journal of Physiology*, **43**, 169–94.

Laszlo, J. I., 1992, Motor control and learning: how far do the experimental tasks restrict our theoretical insight?, in Summers, J. J. (Ed.), *Approaches to the Study of Motor Control and Learning*, Advances in Psychology 84, Stelmach, G. E. and Vroon, P. A., series editors, pp. 47–79, Amsterdam: North-Holland, Elsevier Science Publishers.

Lederman, S. J., 1981, The perception of texture by touch, in Schiff, W. and Fouke, E. (Eds), *Tactual Perception: A Sourcebook*, pp. 130–67, New York: Cambridge University Press.

Lee, D. N., 1976, A theory of visual control of braking based on information about time-to-collision, *Perception*, **5**, 437–59.

Lee, D. N. and Aronson, E., 1974, Visual proprioceptive control of standing in human infants, *Perception and Psychophysics*, **15** (3), 529–32.

Lee, D. N. and Lishman, 1975, Visual proprioceptive control of stance, *Journal of Human Movement Studies*, **1**, 87–95.

Legge, D. and Barber, P. J., 1976, *Information and Skill*, Essential Psychology series A5, Herriot, P. series editor, London: Methuen.

Lennie, P., Trevarthen, C., Van Essen, D. and Wässle, H., 1990, Parallel processing of visual information, in Spillman, L. and Werner, J. S. (Eds), *Visual Perception: The Neurophysiological Foundations*, pp. 103–28, San Diego, CA: Academic Press.

Lisberger, S. G., 1988, The neural basis for learning of simple motor skills, *Science*, **242**, 728–35.

Lishman, J. R. and Lee, D. N., 1973, The autonomy of visual kinaesthesis, *Perception*, **2**, 287–94.

Long, C. L., Conrad, P. W., Hall, E. A. and Furler, S. L., 1970, Intrinsic–extrinsic muscle control of the hand in power grip and precision handling, *The Journal of Bone and Joint Surgery*, **52-A** (5), 853–67.

Lorenz, K. Z. and Tinbergen, N., 1938, Taxis und Instinktbewegung in der Eirollbewegung der Graugans, *Zeitschrift für Tierpsychologie*, **2**, 1–29.

Luria, A. R., 1973, *The Working Brain. An Introduction to Neuropsychology*, Harmondsworth: Penguin Books (reprinted 1978).

MacNeilage, P. F., 1970, Motor control of serial ordering of speech, *Psychological Review*, **77**, 182–96.

Masterton, R. B. and Imig, T. J., 1984, Neural mechanisms of sound localisation, *Annual Review of Physiology*, **46**, 275–80.

Matthews, P. B. C., 1964, Muscle spindles and their motor control, *Physiological Review*, **44**, 219–88.

McBride, E. D., 1942, *Disability Evaluation*, 3rd Edn, Philadelphia: J. B. Lippincott.

McCloskey, D. I., 1978, Kinaesthetic sensibility, *Psychological Review*, **58**, 763–873.

McDonnell, J. R., Solorzano, M. R., Martin, S. W. and Umeda, A. Y., 1990, A head coupled sensor platform for teleoperated vehicles, *Unmanned Systems*, **8** (4), 33–8.

Meijer, O. G. and Roth, K. (Eds), 1988, *Complex Movement Behaviour:'The' Motor-Action Controversy*, Amsterdam: North-Holland.

Meredith, M. A. and Stein, B. E., 1986, Visual, auditory and somatosensory convergence on cells in superior colliculus results in multisensory integration, *Journal of Neurophysiology*, **56**, 640–62.

Meyer, D. E., Abrams, R. A., Kornblum, S., Wright, C. E. and Smith, J. E. K., 1988, Optimality in human motor performance: ideal control of rapid aimed movements, *Psychological Review*, **95** (3), 340–70.

MIL-STD-1472C, 1981, Military Standard Human Engineering Design Criteria for Military Systems, Equipment and Facilities, Department of Defense, Washington, DC, 20301.

Mohler, C. W. and Wurtz, R. H., 1976, Organization of monkey superior colliculus: intermediate layer cells discharging before eye movements, *Journal of Neurophysiology*, **39**, 722–44.

Morgan, M. J., 1977, *Molyneux's Question: Vision, Touch and the Philosophy of Perception*, New York: Cambridge University Press.

Muller, H. J. and Findlay, J. M., 1987, Sensitivity and criterion effects in the spatial cueing of visual attention, *Perception and Psychophysics*, **42**, 383–99.

Napier, J. R., 1956, The prehensile movements of the human hand, *The Journal of Bone and Joint Surgery*, **38B** (4), 902–13.

Navon, D., 1977, Forest before the trees: the precedence of global features in visual perception, *Cognitive Psychology*, **9**, 353–83.

Navon, D. and Gopher, D., 1979, On the economy of the human processing system, *Psychological Review*, **86**, 254–5.

Neisser, U., 1967, *Cognitive Psychology*, New York: Appleton-Century-Croft.

Newell, K. M., Shapiro, D. C. and Carlton, M. J., 1979, Coordinating visual and kinaesthetic memory codes, *British Journal of Psychology*, **70**, 87–96.

Paillard, J., 1960, The patterning of skilled movements, *Handbook of Physiology, I, Neurophysiology*, Vol. III, pp. 1679–708, Bethesda, MD: American Physiological Society.

Pausch, R., Crea, T. and Conway, M., 1992, A literature survey for virtual environments: military flight simulator visual systems and simulator sickness, *Presence*, **1** (3), 344–63.

Penfield, W. and Rasmussen, T., 1950, *The Cerebral Cortex of Man: A Clinical Study of Localization of Function*, New York: Macmillan.

Penfield, W. and Welch, K., 1951, The supplementary motor area of the cerebral cortex, a clinical and experimental study, *Archives of Neurological Psychiatry*, **66**, 289–317.

Perrett, D. I., Rolls, E. T. and Caan, W., 1982, Visual neurones responsive to faces in the monkey temporal cortex, *Experimental Brain Research*, **47**, 329–42.

Pew, R. W., 1966, Acquisition of hierarchical control over the temporal organization of a skill, *Journal of Experimental Psychology*, **71**, 764–72.

Posner, M. I., Nissen, M. J. and Klein, R., 1976, Visual dominance: an information-processing account of its origins and significance, *Psychological Review*, **83**, 157–71.

Posner, M. I., Snyder, C. R. R. and Davidson, B. J., 1980, Attention and the detection of signals, *Journal of Experimental Psychology: General*, **109**, 160–74.

Postman, L. and Egan, J. P., 1949, *Experimental Psychology*, New York: Harper & Row.

Power, R. P., 1981, The dominance of touch by vision: occurs with familiar objects, *Perception*, **10**, 29–33.

Praske, V., Schaible, H. G. and Schmidt, R. F., 1988, Joint receptors and kinaesthesia, *Experimental Brain Research*, **72**, 219–24.

Price, A., 1975, *World War II Fighter Conflict*, London: Purnell.

Reed, E. S., 1982, An outline of a theory of action systems, *Journal of Motor Behavior*, **14**, 98–134.

Revesz, G., 1950, *Psychology and Art of the Blind*, New York: Longmans Green.

Rock, I. and Harris, C. S., 1967, Vision and touch, in *Perception: Mechanisms and Models (Readings from Scientific American)*, San Francisco: W. H. Freeman.

Rock, I. and Victor, J., 1964, Vision and touch: an experimentally created conflict between the senses, *Science*, **143**, 594–6.

Roland, P. E., Larsen, B., Lassen, N. A. and Skinhoj, E., 1980, Supplementary motor area and other cortical areas in the organization of voluntary movements in man, *Journal of Neurophysiology*, **43**, 118–36.

Romanes, G. J., 1976, *Cunningham's Manual of Practical Anatomy*, Vol. 3, *Head and Neck and Brain*, 13th Edn (revised by Romanes, G. J.), London: Oxford University Press.

Rosenbaum, D. A., 1991, *Human Motor Control*, London: Academic Press.

Rosenbloom, L. and Horton, M. E., 1971, The maturation of fine prehension in young children, *Developmental Medicine and Child Neurology*, **13**, 3–8.

Ross, H. E., 1975, Mist, murk and visual perception, *New Scientist*, 19 June, 658–60.

Ross, H. E. and Reschke, M. F., 1982, Mass estimation and discrimination during brief periods of zero gravity, *Perception and Psychophysics*, **31**, 429–36.

van Rossum, J. H. A., 1989, *Motor Development and Practice: The Variability of Practice Hypothesis in Perspective*, Amsterdam: Free University Press.

Rothwell, J. C., Traub, M. M., Day, B. L., Obeso, J. A., Thomas, P. K. and Marsden, C. D., 1982, Manual motor performance in a deafferented man, *Brain*, **105**, 515–42.

Sams, M., Aulanko, R., Hamalainen, M., Hari, Lounasmaa, O. V., Lu, S.-T. and Simola, J., 1991, Seeing speech: visual information from lip movements modified activity in the human auditory cortex, *Neuroscience Letters*, **127**, 141–5.

Schiffman, H. R., 1990, *Sensation and Perception An Integrated Approach*, 3rd Edn, New York: John Wiley.

Schmandt, C.,1983, Spatial input/display correspondence in a stereoscopic computer graphic workstation, *Computer Graphics*, **17 (3)**, 253–61.

Schmidt, R. A., 1975, A schema theory of discrete motor skill learning, *Psychological Review*, **82**, 225–60.

Schmidt, R. A., 1980, Past and future issues in motor programming, *Research Quarterly for Exercise and Sport*, **51**, 122–40.

Schmidt, R. A., 1982, The schema concept, in Kelso, J. A. S. (Ed.), *Human Motor Behavior: An Introduction*, pp. 219–35, Hillsdale, NJ: Lawrence Erlbaum.

Schmidt, R. A., Zelaznik, H. N., Hawkins, B., Frank, J. S. and Quinn, J. T. 1979, Motor-output variability: a theory for the accuracy of rapid motor acts, *Psychological Review*, **86**, 415–51.

Schmidt, R. C., Shaw, B. K. and Turvey, M. T., 1993, Coupling dynamics in interlimb coordination, *Journal of Experimental Psychology: Human Perception and Performance*, **19** (2), 397–415.

Schneider, G. E., 1967, Contrasting visuomotor function of tectum and cortex in the golden hamster, *Psychologische Forschung*, **31**, 52–62.

Schneider, G. E., 1969, Two visual systems: brain mechanisms for localization and discrimination are dissociated by tectal and cortical lesions, *Science*, **163**, 895–902.

Schneider, W. and Shiffrin, R. M., 1977, Controlled and automatic information processing: I, Detection, search and attention, *Psychological Review*, **84**, 1–66.

Sekuler, R. and Blake, R., 1990, *Perception*, 2nd Edn, Singapore: McGraw-Hill.

Shaffer, L. H., 1975, Multiple attention in continuous verbal tasks, in Rabitt, P. M. and Dornic, S. (Eds), *Attention and Performance*, Vol. 5, pp. 157–67, London: Academic Press.

Sherrington, C. S., 1906, On the proprio-ceptive system, especially in its reflex aspect, *Brain*, **29**, 467–82.

Shik, M. L., Severin, F. V. and Orlovsky, G. N., 1966, Control of walking and running by means of electrical stimulation of the mid-brain, *Biophysics*, **11**, 756–65.

Slocum, D. B. and Pratt, D. R., 1946, Disability evaluation for the hand, *Journal of Bone and Joint Surgery*, **28**, 491.

Smith, W. M. and Bowen, K. F. , 1980, The effects of delayed and displaced visual feedback on motor control, *Journal of Motor Behavior*, **12**, 91–101.

Sprague, J. M. and Meikle, T. H. Jr, 1965, The role of the superior colliculus in visually guided behavior, *Experimental Neurology*, **11**, 115–46.

Stein, B. E., 1984, Development of the superior colliculus, *Annual Review Neuroscience*, **7**, 95–125.

Stein, B. E. and Meredith, M. A., 1993, *The Merging of the Senses*, Cambridge, MA: MIT Press.

Stevens, J. C.,1979, Thermo-tactile interactions: some influences of temperature on touch, in Kenshalo, D. R. (Ed.), *Sensory Function of the Skin of Humans*, New York: Plenum Press.

Sullivan, A. H., 1923, The perceptions of liquidity, semi-liquidity and solidity, *American Journal of Psychology*, **34**, 531–41.

Summers, J. J., 1989, Motor programs, in Holding, D. H. (Ed.), *Human Skills*, 2nd Edn, pp. 49–69, Chichester: John Wiley.

Sutherland, I. E., 1965, The ultimate display, *Proceedings of IFIP Congress*, pp. 506–8.

Sutherland, I. E., 1968, A head-mounted three-dimensional display, FJCC Fall Joint Computer Conference, *American Federation of Information Processing Societies*, **33** (1), pp. 757–64.

Taub, E., 1976, Movements in nonhuman primates deprived of somatosensory feedback, in Keogh, J. and Hutton, R. S. (Eds), *Exercise and Sport Science Reviews*, Vol. 4, pp. 335–74, Santa Barbara: Journal Publishing Affiliates.

Tastevin, J., 1937, En partant de l'experience d'Aristote, *L'Encephale*, **1**, 57–84.

Thelen, E., 1987, We think; therefore, we move, *Cahiers de Psychologie Cognitive*, **7**, 195–8.

Thelen, E. and Fisher, D. M., 1982, Newborn stepping: an explanation for a 'disappearing' reflex, *Developmental Psychology*, **18**, 760–75.

Thelen, E., Kelso, J. A. S. and Fogel, A., 1987, Self-organizing systems and infant motor development, *Developmental Review*, **7**, 39–65.

Thompson, R. F., 1967, *Foundations of Physiological Psychology*, New York: Harper and Row.

Thornbury, J. M. and Mistretta, C. M., 1981, Tactile sensitivity as a function of age, *Journal of Gerontology*, **36**, 34–9.

Treisman, A. M. and Gelade, G., 1980, A feature-integration theory of attention, *Cognitive Psychology*, **12**, 97–136.

Trevarthen, C. B., 1968, Two mechanisms of vision in primates, *Pyschologische Forschung*, **31**, 229–337.

Ungerleider, L. G., and Mishkin, M., 1982, Two cortical visual systems, in Ingle, D. J., Goodale, M. A. and Mansfield, R. J. W. (Eds), *Analysis of Visual Behavior*, pp. 549–86, Cambridge, MA: MIT Press.

Vallbo, Å. B. and Hagbarth, K. E., 1968, Activity from skin mechanoreceptors recorded percutaneously in awake human subject, *Experimental Neurology*, **21**, 270–89.

Vertut, J. and Coiffet, P., 1985, *Teleoperation and Robotics: Evolution and Development*, Vol. 3A, London: Kogan Page.

Vickers, D. L., 1973, Sorcerer's Apprentice: head-mounted display and wand, *Symposium on Visually Coupled Systems: Development and Applications*, in Task, J. A. B. and H. L. (Eds), pp. 522–41, Brooks AFB, TX: Aerospace Medical Division.

Vince, M. A., 1948, Corrective movements in a pursuit task, *Quarterly Journal of Experimental Psychology*, **1**, 85–103.

Wade, N. J., 1988, On the late invention of the stereoscope, *Perception*, **16**, 785–818.

Wade, N. J., Swanston, M. T. and de Weert, C. M. M., 1993, On interocular transfer of motion aftereffects, *Perception*, **22**, 1365–80.

Walsch, K. W., 1978, *Neuropsychology: A Clinical Approach*, Edinburgh: Churchill Livingstone.

Weiskrantz, L., Warrington, E. K., Sanders, M. D. and Marshall, J., 1974, Visual capacity in the hemianopic field following a restricted occipital ablation, *Brain*, **97**, 709–28.

Westling, G. and Johansson, R. S., 1984, Factors influencing the force control during precision grip, *Experimental Brain Research*, **53**, 277–84.

Whiting, H. T. A., Gill, E. B. and Stephenson, J. M., 1970, Critical time intervals for taking in flight information in a ball-catching task, *Ergonomics*, **13**, 265–72.

Wiesendanger, M., 1981, Organization of secondary motor areas of cerbral cortex, in Brooks, V. B. (Ed.), *Handbook of Physiology*, Vol. 2(1), pp. 1121–47, Bethesda, MD: American Physiological Society.

Wiesendanger, M., 1987, Initiation of voluntary movements and the supplementary motor area, in Heuer, H. and Fromm, C. (Eds), *Generation and Modulation of Action Patterns*, pp. 3–13, Berlin: Springer-Verlag.

Winstein, C. J. and Schmidt, R. A., 1989, Sensorimotor feedback, in Holding, D. (Ed.), *Human Skills*, 2nd Edn, Chichester: John Wiley.

Woodworth, R. S., 1899, The accuracy of voluntary movement, *Psychological Review*, **3**, Monograph Supplement, 1–119.

Woolsey, C. N., 1958, Organization of somatic sensory and motor areas of the cerebral cortex, in Harlow, H.F. and Woolsey, C. N. (Eds), *Biological and Biochemical Bases of Behaviour*, Madison, WI: University of Wisconsin Press.

Woolsey, C. N., Settlage, P. H., Meyer, D. R., Sencer, W., Hamuy, T.P. and Travis, A. M., 1950a, Patterns of localization in precentral and 'supplementary' motor areas and their relation

to the concept of a premotor area, *Ass. Res. nerv. Dis. Proc.*, **30**, 238–64.
Woolsey, C. N., Settlage, P. H., Meyer, D. R., Sencer, W., Hamuy, T. P. and Travis, A. M., 1950b, Somatotopic organization of 'supplementary motor area' of the monkey, *American Journal of Physiology*, **163**, 763.
Zeki, S. M., 1973, Colour coding in rhesus monkey prestriate cortex, *Brain Research*, **53**, 422–7.

8

Auditory virtual environments

Mark Williams

8.1 Introduction

> . . .human beings enjoy an ability which has given them their most satisfactory appellation, Homo Loquens. (O'Neill, 1980)

Elsewhere in this volume one can read about the purely visual aspects of human communication which throw up their particular challenges to the virtual environment designer, but it seems unlikely that humans will exist comfortably in a synthetic environment without the availability of the 'audio channel'. Shelley says, perhaps a little controversially, in *Prometheus Unbound*:

> He gave man speech, and speech created thought,
> Which is the measure of the Universe.

An indicator of the value of speech, if any were needed, comes from the fact that man has developed the vocal apparatus for communication at some increase in the risk of choking (Borden and Harris, 1984). The ability to swallow whilst breathing is lost early in life as the larynges and tongue move down to a position suitable for speech production.

This chapter presents a brief survey of some of the research relevant to auditory communications in a virtual environment and some of the display technology which makes it possible. Auditory input devices are not covered, these being less well developed, but progressing (Ando *et al.*, 1993).

In the particular case of speech communication, of course, we ought to be aware that there is a visual element in addition to the acoustic signal. No doubt to provide more 'redundancy' in the signal, humans quickly learn to use both auditory and visual cues when using speech communication (from the age of 18 to 20 weeks: Kuhl and Metcalf (1982)) and there is some evidence that mis-registration of these auditory and visual cues causes complete misperception of syllables and words. If one presents filmed examples of a person speaking a word and overdubs another word, the perceived word may be different from both those presented. For example, nearly all observers heard /da/ when /ba/ was dubbed over the lip movements for /ga/, and this has been shown to extend to whole words (McGurk and MacDonald 1976). Such visual interference in auditory perception might be interpreted as 'visual dominance' (see section 7.4.2). But some recent work, using video in the context of an educational tool, seemed to show no enhancement of understanding of a speech-based task (comprehension of English sentences by students with a different mother tongue) with the presence of a video image of the speaker's head and shoulders. Nor did the 'reinforcing' of a sentence's meaning with a supposed salient image (e.g. 'Kate was really down in the dumps' accompanied by an image of girl looking miserable) have any significant affect (Greaves *et al.*, 1993).

8.2   Human auditory system

8.2.1   Physiology

Although humans can perceive other sounds, there is some evidence that they are adapted to perceive the sounds that they make. The human ear consists of three main parts. The first part is the outer ear: this comprises the pinna, or ear flap, and the external auditory meatus or ear canal, a tube some 25 mm long and 7 mm in diameter. The role of the pinna is described more fully below, but the main impact of the auditory meatus is that it enhances sound frequencies around 3 kHz, the region of the audio spectrum that a young adult is typically most sensitive to. The ear canal leads to the middle ear which consists of the tympanic membrane (ear drum), three tiny bones, named the hammer, anvil and stirrup (malleus, incus and stapes) arranged in a 'chain', and some muscles and ligaments. The middle ear is arranged to maximize the transfer of sound energy from air to the fluid within the cochlea (inner ear), with the end of the bony chain, the stapes, acting like a plunger forcing waves through the fluid. In addition, the muscles attached to the bones provide overload protection for extended exposure to loud sounds, they suppress the effects of one's own voice, and help suppress the effects of chewing (the hinge of the mandible is located just below the ear canal and, if the entrance to the outer ear is sealed, it is possible to measure changes of pressure within the ear canal due to jaw movements (Franke *et al.*, 1952)).

Finally we come to the inner ear, which comprises the cochlea and the balance organs. The cochlea is the main transducing system for converting mechanical vibrations to electrical impulses. It is a tapering, coiled tube divided longitudinally by two partitioning membranes and is filled with fluid. One of these membranes, the basilar membrane, has hair sensors embedded in it, and the tops of some touch the other, tectorial, membrane. These outer hair cells, through their interaction with the covering membrane, serve to amplify and tune the processes which stimulate the other, inner, hair cells.

The auditory system has an extraordinary dynamic range of some 120 dB and is capable of detecting sound pressure changes which displace the ear drum by amounts comparable to fractions of the width of a hydrogen molecule (Von Bekesy and Rosenblith, 1951). This seems remarkable, given that the neurons associated with the auditory system have only a dynamic range of 40 dB. One explanation, for which there is some neurological evidence, is that there are several classes of neurons which operate over different parts of this range (Kiang, 1968). The other source of this additional range is the availability of 'out of band' energy from the resonating hair systems within the cochlea. That is to say, the hair sensor systems in the cochlea respond to narrow frequency bands of stimulation, but at high stimulation intensities they are able to react to signals out of their normal response range.

The basilar membrane, mentioned above, provides a frequency analysis of its stimulus because different parts of it experience displacement, depending on the frequency of the stimulus (Von Bekesy, 1960). This is usually termed 'place theory'. Below 1kHz the nerve fibres can fire in synchrony with the pulses of sound pressure, but above this frequency the cells are not able to fire quickly enough, and it is the place that is stimulated on the membrane which gives tone sensation (Crowe *et al.*, 1934).

8.2.2 Identifying a sound source

The perception of tonal quality and the ability to categorize sounds (e.g. the difference between a piano and a violin playing the same note) is dependent on several characteristics. The case of musical instruments can be taken as an example. Firstly, a variation in sound pressure is produced which cannot be characterized by a single frequency but consists of several frequencies related to the base, or fundamental, frequency by simple arithmetic ratios. These are referred to as overtones of the fundamental and together they form a series of harmonics. The relative contribution of these harmonics differs between instruments. Secondly, the excitation and decay curves of the notes from different instruments are different and this also helps to classify them (Berger, 1964).

The perception of these related tones as forming a single source has been investigated by many workers. One important feature of the auditory signal is that all frequency components grow and decay together (McAdams, 1981). Different rates of change in the harmonic structure of speech signals is one feature which seems perceptually important in classifying vowels (Sekuler and Blake, 1994). Bergman and Pinker (1978) found that frequency components which were delayed from the others by more than 50 ms are perceived as being separate sources, and tones which are not harmonically related are more difficult to fuse into one source than those which are (de Boer, 1956). An interesting concept, which has also been developed by Gaver (1994, also section 8.5), is that of auditory texture. Lederman found that subjects could judge quite accurately the relative roughness of a surface simply from the sound it made when it was rubbed (Lederman, 1979).

8.2.3 Auditory localization

The process of localization of sound sources has been the subject of much investigation. There are three 'classical' cues which have been identified as being important in helping to localize sound. Wenzel *et al.* (1988a) cite Lord Rayleigh's 'Duplex Theory' of localization as a first attempt at explaining the process. This includes two important aspects: inter-aural intensity differences and inter-aural time differences. Shaw (1974) used dummy head recordings (that is a physical model of a human head which has small microphones positioned in the entrance to the ear canal) to record the intensity differences and demonstrated that there can be a 20 dB difference when the sound source is located opposite one ear for a 6 kHz tone. Using similar apparatus, Shaw (1974) was also able to measure inter-aural time differences and when the source was again located at 90° to the straight ahead position, the signal was shown to arrive approximately 0·6 ms later at the opposite ear. These two mechanisms of intensity and time assume relatively different degrees of importance depending on the frequency of the stimulus. The first depends on shadowing effects from the geometry of the head, but this mechanism will only work if the wavelength of the stimulus is small compared with the size of the head. For longer wavelengths, the second mechanism of time applies. Here the two signals presented at the ears are discriminable by phase in a systematic way. Simple localization experiments varying these two factors were performed by Naish (1990) in the context of a visual and auditory cueing task both for simple tones and spoken messages.

More sophisticated modelling of the hearing apparatus has been carried out by Wenzel *et al.* (1988a) to include another important localization cue, namely the spectral and temporal modifications introduced by reflections at the pinna. The preceding two 'time-domain' mechanisms prove inadequate at explaining why localization is poorer without the benefit of the pinna. Burger (1958) and Batteau (1967) suggested that the pinna might provide a mechanism for colouring and labelling the sound depending on the source location. Wenzel (1988b) experimented with recordings made using several different pinnas (using a small microphone in the subject's ear canal) and found that playback to a subject using a recording from another's pinna caused the subject difficulties when performing the localization task, particularly when a 'bad localizer's' head-related transfer function (HRTF; see section 8.4.1) was used by a 'good localizer' (see section 8.4.2). This implies that, in virtual environment, it might be possible to improve an individual's localization ability.

A further, second-order, aid to localization is head movement, but clearly this cannot help in the localization of transients because head movements are too slow (Pollack and Rose, 1967). Binaural dummy head recordings give good images which appear external to a headphone wearer's head if the wearer keeps his or her head stationary (Belendiuk and Butler, 1978), but the tendency to 'visualize' a sound source within the head has been a recurrent difficulty with artificial localization systems and is discussed further below (section 8.4).

In rooms, unsurprisingly, reflections have been found to have a significant impact on localization performance. Taking an 'ecological view' of perception (see also sections 3.2.4 and 9.3), Guski (1990) points out that human hearing has evolved in the context of an upright-standing animal, and ground reflections are important for elevation cues. Experiments conducted within an anechoic chamber where one surface could be made reflecting (in this case, a somewhat 'non-ecological' aluminium sheet) showed that elevation judgements were enhanced in the presence of a floor reflector but could be degraded if the reflecting surface was placed as a ceiling. The stimulus used was that of a male voice saying 'hello'.

Another study of localization effects in reverberant rooms explicitly tried to isolate depth perception (Nielsen, 1993). Nielsen notes that there are three factors which affect distance perception: overall loudness, which is most effective in anechoic conditions, the frequency spectrum (sounds heard from a distance tend to have the high frequency content of their spectra attenuated), and the ratio between the intensity of the direct and reverberant portions of a sound (the latter arrives later, of course). The study cited looked at three types of room with different reverberation characteristics, and it was found that changes in absolute loudness had different effects in different rooms. In an anechoic chamber, the physical distance had no effect on perceived distance if the absolute loudness was kept constant at the listener's position. In reverberant rooms, the perceived distance is almost independent of playback level, and sources that were located to one side were always perceived to be further away than their true distance. These and other issues, such as how to enhance localization, are discussed by Durlach *et al.* (1993). Accurate modelling of reverbation could thus have similar importance in virtual environments to the accurate modelling of light transmission as discussed by Christou and Parker (section 3.7).

As has been mentioned earlier in the context of speech perception (section 8.1), visual cues interact with auditory cues, and they can have a dominant effect on localization. The simple observation that sound sources in the cinema are usually perceived to be at the location of the visual stimulus is one example (Welch and Warren, 1980).

8.3 Virtual speech

8.3.1 Natural speech

Lofqvist (1990) refers to speech as audible gestures, and as such speech is an important capacity of many virtual 'actors' (see section 2.1.4). In order to help in the understanding of speech synthesis (described below), it is worth describing very briefly the nature of speech and the vocal apparatus used to produce it.

The vocal system uses a reservoir of moist air within the lungs, usually above atmospheric pressure (pressure is controlled by the pulmonary musculature), which may be expelled through the glottis, consisting of a gap between two fleshy folds (the vocal cords). These folds may be tensed and stiffened by a complex muscle system around the larynx. The folds are forced apart by the pressure and then drawn back together again by the Bernoulli forces exerted by the accelerated air passing between them (Van den Berg, 1958). A simple analogy is to blow between two sheets of paper held apart a little way with their surfaces parallel: perhaps counter-intuitively, the sheets are drawn together.

The source of the vocal signal, then, is the opening and closing of the glottal folds, allowing a train of pulses of air into the vocal tract above it, and then into the mouth. A graph plotting pressure against time reveals a 'saw-tooth' waveform. The spectrum of this type of waveform consists of a set of harmonically related frequencies, decreasing in magnitude with increasing frequency. The fundamental frequency of these spectra usually lies between about 120 and 500 Hz, depending on the age and sex of the speaker. The vocal tract above the glottis is a tube approximately 22 cm long, again this varies with the age and sex of the speaker, and the diameter of the tube varies along its length (partly because of skeletal construction and partly because the tongue and jaw position vary as the speaker speaks). The vocal tract acts as a complex bandpass filter, modifying the envelope of spectrum of the glottal wave so that some parts are emphasized and others suppressed. The broad peaks produced in the spectrum are usually referred to as formants, and these vary according to tract configuration. With some configurations of the vocal tract, constrictions produce turbulent flow which generate bandlimited gaussian noise sources at, or downstream of, the constriction (Shadle, 1990). These are manifested as unvoiced sounds, such as 's' and 'f' and other hissing sounds.

8.3.2 Speech synthesis

This topic has a surprisingly long history, dating to von Kemplen, who, in 1769, developed a bellows-operated device capable of producing spoken phrases in French and Italian, according to the inclination and dexterity of its operator (Linggard, 1985). Subsequently, various analogue-computing and electrical-circuit synthesizers have been developed (Flanagan, 1983), but the advent of digital computers enabled the production of synthesizers of great flexibility.

Leaving aside crude systems which store representations of individual phones and concatenates them according to stored rules, there are two basic approaches to speech synthesizers (Linggard, 1985). They are distinguished by the type of information which is required to drive the synthesizer.

One type of synthesizer models the speech process in terms of its acoustic properties. This can be the record of whether a sound is voiced or unvoiced, together with the record of the formant frequencies, usually termed a formant synthesizer. Alternatively, 'linear predictive analysis and resynthesis' (sometimes referred to as 'linear predictive coding', or LPC) can be used. Here the speech signal is stored in an approximate form, which is a value for the frequency of the periodic excitation and a value for its amplitude at discrete time intervals. These data are then supplied as the input to a digital filter, consisting of a set of weighting factors applied to a number of past samples of the data. These weights adapt with time according to some statistical constraints, usually of the form of a least mean squared error (see Rabiner and Shafer (1980) or Haykin (1991) for a detailed description). Speech synthesis was one of the first major successes of the application of adaptive, predictive filter techniques, and it works well for non-nasalized, voiced sounds. Other sounds require modelling approximations. The main reason for this is that it is assumed the spectrum of the signal being synthesized contains only poles (roughly, 'peaks'), and this is only true for the types of sound mentioned.

A rather different strategy is to attempt to model the physiology of the human vocal system (e.g. Ishizaka and Flanagan, 1972; Rice, 1987). The input to this sort of synthesizer is composed of approximations of articulatory parameters, such as sub-glottal pressure, vocal fold tension and vocal tract shape. The raw data for the latter are derived from X-ray studies of humans phonating (usually in Russian, e.g. Fant, 1963). This type of synthesizer has found most application with those interested in speech production motor issues (e.g. Howell *et al.*, 1987). These models have been extended so that they can produce nasalized and unvoiced sounds well in text-to-speech systems. They tend, however, to be used to generate speech 'off-line', because of their computationally expensive nature (Flanagan *et al.*, 1975).

Recommendations exist for when it is appropriate to use synthesized (usually based on an LPC system), and when to use pre-recorded digitized real speech (see Cowley and Jones (1992) for a survey). For some applications there may be advantages in having speech which is not 'natural'. For example, in the case of aircraft pilots, it may be advantageous to discriminate between messages from (human) ground controllers and from on-board voice-generating systems. The main obstacle to successful synthetic speech currently is the difficulty in producing the prosody of normal speech: that is, how we connect up all the pieces of speech.

8.4   *Creating an auditory virtual environment*

8.4.1   Audio hardware

Two of the most widely known 3-D audio display systems have been developed in parallel. One is the commercially available 'Convolvotron' (and its cheaper version, the 'Beachtron') system, and is associated with the NASA-Ames Laboratory. The other system is under development at the Armstrong Laboratory of the Wright-Patterson Airforce Base. From published descriptions, they both function in similar ways.

As mentioned earlier in this chapter (section 8.2.3), it has been shown that the sound incident upon a listener is modified by reflections in the pinna and by diffraction effects about the head and torso. These modifications affect the spectrum of the incident sound, and they depend on the relative position and direction of the sound's source.

The approach taken by both systems mentioned is to measure these HRTFs at many points in azimuth and elevation for each ear, by insertion of miniature microphones within the ear canal. Typically, then, there are pairs of frequency responses at 10° to 15° intervals. Wenzel *et al.* (1988a) cite informal results that there is no perceptual difference between spectra interpolated between measured values as much as 60° apart, and those which have been measured at the interpolated points.

The spectra obtained are then represented as a filter in the digital domain. There are two major types of digital filter: infinite impulse response (IIR) and finite impulse response (FIR). Some of the terminology in digital filter design carries over from analogue filter design, and in essence the filter may be characterized by how it behaves when a very short 'spike' signal is applied to its input. In the digital domain this usually means the observed output values when there is a single input sample of unit amplitude. As is implied by its name, an IIR filter would be expected to produce an infinite train of values which were not all zero. An FIR filter, in contrast, would only produce a finite number of output samples, the rest being zero. Generally, an IIR design is recursive: its current output value depends on previous output and, perhaps, input values. FIR designs are usually, but not exclusively, non-recursive so that the current output value depends solely on past and current input values.

3-D audio systems are implemented as FIR filters. In the Convolvotron the FIR filter stores pairs (one for each ear) of 256 filter coefficients which are then applied to 256 past input values of some monophonic sound, in order to produce a single output sample of the 'spatialized' equivalent sound. This description is rather simplified because it is also necessary to use various interpolation techniques to position sound which is to appear in directions between those for which the measured responses exist. The most recent versions of this equipment use more subtle methods of combination than simple linear interpolation (Foster *et al.*, 1992).

8.4.2   Assessments of 3-D audio hardware

One recent study (Pellieux *et al.*, 1993) assessed the Convolvotron. In their experiment they measured the head-related transfer function for four subjects, and assessed their ability to localize a virtual sound source by tracking the users' head movement and asking them to face the apparent location. They found that the subjects demonstrated different types of response. Two of the subjects could localize sound sources relatively well, to within 4° or 5° r.m.s. error. The other two subjects showed errors in the order of 25° to 30° r.m.s. error. This trend for two types of response has also been observed by Wenzel *et al.* (1988a), where goodness-of-fit factors to real and apparent position functions varied from 0·89 to 0·97 in the azimuth but from 0·69 to 0·98 in elevation.

Another assessment recently published, this time using hardware developed at the Armstrong Laboratory, was reported by McKinley *et al.* (1994). It was found that the minimum audible angle in azimuth was 4–5° in quiet conditions and 6–7° in a noisy environment (a simulation appropriate to an AV-8B or Harrier aircraft). The conclusions of this paper were that, during flight trials, the auditory information presented using this system gave an increase in 'situational awareness' reported by pilots, and improvements in many other performance indices. Also, although further improvements in technology were expected, the current technology was deemed adequate for their particular application. Similar results were obtained by Calhoun *et al.* (1988), showing an improvement in left/right cueing by use of '3-D speech' as opposed to non-localized speech.

A common problem experienced with 3-D sound systems is that the sound source often appears localized within the head. Bergault and Wenzel (1993) found that up to 46% of distance judgements put the source within the head when using generic HRTFs. In this study, the listeners were inexperienced and the HRTFs being used were not their own (usually generic HRTFs belong to a 'good localizer'). This localization effect can be suppressed to some extent by adding artificial reverberation to the sound source (Bergault, 1992).

8.5   *Auditory icons and virtual environments*

There are many other speech and auditory issues which are of interest to the inhabitants of virtual environments, including the effects of external noise, the precedence effect, and binaural noise suppression. These are touched on by Durlach *et al.* (1993). This chapter will conclude by discussing an issue central to the human–computer interface design of auditory virtual environments, and about which there is clearly a need for much research.

Gaver (1994) discusses the application of the auditory icon or 'Earcon' (Blattner *et al.*, 1989) in the generation of human–computer interfaces which use both visual and auditory symbology. He stresses the concept of hearing events, rather than using musical analytical techniques. This requires metaphors relevant to 'everyday listening', e.g. representations of heavy and light or hard and soft. It is thus possible to have what Gaver calls 'parameterized auditory icons'. In the context of the familiar desktop metaphor for computer interaction for example, when a large file is dragged across different parts of a screen, we hear firstly that the file is large and heavy, and secondly that the different parts of the screen may have different auditory textures. In this way, we build in multidimensional, mutually reinforcing cues into an interface.

Two examples of this approach are the 'Sonic Finder' (Gaver, 1989), in which copying files is accompanied by the sound of a filling vessel, and the 'ARKola plant' (Gaver, 1994). This latter was an application which allowed several users to operate a virtual bottling plant. Associated with various parts of the plant were distinct sounds, so that as many as twelve sounds might be playing simultaneously! Two important results came from a study of this application, which should be noted by designers of virtual environments. Firstly, it helped in collaborative working: one could still hear colleagues even when one could not see what they were doing. Secondly, the concept of a 'healthy plant' emerged, as an integral entity and not just as a collection of parts. This collective concept was presumably identifiable by a regular pattern of sounds.

Blattner *et al.* (1989) discuss various techniques that might be used for designing 'Earcons', including musical ones. Musical techniques lead quite naturally to the Wagnerian concept of *leitmotiv*: small 'snippets' of music in an opera which are easily recognizable and which may be modified depending how the characters interact. In their concluding remarks, Blattner *et al.* (1989) say: 'We found that playing earcons concurrently created unanticipated problems that could not be resolved until the structure of individual earcons had been thoroughly studied . . .'. It is perhaps as well to remember that in mediaeval times certain combinations of notes were forbidden in music by papal decree: they were the *diabolo in musica*!

References

Ando, S., Shinoda, H., Ogawa, K. and Mitsuyama, S., 1993, A three-dimensional sound localization sensor system based on the spatio-temporal gradient method, *Transcripts of the Society of Instrument and Control Engineers*, **29** (5), 520–8.

Batteau, D. W., 1967, The role of the pinna in human localization, *Proceedings of the Royal Society of London, Series B*, **168**, 158–80.

Belendiuk, K. and Butler, R. A., 1978, Directional hearing under progressive impoverishment of binaural cues, *Sensory Processes*, **2**, 58–70.

Bergault, D. R., 1992, Perceptual effects of synthetic reverberation on three-dimensional audio systems, *Journal of the Audio Engineering Society*, **39**, 864–70.

Bergault, D. R. and Wenzel, E. M. , 1993, Headphone Localization of Speech, *Human Factors*, **35** (2), 361–76.

Berger, K. W., 1964, Some factors in the recognition of timbre, *Journal of the Acoustical Society of America*, **36**, 1881–91.

Bergman, A. S. and Pinker. S., 1978, Auditory streaming and the building of timbre, *Canadian Journal of Psychology*, **31**, 151–9.

Blattner, M. M., Sumikawa, D. A. and Greenberg, R. M., 1989, Earcons and icons: their structure and common design principles, *Human Computer Interaction*, **4** (1), 11–44.

de Boer, I. 1956, Pitch of inharmonic signals, *Nature*, **178**, 535–6.

Borden, G. and Harris, K., 1984, *Speech Science Primer: Psychology, Acoustics and Perception of Speech*, 2nd Edn, Baltimore, MD: Williams and Wilkins.

Burger, J. F., 1958, Front back discrimination of the hearing system, *Acustica*, **8**, 302–10.

Calhoun, G., Janson, W. P. and Valencia, G., 1988, Effectiveness of three-dimensional auditory directional cues, in *Proceedings of the Human Factors Society's 32nd Annual Meeting*, pp. 68–72, Santa Monica, CA: Human Factors Society.

Cowley, K. C. and Jones, D. M., 1992, Synthesized or digitized? A guide to the use of computerized speech, *Applied Ergonomics*, **23** (3), 172–6.

Crowe, S. J., Guild, S. R. and Polvost, L. M., 1934, Observations on the pathology of high tone deafness, *Bulletin of the Johns Hopkins Hospital*, **54**, 315–79.

Durlach, N. I., Shinn-Cunningham, B. G. and Held, R. M., 1993, Supernormal auditory localization, *Presence*, **2** (2) 89–103.

Fant, G., 1970, *Acoustic Theory of Speech Production*, The Hague: Mouton.

Flanagan, J. L., 1983, *Speech Analysis Synthesis and Perception*, 2nd Edn, Berlin: Springer-Verlag.

Flanagan, J. L., Ishizaka, K. and Shipley, K. L. 1975, Synthesis of speech from a dynamic model of the vocal cords and vocal tract, *Bell System Technical Journal*, **54** (3), 485–505.

Foster, S. H., Chapin, W. and Longley, L., 1992, Minimum phase HRTF convolution technology, Technical Memo from Crystal River Engineering Inc., Groveland, CA.

Franke, E. K., von Gierke, H. E., Grossman, F. M. and von Wittern W. W., 1952, The jaw motions relative to the skull and their influence on hearing by bone conduction, *Journal of the Acoustical Society of America*, **24**, 142.

Gaver, W. W., 1989, The Sonic Finder: an interface using auditory icons, *Human Computer Interaction*, **4**, 67–94.

Gaver, W. W., 1994, Sound effects for computational worlds, in *Proceedings of the Ergonomics Society/IEE Conference 'Designing Future Interaction'*, Warwick University UK, April 8th, Loughborough, UK: The Ergonomics Society.

Greaves, C., Warren, M. and Ostberg, O., 1993, Enhancing speech intelligibility using visual images, *Advances in Human Factors Ergonomics*, **19** (B), 1097–102.

Guski, R., 1990, Auditory localization: effects of reflecting surfaces, *Perception*, **19**, 819–30.

Haykin. S., 1991, *Adaptive Filter Theory*, 2nd Edn, Englewood Cliffs, NJ: Prentice Hall.

Howell P., Williams, M. and Vause, L., 1987, Acoustic analysis of repetitions in stutterers' speech, in Peters, H. F. M. and Hulstijn, W. (Eds), *Speech Motor Dynamics in Stuttering*, pp. 371–80, Vienna: Springer-Verlag.

Ishizaka, K. and Flanagan, J. L., 1972, Synthesis of voiced sounds from a two mass model of the vocal cords, *Bell Systems Technical Journal*, **51** (6), 1233–68.

Kiang, N. Y. S., 1968, A survey of recent developments in the study of auditory physiology, *Annals of Otology, Rhinology and Laryngology*, **77**, 656–75.

Kuhl, P. K and Meltzoff, A. N., 1982, The bimodal perception of speech in infancy, *Science*, **218**, 1138–41 (reported in Sekuler, R. and Blake, R., 1994, *Perceptions*, 3rd Edn, Singapore: McGraw-Hill).

Lederman, S. J., 1979, Auditory texture perception, *Perception*, **9**, 93–103.

Linggard, R., 1985, *Electronic Synthesis of Speech*, Cambridge: Cambridge University Press.

Lofqvist, A., 1990, Speech as audible gestures, in Hardcastle, W. J. and Marchal, A. (Eds), *Speech Production and Speech Modelling*, Vol. 55, Behavioural and Social Sciences Series, NATO ASI series D, pp. 289–322, Dordrecht: Kluwer.

McAdams, S., 1981, Spectral fusion and the creation of auditory images, in Clynes, M. (Ed.), *Music, Mind and Brain: the Neuropsychology of Music*, pp. 279–98, New York: Plenum Press.

McGurk, H. and McDonald, J., 1976, Hearing lips and seeing voices, *Nature*, **264**, 746–8.

McKinley, R. L., Erickson, M. A. and D'Agnelo, W. R., 1994, 3-Dimensional auditory displays:

development, applications and performance, *Aviation, Space and Environmental Medicine*, **65** (5), Suppl., A31-8.

Naish, P. L. N., 1990, Simulated directionality in airborne auditory warnings and messages, in Life, A., Narborough-Hall, C. S. and Hamilton, W. I. (Eds), *Simulation and the User Interface*, pp. 127-41, London: Taylor & Francis.

Nielsen, S. H., 1993, Auditory distance perception in different rooms, *Journal of the Audio Engineering Society*, **41** (10), 755-70.

O'Neill Y. V., 1980, *Speech and Speech Disorders in Western Thought before 1600*, Westport, CT: Greenwood Press.

Pellieux, L., Gulli, C., Leroyer, P., Piedecocq, B., Leger, A. and Menu, J. P., 1994, 'Approche experimentale du son 3D: methodologie et resultats preliminaire, in *AGARD Aerospace Medical Panel Symposium on 'Virtual Interfaces: Rsearch and Design'*, Lisbon, Portugal, 18-22 October 1993. AGARD Conference Proceedings No. 541, pp. 19-1-6, Washington, DC: NASA Scientific and Technical Information Branch.

Pollack, I. and Rose, M., 1967, The effect of head movements on the localization of sounds in the equatorial plane, *Perception and Psychophysics*, **2**, 591-6.

Rabiner, L. R. and Schafer, R. W., 1980, *Digital Processing of Speech Signals*, Englewood Cliffs: Prentice Hall.

Rice, D. L., 1987, A twisting one mass glottal model, paper presented at session H, 114th Meeting of the ASA, Miami, FL, 17 November 1987.

Sekuler, R. and Blake, R., 1994, *Perception*, 3rd Edn, Singapore: McGraw-Hill.

Shadle, C. H., 1990, Articulatory-acoustic relationships in fricative consonants, in Hardcastle, W. J. and Marchal, A. (Eds), *Speech Production and Speech Modelling*, Vol. 55, Behavioural and Social Sciences Series, NATO ASI series D, pp. 187-209, Dordrecht: Kluwer.

Shaw, E. A. G., 1974, Transformation of sound pressure level from the free field to the ear drum in the horizontal plane, *Journal of the Acoustical Society of America*, **56**, 1848-61.

Van den Berg, J. W., 1958, Myoelastic-aerodynamic theory of voice production, *Journal of Speech and Hearing Research*, **3**, 227-44.

Von Bekesy, G., 1960, *Experiments in Hearing*, New York: McGraw-Hill.

Von Bekesy, G. and Rosenblith, W. A., 1951, The mechanical properties of the ear, in Stevens, S. S. (Ed.), *Handbook of Experimental Psychology*, pp. 1075-115, New York, John Wiley.

Welch, R. B. and Warren, D. H., 1980, Immediate perceptual response to intersensory discrepancy, *Psychological Bulletin*, **88**, 638-67.

Wenzel, E. M., Wightman, F. L. and Foster, S. H., 1988a, A virtual display system for conveying three dimensional acoustic information, in *Proceedings of the Human Factors Society 32nd Meeting*, pp. 86-90, Santa Monica, CA: Human Factors Society.

Wenzel, E. M., Wightman, F., Kistler, D. J. and Foster, S. H. 1988b, Acoustic origins of individual differences in sound localisation behaviour, *Journal of the Acoustical Society of America*, **84**, 579.

9

Designing in virtual reality: perception–action coupling and affordances

Gerda J. F. Smets, Pieter Jan Stappers, Kees J. Overbeeke and Charles van der Mast

'We all agreed,' he said, 'that your theory is crazy. The question which divides us is whether it is crazy enough to have a chance of being correct. My own feeling is that it is not crazy enough.' Bohr to Pauli.(F. J. Dyson, Innovation in physics, *Scientific American*, September 1958, 74–82)

9.1 Introduction

Simulations have always been important to engineers who use preliminary models and test rigs to find out whether the designs they have thought up will actually work. It is important, however, to know when such simulations are useful to the design engineer and when their similarity to reality, and hence their usefulness, ends. This is particularly true of computer simulations, where there is not even a physical model, but only a computer program.

Computer simulations are used in two ways (see also sections 2.3 and 10.3 of this book). On the one hand, they can be an aid to learning how a complex system works and are particularly useful when the system will have to work in a variety of conditions. For example, a computer simulation of the way a vehicle behaves on the road can help predict at what point it will start to become unstable (Pacejka, 1991). On the other hand, computer simulations can also be used in working, or learning to work, with something (see also Rheingold, 1991, p. 212). An aircraft simulator, for example, helps a pilot to learn to fly better. For it to work properly, a computer simulation of this kind has to be capable of convincing us that we are somewhere else: high up over the ocean, with engines faltering, for instance. This is what Sheridan (1989) calls 'telepresence'. With the passage of time, computer systems have been developed which are capable of creating telepresence not just in a single virtual world but in a large number of virtual worlds. Such systems are often termed virtual reality (VR) systems.

We should like here to take a closer look at the latter kind of computer simulation, and in particular to discuss the question of whether it can be useful in the design process. At the same time, we shall illustrate the relevance of perceptual psychology research to the development of a computer-aided design (CAD) program for use in VR. We shall take

189

the risk of outlining on the basis of only one example where we might be able to go with VR in the near future. Although VR systems are still limited (as limited as the first generation of telephones, radios, television sets and computers), they do already give us the promise of what they might be able to offer in the future.

We shall start by defining the terms VR and CAD, and then describe our CAD program and discuss its technological and scientific relevance.[1]

9.2   Definitions

9.2.1   Virtual reality (VR)

A VR system is a computer you can no longer see.[2] There are two sorts: systems which completely surround you and offer a complete virtual environment, and systems which only offer a window onto a virtual reality. Both are important, and we carry out research using both, but here we shall describe work with a system of the first kind only.

The immersive system that we use is a *Virtuality 1000 SU* made by W Industries (Figure 9.1). Naturally, it has a number of limitations. The helmet is heavy (2·9 kg), the field-of-view is relatively small (90°) and there is a relatively large time lag (up to 0·5 s for complex scenes). Image resolution is low (276 × 372 pixels with a refresh rate of 25 Hz), but research has shown that with moving images this is not an incapacitating constraint (Smets and Overbeeke, 1993).[3]

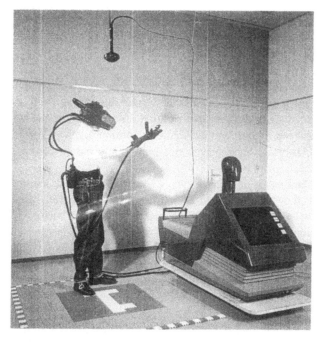

Figure 9.1   Virtuality 1000 SU virtual reality simulator. The system comprises a computer which drives visual, auditory and tactile outputs using data derived from sensors fixed to the operator. This sensor monitors the position and attitude of the operator's head and the position of his right hand. When the system is working the operator receives feedback in the helmet and Feedback Glove. The helmet contains two small screens and four loudspeakers, so the overall effect is stereo information, both visual and auditory, about the observable world surrounding the operator. The Feedback Glove contains tiny balloons which are inflated when a virtual object is touched, and thus transmit slight pressure to the fingers and palm of the hand.

Another disadvantage is that the 'Feedback Glove' provides tactile feedback but no force feedback. If you touch something you feel it (because the balloons in the Feedback Glove are inflated), but it does not 'push back' or resist you as it would in the real world. There are, however, aids which can provide such force feedback. One, shown in Figure 9.2, is a concept of a 'scooter' with which it is possible to move through virtual space, fast or slow, leaning into corners or not, until you hit a wall, when you feel the counter pressure by the blow against the handlebars and the stopping of the band under your foot.

We bought our VR simulator in 1989 to aid research into the importance of body motion in visual perception. This research is relevant to the design of technical products such as teleoperation systems (Smets, in press). (A new version of the *Virtuality VR* system has since been launched with enhanced capabilities.)

9.2.2 Computer-aided design (CAD)

Computer-aided design is a way of designing in which the computer is integrated into the design process. This confers a number of important benefits. One benefit it offers is the easy creation and correction of working drawings, with a complete datafile which allows the rapid and accurate calculation of various alternatives. This is the first kind of computer simulation to which we referred in section 9.1, when the computer is used as an aid to understanding how a complex system works. CAD is also used to visualize product design, which enables aesthetic (Figure 9.3) and ergonomic (Figure 9.4) evaluations of the design at an early stage in the design process.

But there are still disadvantages with this kind of simulation. Two of the most important ones can be eliminated with a VR system because it offers telepresence. However

Figure 9.2   Scooter (C. Hummels, 1993).

Figure 9.3   Aesthetic evaluation using computer visualization. Street furniture (dipl. des. C. Maurer).

sophisticated computer visualizations may have become, they still have the disadvantage of being flat pictures of what is generally a three-dimensional design. For example, with an ordinary computer the representation of a newly designed car is restricted to, at most, a detailed picture that can be viewed from various angles. When using VR to design a car, you can 'virtually' walk around it, open its doors, and experience and evaluate its spatiality without having to build a physical model.

The other disadvantage is even more important, if less obvious. There are gaps in the conceptual design of CAD packages (see also Neuckermans, 1992), as most CAD packages do not allow what a designer needs most: direct manipulation. Generally, CAD programs work with menus or keys with which the user selects commands. There is no link whatever to the designer's perceptual and motor skills. In contrast, the CAD program to be described here, which is still in the process of being developed, allows virtual objects to be manipulated directly. The program offers a view of CAD which, given the rapid development of VR systems, will be capable of being used in practical design in the not too distant future.[4]

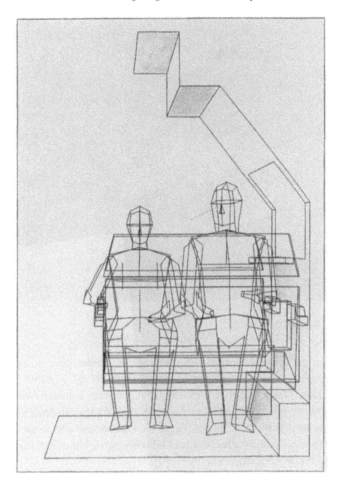

Figure 9.4 Swift ergonomic evaluation is carried out using computer visualization by ADAPS (Anthropometric Design Assessment Program System, Faculty of Industrial Design, Delft University of Technology). In this anthropometric evaluation of upper deck seats on a double-decker train operated by Netherlands Railways, we see on the left a model of a woman with average height (50th percentile for The Netherlands) and on the right that of a large man (95th percentile for The Netherlands). There are some problems. When he rises from his seat the man has too little headroom. There is too little elbow-room. A large section of the population will be unable to reach the floor with their feet. There is insufficient leg-room by the window.

9.3   CAD in VR

9.3.1   The program

Figure 9.5 provides a first impression of what the VR CAD program has to offer. The designer can copy three-dimensional objects from a three-dimensional pin-board, place them above his drawing board, reshape them using a number of tools available to him, and then put them away on a more distant memo-board. To do this he pushes his hand through a telewindow, which stretches his arm to roughly six times its normal length.

If the designer grasps a tool with his hand, the function of his hand itself changes. Thus if he takes the stapler he can staple two volumes to each other by pinching them, even if the volumes are not adjacent in space. He can then rescale a bundle of volumes stapled to each other using the pliers to manually stretch the forms by pulling on one of the handles

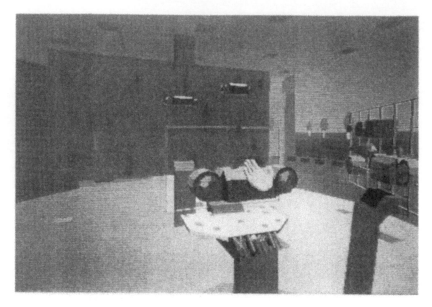

Figure 9.5 The virtual design space.

which appear. He can cut shapes loose with the scissors, throw them away by depositing them in the wastepaper basket, or put them away through the telewindow. He can distort surfaces like a kind of plastic clay by pressing on them with a smaller or larger stamp. But with this kind of clay he can choose his own plasticity and he can shape surfaces without causing undesirable distortion of adjacent parts. He can choose the most suitable colour scheme from the Munsell colour atlas and immediately see what effect his choice has on the volume he has made (Figure 9.6).

The program was to some extent inspired by the so-called 'ecological perception theory' propounded by James Gibson (1904–1978) (Gibson, 1966, 1979; Flach *et al.*, in press).[5] (See also section 3.2.4.) The term 'ecological' refers to the importance of the relationship between the observer and his surroundings. We shall discuss the program in the context of two implications of this theory. The first concerns the importance of perception–action coupling in achieving direct manipulation. The second concerns the types of information which an organism uses to guide his behaviour; these are formal characteristics of the environment which are scaled relative to the perceptual motor skills of the organism (see also Flach and Warren, 1993). This is sometimes referred to as 'form semantics' in current literature on design engineering (semantics is the theory of the meaning of words; form semantics deals with the meaning of forms).

9.3.2 The importance of perception–action coupling for direct manipulation

9.3.2.1 Spatial perception and telepresence

The virtual workplace in the VR CAD program has been designed to give the best possible impression of three-dimensionality. In the VR system, images presented to the left and right eyes are slightly different and provide stereoscopic information, which is known to enhance the perception of space. The virtual environment we designed also provides texture, a horizon, and motion parallax; the importance of these types of information for spatial perception is less well known.

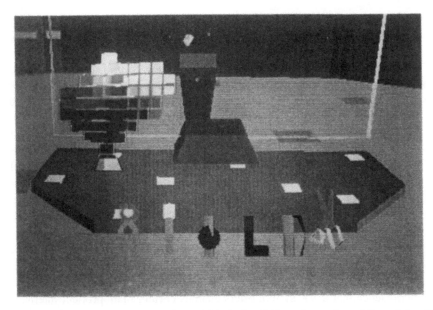

Figure 9.6 Overview of tools currently possible in the CAD package, from left to right: Munsell colour tree, tongs, brush, stamp, angle bar, stapler and a pair of scissors. (Based in part on the final degree projects of T. Hoek (1992) and P. Brijs (1992)).

Figure 9.7 The horizon specifies the height of an object because it always bisects objects of the same height in the same proportions, unless objects are at different heights above the ground, or the ground itself slopes. Explicit horizon left, implicit horizon right (terms used by Gibson (1979)).

The horizon always 'cuts' objects of the same height according to the same proportions, regardless of their distance, and probably specifies the perception of the height of an object (e.g. Sedgwick, 1973; Haber and Levin, 1992). There is thus no need for distance estimates (Figure 9.7).

Even if the horizon is not explicitly indicated, it is still implicitly given as the limit of the textural gradation of the ground. The space in which we live is always characterized by a ground surface which always has a certain roughness or 'grain', i.e. a more or less regular alternation of minute hills and valleys. This grain can be fine, as in linoleum, or coarse, as in a ploughed field. The perceived grain and texture of the surface corresponds to the structure of the light arriving at an observation point. This optical structure specifies the surface and the objects upon it very precisely, and there is no need to learn to understand it. Gibson calls this optical structure the 'optic array'. If the observation point moves, the changing optic array is called 'optic flow'.

The relationships of the objects in our environment to the ground surface provide crucial perceptual information. Consider Figure 9.8. The pictures of the two wastepaper baskets occupy areas of different size, but because they occlude the same amount of texture on the ground surface, they are perceived as equally large.[6]

The horizon is one means by which perception of the environment is coupled to our experience of our bodies. In particular, the horizon is coupled to eye-height. But if the ground surface slopes or the perceived object hangs above it, this perceptual coupling is too limited to be informative. If we can move about, however, the perceptual coupling is again strong enough to allow reliable spatial judgements. Moving gives more information about the position of the body (or parts of the body) relative to the surroundings. It allows us, by the coupling between our own movements and shifts in the optic flow, to estimate where things are in space relative to ourselves. Smets (Smets, in press) has argued, on the basis of experimental research, that it is precisely this coupling that causes a sense of 'telepresence'.

The telepresence that a VR system offers enables certain man–product interactions to be evaluated before a physical model is built. A virtual model is three-dimensional and

Figure 9.8   Use of texture to influence perceived size.

life-size, so that we can walk round it and subject it to visual evaluation. We can do more than this. For example, we can put a radio cassette-recorder on our shoulders and decide whether or not we have to stretch our arms too far. We can see whether the handle is large enough, and whether we have a tendency to use it or to hold the recorder in some other way. But we only feel the recorder in our hands, nowhere else, so that our judgement still remains principally a visual one.

The VR system can be coupled to a robot so that the design can be machined. In this way, it is possible to make a virtual prototype of a design very early on in the design process in greater detail than is possible now, and then to convert the virtual prototype into a simplified physical model should the need arise (Figure 9.9).

9.3.2.2 Direct manipulation

All the tools in the CAD program are designed so as to elicit the behaviours for which they are intended, rather as a hammer or a saw elicits the appropriate behaviour in the real world. As you take hold of the tools, the function of your hand changes. It is not the intention that these new tools should be imitations of those in the real world; on the contrary, they can offer new functions and new behaviours. They allow the designer to

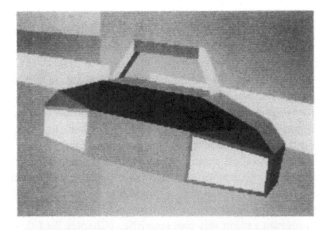

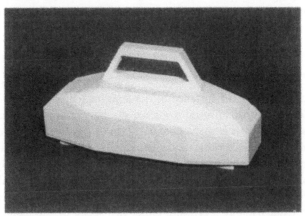

Figure 9.9   Some man–product relationships can be evaluated by reference to the virtual model (above), but if necessary the model can also be machined (below). (In collaboration with J. Vergeest.)

work with the environment in a new way: a way which is beneficial to the design process. The designer can now link forms, uncouple them, rescale, colour, and distort them, and so on, all by means of direct manipulation. This is accompanied by visual, auditory and tactile feedback: the designer can see, hear and feel what he is doing.

Human–computer interaction is most often described in conversational, rather than behavioural, terms. We, however, share Walker's opinion (quoted in Rheingold, 1991, p. 183), when he says:

> Conversation is the wrong model for dealing with a computer – a model which misleads inexperienced users and invites even experienced software designers to build hard-to-use systems. Because the computer has a degree of autonomy and can rapidly perform certain intellectual tasks we find difficult, since inception we've seen computers as possessing attributes of human intelligence ('electronic brains'), and this has led us to impute to them characteristics they don't have, then expend large amounts of effort trying to program them to behave as we imagine they should. When you're interacting with a computer, you are not conversing with another person. You are exploring another world.

Walker (1988) used the way we explore the world of the computer as a basis for a computer revolution taxonomy, arranged in terms of generations of interaction techniques rather than generations of software of hardware components (Figure 9.10). In the 1940s we controlled computers with plugs, ten years later with punched cards, another decade later with a keyboard; and now most of us use computer menus. The next step, according to Walker, ought to be a form of direct manipulation. But Walker, as is the usual case, visualizes direct manipulation as a 'point and click' manipulation, perhaps using a mouse-controlled cursor, or possibly a finger on a touch-sensitive screen. But manipulation of this kind does not allow genuine convenience and manoeuvrability. In the case of an ordinary computer system this limitation on manipulation is understandable; but in VR systems the same kind of convention has arisen. The same 'point and click' mechanism is used, except that it is controlled by a Feedback Glove, or equivalent, pointing at a menu with three-dimensional text floating somewhere in virtual space (Figure 9.10). This convention may be suited to verbal and logical communication, but it is not suitable for the perceptual motor tasks for which VR systems were developed in the first place, and which are also essential to the design process. A VR system allows the movements of the user, whether hand or whole body movements, to control the computer; the tools that we have designed for our VR CAD program exploit this characteristic. (Chapter 7 of this book considers the human requirements and capabilities for direct manipulation; 7.5.3 describes 'ecological' motor control theories which identify perception–action coupling.)

9.3.2.3 Affordances

So far we have discussed the importance of perception–action coupling. But what kind of information does an organism need to create the perception–action coupling?

Most people assume that we perceive the world on the basis of the detection of simple or lower-order variables, such as distance and speed, which then, in the brain, are combined to make something more complex. For example, the prediction of exactly when I would bump into a wall directly ahead of me would be based upon my speed of walking, combined with the distance from the wall. But in fact the time left to me before I collide with the wall is easily predicted without specifying either my speed or the distance, or the need to perform some complex calculation. This 'time-to-contact', generally known as 'tau', is given in the expansion gradient on my retina (Figure 9.11). It is given by the speed with

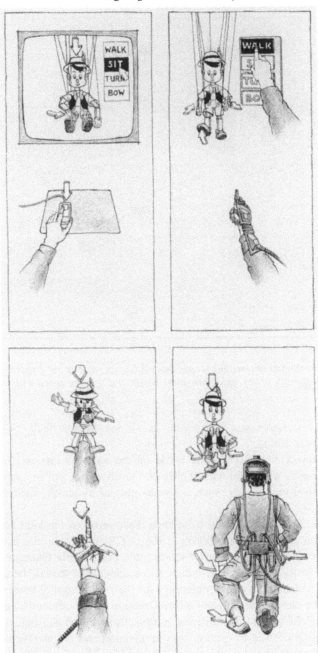

Figure 9.10   Taxonomy of the man–computer interaction in VR (after Walker, 1988). Top left: menu-driven system by the use of a mouse; top right: 'point and click' manipulation (Walker's direct manipulation); bottom: direct manipulation by movements of either the hand or the entire body.

which a texture point at a particular distance from the expansion centre of the optic flow moves away from that centre. This variable, which has been shown to be used by birds diving towards their prey, and which also plays a part in playing table tennis (particularly in deciding when precisely a smash should be initiated; Bootsma, 1988), cannot convincingly be simulated on an ordinary computer which does not respond to the observer's head

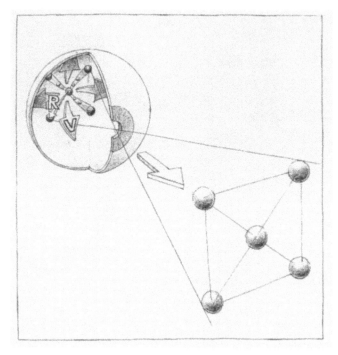

Figure 9.11   Time to contact at constant speed. Tau=R/V, where R is the distance from a texture point to the expansion centre of the optic flow on the retina and V is the speed with which this point moves (Lee, 1976).

movements; but it is possible on a VR system. This is why it is perfectly possible to play table tennis in virtual reality.[7]

Biological systems do not perceive the world on the basis of extrinsic units, such as centimetres or seconds, which are meaningless to the observer, but in terms of intrinsic units which are linked to the body, such as eye-height, or to action, such as the time to contact.[8]

In recent years several 'affordances' have been discovered of the kind just described. Affordances (a term introduced by Gibson (1966, 1979) to denote 'the actions a given environment affords to a given acting observer') are perceivable characteristics of the environment relative to the organism, which allow adequate (spatial) behaviour. Thus Warren (1984) showed that the visual preference for the riser height of the steps in a flight of stairs can be quite closely predicted as a fixed fraction of the climber's leg length. You choose that step height which you can most easily climb, in physiological terms, even if you have only been allowed to see the steps in question and not use them. People can see which particular riser height calls for the least effort (Figure 9.12).

Analogous research has been carried out into the 'sitability' of chairs (Mark, 1987) and the 'passability' of doors. The results are always expressed in terms of dimensionless variables, not bound to any particular external units but to the relation between mover and environment, such as the width ratio between the door and the body of the person moving through it (where 'body' includes the vehicle, in the case of driving). Thus door passability calls for:

$$(\text{door width})/(\text{body width}) \geq 1 \cdot 3$$

whether it is for frogs, people, or even people in a car (Shaw *et al.*, in press; Ingle and Cook, 1977; Warren and Whang, 1987).

Figure 9.12 The visual preference for the riser height of a stair step (h) *depends on leg length* (L), *where the chosen height* h=0·25L. *By measuring energy consumption while climbing on a variable speed treadmill it is found that energy is most efficiently used at* h=0·26L *(Warren, 1984).*

All these dimensions and sources of information can help in designing a virtual environment that provides as three-dimensional an experience as possible. But that is not enough for our virtual workplace. We should also consider the affordances offered by instruments and tools. Although this is more difficult, we think it is essential that the tools in the VR design environment should be developed in such a way as to express their functions as clearly as possible, and just as obviously as the tools we are familiar with in everyday life. The very shape of a pair of scissors, for example, is a clear indication of how to hold them, and the possibilities for movement which they offer suggest what this tool may be used for (see section 6.2.3 for comments on the role of affordances in object recognition). Our first step in this direction was to choose specialized tools, each of which looks different, instead of a universal tool which is controlled by a keyboard or some point-and-click mechanism (see also Norman, 1988). In this way we immediately make clear how many functions are available, and what they are. Thus the electronic clay is modelled using a large spatula for major distortions and a smaller one for making more localized changes. Although it is difficult to develop tools which clearly express how they are to be used, we think this is an important goal. For help in achieving this, we can look at recent developments in the artificial intelligence debate.

9.3.3 Artificial intelligence and the evolution of affordances

The aim of artificial intelligence (AI) is to make a computer achieve results that would be regarded as intelligent if they had been produced by a human.[9]

In the 1960s the idea arose that all intelligence, including human intelligence, looked so much like what a computer does that it could be defined in terms of calculations using symbols. The content of the symbols (the semantics) was irrelevant. It was the formality (the syntax) which counted, and only the formality. Meaning resided only in the external user of the program, and the way he interpreted its output. The human perceptual system was seen as a program which in principle could run on any machine and which merely happened to be confined to a single organism. This approach was successful in the development of expert systems,[10] robotics and image processing systems. Computer simulation of the way we process information came to be regarded by many, partly as a result of this, as a duplication of our perceptual system. This was overstating things: since then researchers have found that they could develop more practical applications if they simulated human knowledge processing by dealing with the semantics, especially the form semantics (i.e. the affordances).

The first computers, which were slow because they processed all information sequentially, and were inflexible because all information was stored locally, are being surpassed by computers which can also process information in parallel, and in which the information is not stored locally but in a distributed manner ('Parallel Distributed Processing' or PDP; see Rumelhart and McClelland (1986)). This makes it possible to treat meaning as something that arises in the interplay between the overall system and its environment.[11] Such interplay is possible only in computers whose operations are not deterministically fixed in every detail, and which organize themselves as they carry out a particular task. Computers operating in this fashion are now in use.

As time proceeds, computers are increasingly regarded as a network of links between simple processors. To a large extent this network can evolve solutions to problems by varying connections in the network until it develops a stable interaction with its environment. A central control mechanism telling all parts exactly what to do and how to do it is no longer necessary. Thus mobile robots can learn how to move through their environment without bumping into things. Ultimately they are found to incorporate rules which are analogous to the Gibsonian rules for perceptual behaviour control (Shaw, 1993). In the VR context a neural network of this kind was set up by Fels and Hinton (1990) to convert a sign language used by deaf people, the movements of which were recorded using a Feedback Glove, into spoken language (Figure 9.13). The computer itself learnt which gestures should be matched to which words.[12] The usefulness of such computational networks for the study of object recognition is discussed in section 6.2.6 of this book.

The research we would like to move on to is analogous to this. The intention, briefly, is to program human–computer interaction as an open system which organizes itself as it adapts to its user's behaviour.

This can be done by starting from an initial state of given interactions by means of a collection of given tools, as in Figure 9.14. They inherently imply an interaction. They tell us something about the movement which they afford, but also indicate which parts of the objects in the environment they will affect. Suppose we ask someone to copy a particular complex shape as efficiently as possible. As soon as a particular combination of actions occurs frequently to achieve a particular intention (distortion, for example) this becomes the command for that behaviour. The interaction is then examined to see whether it can be improved by each time giving it a little nudge (a mutation, let us say) either to the form of the tool, to the way it is used, or to its interaction with the environment, and to record what effect this has on efficiency. In this way it is possible to indicate where the interaction is relatively stable, where it becomes unstable, and where it changes into another behaviour (another regime). Thus we can program tools which, in the very act of being used, can evolve; just as over the centuries the plough through a process of constant readjustment gradually acquired its present efficient shape.

Figure 9.13   The speaking glove which enables a deaf-mute to speak to people who do not understand sign language. She wears a right glove, a loudspeaker round her neck and an alphanumeric LCD round her left wrist, for visual feedback. Her gestures (provided they are slow enough) are converted into words and played by the loudspeaker.

9.4   Conclusions

Ever since Descartes, visualization has progressively disappeared from the scientific literature. The instruments that have been developed have been aimed mainly at analysis, supporting the rationalistic cartesian paradigm. Now, however, this approach is receiving more and more criticism and the importance of the observer and of sensory experiences has been recognized in many disciplines, including physics, AI, and technological development, such as teleoperation systems. Scientific visualization is thus regaining a prominent role in science and engineering, just as it did before Descartes. Computer simulations are being developed which pay close attention not only to the datafile and calculating design alternatives, but also to the visualization of these. VR systems go even further: they offer telepresence. Visualization and, where necessary, telepresence help us to synthesize our knowledge. We can take similar examples to Ellis in sections 2.3.4 and 2.3.5 of this book. A surgeon who has a datafile of a human brain, obtained by computer tomography, cannot read the information. He will have difficulty seeing the three-dimensional structure of that brain if he has to rely solely on two-dimensional cross-sections. Telepresence techniques can help him literally to see through his data. The same applies to a chemist designing a new molecule (perhaps one which uses force feedback to sense how large the attraction or repulsion between different particles is; Brooks *et al.* (1990)), an astronomer trying to visualize the surface structure of Venus on the basis of telemetric data received from space, an aircraft builder who can evaluate and optimize a virtual aircraft or space vehicle in a virtual wind tunnel (Kalawsky, 1993, p. 314), or an industrial designer who designs a product using a computer. Sometimes immersive VR is better (as in CAD), but often virtual window systems are sufficient (e.g. for the inspection of baggage at an airport, where security and safety would be improved if a good three-dimensional impression of the contents of cases were available). Sometimes it is useful to use semi-transparent visual displays, so that the virtual world merges into the real world. Thus an echoscopist would be able to see the unborn baby both through and in the mother's body (Kalawsky, 1993, p. 319).

Simulated and virtual realities

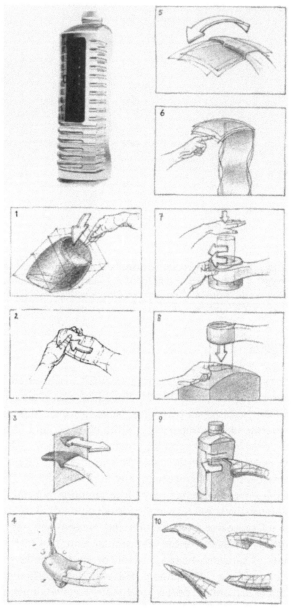

Figure 9.14 Our aim: a natural way of designing which ties in with perceptual motor skills and in which we have to learn as few extrinsic commands as possible. The emphasis is on sketching. You can make the product on the left (a mineral-water bottle) with your hands. This can be done by first dipping your right virtual hand (left for left-handed) into a container to give it a viscous envelope (1) which you can use your left virtual hand to shape until you are satisfied with it (2). The form of the hand may be saved by making a mould of it (3). Then you provide your hand with some material (4) after which you use your right hand to make three-dimensional sketches. During sketching several transparent layers are formed (5). When you are satisfied with one of these three-dimensional shapes, you can select it (6) and the others will disappear. To create a different kind of layer your right hand must be reshaped for which eventually you may use tools (7). With your left hand, you can then indicate on the product itself which part of it is to be affected by your next treatment of it (8). Besides creating it is also possible to manipulate the product—cutting, for example—by giving your hand another function. This is possible by changing the new covering of your hand (9). In this way many functions are possible. The form of the virtual right hand indicates what you can do with it.[13]

Although the development of this CAD program is still at an early stage, we believe it nevertheless clarifies some advantages of designing with VR. These advantages can be best understood against the background of the ecological theory of perception.

The possibility of coupling perception and action permits the development of an interaction with the computer which is based on direct manipulation: a true direct manipulation interface, which is difficult to achieve with other computer systems. In this way it is possible to design products which can be manipulated before a physical model is made. If necessary, however, a physical model can quickly be made on the basis of the VR data.

VR allows us to develop new tools which elicit behaviours in a way that is comparable with ordinary tools in everyday life, even though they are not limited to the functions of everyday life. The development of these tools is only just beginning. The user-friendliness that can be achieved in this way is technologically relevant because it is crucial to the development of highly interactive computer systems. The approach is scientifically interesting because it ties in with recent scientific debates on perception and artificial intelligence.

Both advantages make virtual reality a challenge to designers and will, in time, lead to implementations for the simple reason that they can accelerate and improve the design process.

Notes

1. The drawings are by C. Hummels. Much of this chapter reflects joint work and thinking at the Laboratory for Form Theory, Delft. The project is supported by a grant from the Delft University Encouragement Fund.
2. A brief introduction to VR and VR technology is given by Biocca (1992a, 1992b); a rather more penetrating overview is provided by Kalawsky (1993), and a survey of research with and into VR appears in Earnshaw *et al.* (1993). In 1992 a scientific journal of VR technology and research was founded under the title *Presence*. In the same year a whole issue of the *Journal of Communication* was devoted to VR.
3. This can be experienced in everyday life. The advantages of a high-definition television set, i.e. one with a relatively high picture resolution, are large when the picture remains static but small when the picture is moving. The old super-eight movies were very fuzzy if you projected them frame by frame, but again, this was not a problem when the film was run at its intended speed.
4. The proposed software package is the result of multidisciplinary teamwork within the Laboratory for Form Theory in the Faculty of Industrial Design, led by Gerda Smets. Software development is carried out in conjunction with the department of Information Systems of the Faculty of Technical Mathematics and Information Technology (to which Charles van der Mast belongs), together with a number of students from both faculties (including Tjeerd Hoek and Peter Brijs). Other research with and into VR in the Laboratory for Form Theory relates to validity research (Fons Blankendaal and Kees Overbeeke), research into the importance of head movements for depth perception (Stappers, 1992), on the possible causes of telepresence (Smets, in press) and multimedia communication research (in conjunction with Bill Gaver, Xerox EuroPARC, Cambridge). Most research, however, concerns non-immersive VR (the Delft Virtual Window System) (e.g. Overbeeke and Stratmann, 1988).
5. This is a relatively recent perceptual theory to which, however, there are frequent references in the VR literature. The theory's relevance in connection with human–computer interaction (HCI) in general is discussed by Wallace and Anderson (1993) and Gaver (1991).
6. This is also the only internally consistent and empirically sound explanation for the difference in perceived size of two known adults in Ames' distorted room (Stoper, 1990; Warren, in press).
7. This was part of a validity study (about perceptual calibration) in which the importance of a time lag in the performance of this task was monitored. It turned out that the start of the smash in VR was exactly the same as in reality (averaging 250 ms before hitting the ball), but that

variability was significantly higher (compared with the values obtained in Bootsma's (1988) study). So far, VR has been used primarily to explore future applications. Extensive reporting on validity studies for various tasks is still scarce, perhaps owing to the newness of the medium. Some work, however, has been done on three-dimensional (physical) calibration, including that by Ellis (e.g. 1993) and ours (Stappers, 1992). Ellis' is as follows: suppose a real space is imitated in VR and that a person is in the same place both virtually and in real life. Suppose also that this observer is then asked to walk towards a cross drawn on the virtual and on the corresponding real wall. If he or she touches the virtual cross, is the three-dimensional calibration good enough to ensure that the real cross is also touched?

8. An observer can see the lower-order variables if he is asked for them. But in most perceptual tasks, i.e. those performed outside the laboratory, they are passed over and the higher-order variables are seen straight away. Several authors (e.g. orientation and curvature of surfaces; see Koenderink, 1990a, 1990b) have mathematically described some of these variables, and have suggested mathematical and physiological methods and mechanisms to detect them. A metaphor for this might be a slide-rule: you can multiply with it without carrying out a single 'calculation'.

9. On cognitive science in general see Varela (1992); on artificial intelligence, see Searle (1990) and Churchland and Churchland (1990); and on computer-aided design, see Coyne and Snodgrass (1993).

10. Expert systems are computer systems into which the knowledge of experts, relating to the diagnosis of a problem by reference to a combination of symptoms, has been fed, so that non-experts, by entering the symptoms, can arrive at a diagnosis of the problem.

11. This is called connectionism. The importance of experience (the interplay of system (human or computer) and context in the origination of meaning), much more than the importance of a formal system of rules (in which observer and environment are separate from each other), has long been propounded by philosophers such as Heidegger, Merleau-Ponty and Foucault. Heidegger's ontology, for example, is a complete reversal of the classical, cartesian ontology (which is implicitly the basis of most CAD programs). Recently Heidegger's ontology has been applied to the development of CAD programs (see, for example, Coyne and Snodgrass, 1993). The CAD program described here also ties in with this.

12. Vänäänen and Böhm (1993) have investigated the use of neural networks in the recognition of gestures in a more general context.

13. In many tasks, including computer tasks, the use of two hands is an advantage. The left hand is best used as it is for ordinary tasks: first the left, then the right (first place the nail, then strike it with the hammer); the left hand supports, the right hand does (the left hand holds the nail, the right hand does the hammering); see also Buxton (1993).

References

Biocca, F., 1992a, Communication within virtual reality: creating a space for research, *Journal of Communication*, **42** (4), 5–22.

Biocca, F., 1992b, Virtual reality technology: a tutorial, *Journal of Communication*, **42** (4), 23–72.

Bootsma, R. J., 1988, The timing of rapid interceptive actions, PhD thesis, Free University of Amsterdam.

Brooks, F. P., Jr, Ouh-young, M., Batter, J. J. and Kilpatrick, P. J., 1990, Project GROPE – haptic displays for scientific visualization, *ACM Computer Graphics*, **24** (4), 177–85.

Brijs, P. A. J., 1992, MOVE: modelling objects in a virtual environment, degree paper, Faculty of Mathematics and Informatics, Delft University of Technology.

Buxton, B., 1993, Two-handed input in human–computer interaction, paper presented at the Laboratory of Form Theory, Department of Industrial Design Engineering, Delft University of Technology.

Churchland, P. M. and Churchland, P. S., 1990, Could a machine think?, *Scientific American*, **262** (1), 26–31.

Coyne, R. and Snodgrass, A., 1993, Rescuing CAD from rationalism, *Design Studies*, **14** (2), 100–23.

Earnshaw, R. A., Gigante, M. A. and Jones, H. (Eds), 1993, *Virtual Reality Systems*, London: Harcourt Brace.

Ellis, S., 1993, Nature and origins of virtual environments, paper presented at the Applied Vision Association Annual Conference, Simulated and Virtual Realities, Bristol, 29–31 March 1993, London: Applied Vision Association, British College of Optometrists.

Fels, S. S. and Hinton, G. E., 1990, Glove-talk: a neural network interface between a data-glove and a speech synthesizer, *IEEE Transactions on Neural Networks*, **4** (1), 2–8, cited in Kalawsky, R. S., 1993, *The Science of Virtual Reality and Virtual Environments*, p. 304, Wokingham: Addison-Wesley.

Flach, J. M. and Warren, R., 1993, Active psychophysics: the relation between mind and what matters, in Flach, J. M., Hancock, P. A., Caird, J. and Vincente, K. J. (Eds), *The Ecology of Human Machine Systems*, Hillsdale, NJ: Erlbaum.

Flach, J. M., Hancock, P. A., Caird, J. and Vincente, K. J. (Eds), in press, *The Ecology of Human Machine Systems*, Hillsdale, NJ: Erlbaum.

Gaver, W. W., 1991, Technology affordances, in *Proceedings CHI'91*, pp. 79–84, New York: Association for Computing Machinery.

Gibson, J. J., 1966, *The Senses Considered as Perceptual Systems*, Boston, MA: Houghton Mifflin.

Gibson, J. J., 1979, *The Ecological Approach to Visual Perception*, Boston, MA: Houghton Mifflin.

Haber, R. N. and Levin, C. A., 1992, The perception of object size is independent of the perception of object distance, paper presented at the 33rd Annual Meeting of the Psychonomic Society, St Louis, MO, 13–15 November 1992, Austin, TX: The Psychonomic Society.

Hoek, T. S., 1992, CAD in virtual environments, degree paper, Faculty of Industrial Design Engineering, Delft University of Technology (in Dutch).

Hummels, C., 1993, A new perspective in an immersive virtual environment, degree paper, Faculty of Industrial Design Engineering, Delft University of Technology.

Ingle, D. and Cook, J., 1977, The effect of viewing distance upon size preference of frogs for prey, *Vision Research*, **17**, 1009–19.

Kalawsky, R. S., 1993, *The Science of Virtual Reality and Virtual Environments*, Wokingham: Addison-Wesley.

Koenderink, J. J., 1990a, Some theoretical aspects of optic flow, in Warren R. and Wertheim, A. H. (Eds), *Perception and Control of Self-motion*, pp. 53–68, Hillsdale, NJ: Lawrence Erlbaum.

Koenderink, J. J., 1990b, *Solid Shape*, Cambridge, MA: MIT Press.

Lee, D. N., 1976, A theory of visual control of braking based on information about time-to-collision, *Perception*, **5**, 437–59.

Mark, L. S., 1987, Eyeheight scaled information about affordances: a study of sitting and stair climbing, *Journal of Experimental Psychology: Human Perception and Performance*, **13**, 361–70.

Neuckermans, H., 1992, Ontwerpmethodiek, deel CAAD, Coursebook for students, Faculty of Applied Sciences, University of Louvain.

Norman, D. A., 1988, *The Psychology of Everyday Things*, New York: Basic Books.

Overbeeke, C. J. and Stratmann, M. H., 1988, Space through movement, PhD thesis, Delft University of Technology.

Pacejka, H. B., 1991, Voertuig in beweging, Foundation Day address, Delft University of Technology.

Rheingold, H., 1991, *Virtual Reality*, London: Secker & Warburg.

Rumelhart, D. E. and McClelland, J. L., 1986, *Parallel Distributed Processing*, Vol. 2, Cambridge, MA: MIT Press.

Searle, J. R., 1990, Is the brain's mind a computer program?, *Scientific American*, **262** (1), 20–5.

Sedgwick, H. A., 1973, The visible horizon: a potential source of visual information for the perception of size and distance, PhD thesis, University Microfilm no. 73-22530, Cornell University.

Shaw, R. E., 1993, Ecological psychology and the new artifical intelligence: seeking common ground, in Pittinger, J. B. (Ed.), *Proceedings of the VIIth International Conference on Event Perception and Action*, 8–13 August 1993, p. 45, Vancouver, BC: University of British Colombia.

Shaw, R. E., Flascher, O. M. and Kadar, E. E., in press, Dimensionless invariants for intentional systems: measuring the fit of vehicular activities to environmental layout, in Flach, J. M., Hancock, P. A., Caird, J. and Vincente, K. J. (Eds), *The Ecology of Human Machine Systems*, Hillsdale, NJ: Lawrence Erlbaum.

Sheridan, Th. B., 1989, Telerobotics, *Automatica*, **25** (4), 487–507.

Smets, G. J. F., in press, Designing for telepresence, in Flach, J. M., Hancock, P. A., Caird, J. and Vincente, K. J. (Eds), *The Ecology of Human Machine Systems*, Hillsdale, NJ: Lawrence Erlbaum.

Smets, G. J. F. and Overbeeke, C. J., 1993, Trading off spatial versus temporal resolution: about the importance of actively controlled movement for visual perception, in Brogan, D., Gale, A. and Carr, K. (Eds), *Visual Search 2*, pp. 389–400, London: Taylor & Francis.

Stappers, P. J., 1992, Scaling the visual consequences of active head movements, PhD thesis, Delft University of Technology.

Stoper, A. E., 1990, Pitched environments and apparent height, paper presented at the 31st Annual Meeting of the Psychonomic Society, New Orleans, LA, cited in Warren, W. H., in press, Environmental design as the design of affordances, in Flach, J. M., Hancock, P. A., Caird, J. and Vincente, K. J. (Eds), *The Ecology of Human Machine Systems*, Hillsdale, NJ: Lawrence Erlbaum.

Vänäänen, and Böhm, K., 1993, Gesture drive interaction as a human factor in virtual environments — an approach with neural networks, in Earnshaw, R. A., Gigante, M. A. and Jones, H. (Eds), *Virtual Reality Systems*, London: Harcourt Brace.

Varela, F. J., 1992, Whence perceptual meaning? A cartography of current ideas, in Varela, F. J. and Dupuy, J. P. (Eds), *Understanding Origins; Contemporary Views on the Origin of Life, Mind and Society*, pp. 235–63, Dordrecht: Kluwer.

Walker, J., 1988, *Through the Looking Glass: Beyond 'User Interfaces'*, Sausalito, CA: Autodesk Inc.

Wallace, M. D. and Anderson, T. J., 1993, The stagnation of theory in HCI, Research Report, Department of Information Systems, University of Ulster.

Warren, W. H., 1984, Perceiving affordances: visual guidance of stair climbing, *Journal of Experimental Psychology: Human Perception and Performance*, 10 (5), 683–703.

Warren, W. H., in press, Environmental design as the design of affordances, in Flach, J. M., Hancock, P. A., Caird, J. and Vincente, K. J. (Eds), *The Ecology of Human Machine Systems*, Hillsdale, NJ: Lawrence Erlbaum.

Warren, W. H. and Whang, S., 1987, Visual guidance of walking through apertures: body scaled information for affordances, *Journal of Experimental Psychology: Human Perception and Performance*, 13 (3), 371–83.

10

Social dimensions of virtual reality[1]

Deborah Foster and John F. Meech

10.1 Introduction

Virtual reality has become a widely publicized technology and as such has provoked a great deal of speculation about the effects it might have on both individuals and society as a whole. Schroeder (1993) highlights the fact that the speculations (good and bad) about virtual reality – that it may threaten our sense of reality and change people for better or for worse – are as much a reflection of current beliefs about society as a whole as they are about virtual reality in particular.

This chapter seeks to evaluate the social dimensions of virtual reality. In doing so it is proposed that any form of new technology requires analysis within a social context: the context in which the technology is developed and used. Such an approach acknowledges the interactive relationship between technology and society which is both complex and inseparable, and seeks to open a forum for debate between technological innovators and social scientists. Given that virtual reality, the technology under discussion, is relatively new and its application limited, it might be supposed that such a discussion is premature. Such an argument, however, would simply serve to illustrate the point that society is often reactive to technological developments and that the technological community is introspective. Whilst an element of truth may exist in such a perception, ultimately the ways in which individuals, groups or society at large mediate technology and the impact and experiences from that process are highly important for both the 'technicist' and the consumer or user. A debate of this kind at this stage of the technological process aims therefore to open up a dialogue around issues associated with virtual reality which are both technical and social.

This chapter begins by addressing the nature of the reality representation which virtual reality seeks to achieve by means of simulation, and goes on to highlight philosophical and socio-cultural questions related to the construction of consensual meaning around the concept of reality. The discussion then proceeds to evaluate the social dimensions of technology and how technology shapes and is shaped by society. In doing so, the place or construction of meaning through technology such as virtual reality is considered. Following this, the role and impact of other forms of media are analysed and comparisons with virtual reality are drawn when relevant. Evidence is presented from case studies of cinema, television and computer games which attempt to illustrate how images and ideologies help construct dominant cultural discourse – a discourse which is by no means value-free in its various representations of reality. The possible implications of a move away from spectatorship to immersion and active participation with the development of virtual reality in the sphere of entertainment is then explored. Finally, issues relating to ethical dilemmas that may arise from virtual reality are considered and further directions for research described. In short, we aim to discuss some of the issues embedded in the question:

> How will any individual or group, carefully and sensitively, with a deep appreciation for cultural, racial, religious and gender bias, create virtual reality systems? (Helsel and Roth, 1991)

10.2 A technological disclaimer

Current virtual reality systems allow one (or more) participants to become actively immersed in a tailored environment. In domains such as manufacturing, training, teleoperation and so on these environments are generally simulations of a particular environment. In other domains such as entertainment, completely artificial worlds can be created by those designing the virtual reality.

The meaning of virtual reality in the context of this chapter is taken to be some form of immersive,[2] synthetic environment which creates a feeling of presence or suspension of disbelief which is sufficient to make the user feel that the artificial world which they appear to inhabit is 'real'. We use this description to avoid questions associated with how real a virtual world appears owing to the particular technologies used in rendering displays or producing tactile feedback etc. In our view, virtual reality is a simulative device which is embodied by current technological development. We may expect resolution and fidelity to improve and the effects of sensor lag to decrease as the technology is improved. In addition it is not clear to what extent the impression of reality gained by a participant depends solely upon such factors such as photo-realism. In one of the first public viewings of cinema the audience ran from the projected image of a steam locomotive apparently arriving in the theatre, even though the image was black-and-white and there was no sound.

One of the reasons that there is so much publicity surrounding virtual reality is that virtual reality seems to combine all the elements of other media forms (e.g. television, film, video games) in a way which makes interaction with the media intuitive to the participant, and therefore the experience may be very direct and personal. For example, in film and television there is a substantial learning process which must be undergone in order to enable one to understand the language of cinematography and editing used. Cuts, pans and zooms do not happen in real life! In video games, the player of the game relates to a viewpoint or character which participates in the action. In immersive virtual reality, the senses are provided with naturalistic data in a form which requires little learning or interpretation. The experience provided to the user appears far more immediate than other forms of media and consequently may have more impact. The first part of this chapter examines the way in which virtual reality combines simulation with participation and the consequent effect which this may have on users.

It seems likely that virtual reality will become a pervasive technology with many users, perhaps eventually dominating our lives in a similar way to television and video games. It is therefore not unreasonable to speculate that virtual reality may have a similar capacity for affecting our lives. Marshall McLuhan (McLuhan, 1965) wrote, 'The medium is the message': the personal and social consequences of any form of media result from the new scale that is introduced into society by each extension of ourselves or by any new technology. In the light of virtual reality combining elements of television and video games and possessing the capacity (at least in theory) to become as widespread, we shall use these as examples and attempt to extrapolate the effects of virtual reality from them.

10.3 Simulation and reality

Virtual reality may be viewed as a high-fidelity simulation of a model world or environment. Simulation, whether it is an aircraft simulation or a mathematical model of chemical molecules, is generally a synthetic construct which is intended to represent a particular environment (see section 2.3 for an overview of different simulation applications, and 9.2 for a specific simulation requirement). The degree of accuracy that the model achieves may be thought of as the degree of realism, or how closely it represents the accepted reality. Flight simulators, for example, model a well-bounded existing environment of the inside of a particular aircraft and an outside world which corresponds to existing geography, and generally achieve a high degree of accuracy in representing a real, flying aircraft.

Virtual reality has the ability not only to simulate existing environments but also to represent environments which may not be consensually agreed as real. Is there a difference between simulation and representation? Sections 6.1.1 and 6.1.3 consider this difference in the context of visual recognition. Baudrillard (1983, p 11) draws the following distinction: 'Whereas representation tries to absorb simulation by representing it as false representation, simulation envelops the whole edifice of representation as itself a simulacrum'. That is, simulation encompasses representation (in which the sign and the real are taken to be equivalent), but simulation provides more than just representation. According to Baudrillard simulation can:

1. be a reflection of a basic reality;
2. mask and pervert a basic reality;
3. mask the absence of a basic reality;
4. bear no relation to any reality whatever: it is its own pure simulacrum.

Simulation may reflect reality in terms of a model of an existing environment, for example, the teleoperation of a device. Masking the presence or absence of an aspect of reality may be thought of as constituting a partial simulation. For example, no matter how realistic an aircraft simulator is, if it is crashed by the pilot, no-one will be killed. A pure simulacrum allows the possibility of wholly artificial environments to be produced which may be thought of as the simulation equivalent of an surrealist painting, or 'virtualization' of another domain such as manipulating chemical molecules (McGuinness and Meech 1992). ('Virtualization' is defined here as taking data not normally directly accessible to the human senses and creating from them an immersive environment. This is in contrast to the sense adopted by Ellis (section 2.2).) Virtual reality systems are obviously capable of all these simulative abilities. We can both simulate (feign to have what we do not have) and dissimulate (feign not to have to have what we have) using virtual reality in a form which may be pictorially illustrated as in Figure 10.1 (after Pimentel and Teixeira, 1993). In Figure 10.1, 'there' may be thought of as a simulated environment (Baudrillard's simulation type 1), and 'elsewhere' as an environment which may be entirely imaginary, such as a role-playing game (Baudrillard's 2, 3, and 4). Virtual reality has the holding power to provide a high level of engagement depth across a wide range of abstractions, but can simulation, through virtual reality, affect an individual's conception of what is real and what is not?

10.4 Reality control and hyperrealism

From a postmodern concept of reality, Baudrillard views culture as being dominated by simulation (Baudrillard, 1983, p. 2): 'Simulation is no longer that of a territory, a referential

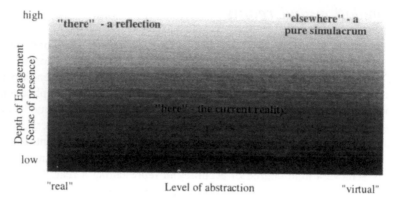

Figure 10.1   Virtual reality and simulation.

being or substance, it is the generation by models of a real without origin or reality–
hyperreal.' In Baudrillard's view, reality itself may be viewed as hyperrealistic. The whole
of everyday political, social, historical and economic reality is incorporated into the
simulative dimension of hyperrealism; what actually happens becomes less important than
what appears to have happened. Consider the picture of the world which is portrayed by
conventional media: 'news' is selected if it is considered significant, interesting, liable to
attract readers or viewers. News, and similar reporting media involves selection on the
part of the reporter, cameraman and editor. Film and video is *directed* and a perspective
is introduced into a portrayal of fact or fantasy. This in turn raises questions about simulation
and virtual reality in that at some point a decision is taken as to what will be included
in the simulation and what will be left out.

10.5   How many realities are there?

The phrase 'virtual reality' is a contradiction in terms: how can something be both virtual
and real? In discussing the effects of virtual reality, we implicitly assume that it is possible
to have more than one reality, and by means of various technical apparatus, move between
these realities. At a more fundamental level, we may ask ourselves if there is a single
reality which is the same for everyone? What exactly is reality? This is a difficult question
to answer as it addresses many philosophical issues of objectivity and subjectivity. Perhaps
it is more profitable to ask if there is a single, objective, 'external' reality which is
independent of an individual's perception of it.

The issue of reality is a philosophical puzzle, but is also related to the social environment
in which we live. It seems difficult to prove that there is an objective reality which is
the same for everyone – independent of observer or observation. Jean-François Lyotard
(see Rendings, 1991, pp. 99–100) identifies the concept of a 'metanarrative' as a sweeping
consensual explanation which claims to objectively explain why things are the way they
appear to be (see Woolley (1992) for a discussion). In this way, the accepted picture of
an objective reality is constructed through social interaction. Accepted scientific theories
are produced in exactly this way, through publication of results and discussions until a
particular theory gains acceptance. Interestingly, Gibson's description of cyberspace as
a 'consensual hallucination' (Gibson, 1986) parallels this: a virtual reality may also be
constructed by social consensus.

Through social interactions each individual constructs his or her own subjective reality which may be viewed as mediated into an 'objective' reality through further interaction in the social group. Each person has his or her conception of reality, shaped by his personal and cultural upbringing, which emphasizes not only the physical characteristics of the world (when you drop an object it always falls to the ground) but also what is acceptable within society: what constitutes a criminal action, how to interact with others, etc. This learning process is not static; we are all equipped with evolutionary mechanisms which allow us to incorporate new experiences into our reality, and in this way our experiences shape our expected reality and consequently our behaviour. We have a constantly changing conception of reality, and this ongoing learning process has implications concerning the generation of constructed realities which may change someone's perception of consensus and therefore modify behaviour.

If our personal idea of reality changes, what might be the effect of multiple realities which we can enter and leave by choice? Can we tell the difference between different virtual realities and can we tell whether an experience is 'real' or not? Heim (1991) suggests that we have three hooks on reality as an 'anchor' which enables us to tell what is real and what is not:

1. *Mortality/natality:* We are each born at a definite time and grow up within distinct social groups. These limits impose external parameters on reality, providing us with a sense of rootedness in the earth; a finite planet with fragile ecosystems.
2. *Temporality:* (Memory, history) uniqueness and irretrievability.
3. *Fragility:* Physical injury, suffering, death.

A central question is how well our hooks on reality can hold when using virtual reality. Even if we can distinguish fact from fiction, is it possible that elements of a virtual reality may be assimilated into an individual's own 'personal' reality? This question becomes particularly important when dealing with users whose sense of reality may already be distorted:

> Careful thought is required before virtual reality-based care or clinical investigation is offered to patients, especially those who are mentally ill. Underlying concerns include the capacity of virtual reality to distort reality-testing in patients whose judgement is already impaired, the loss of freedom of choice of experience when in virtual reality. (Anon., 1991)

Descartes wrote a passage concerning the testing of reality which provides some advice:

> I will suppose, therefore, that there is no God of goodness, the sovereign source of truth, but a malignant genius, as powerful as he is cunning and deceitful, who has used all his zeal to deceive me; and I will make myself think that the sky, the air, the earth, colours, shapes and sound, indeed every external thing we perceive, are all no more than the illusions of dreams, by which this Demon has laid traps for my credulity. I will conceive myself as having neither hands nor eyes, neither flesh nor blood, and no senses at all, and yet as falsely believing myself to be possessed of all these things. I shall cling obstinately to this notion and though with its help I may not be able to acquire the knowledge of any truth, it will at least enable me to suspend my judgement. (Translation by Wollaston, 1960)

Although some researchers have pointed out that it is unlikely that virtual reality will attain a level where no cues are available to determine whether the current environment is real or virtual, the effects of stress or time pressures may make this less critical than it appears in convincing the user that the experience is real at that moment (Shapiro and McDonald, 1992), and this would seem to suggest that virtual reality can, under certain circumstances, affect an individual's behaviour.

10.6   Who controls the virtual reality?

It is important to conceptualize the systems that create virtual reality as a form of technology within a social context, because it is within this context that it produces meanings of a social, cultural and political nature which need to be analysed. Underlying such an analysis is the rejection of the view that technology is somehow independent of society: a force over which we have no control and are helpless in its advance. Such a viewpoint is encapsulated in the technological determinist perspective, once dominant within sociological discourse, which focused upon the effects of technology over society without questioning the design origins, or applications of technologies, or the power of human intervention in these processes. A rejection of this viewpoint is central to our discussion, in so far as an emphasis on the social dimensions of virtual reality involves an acknowledgement of the interactive process between technology and society. Crucially, a recurrent theme we raise is how does the kind of society we live in affect the technology we produce and its application?

10.7   Is virtual reality a 'neutral' technology?

Technology at its most basic level of interpretation is simply a collection of physical objects which present themselves as an objective and tangible reality. In this sense, technology appears to be neutral and value-free. Such a view, however, disguises the 'meanings' inherent within technology which are derived from subjective human intent and knowledge embedded within the assumptions underlying its development and application. The mystique of technology surrounded by a 'culture of expertise' (Pacey; 1983) hides the essential social content of meaning and representation that technologies embody. This is particularly relevant to virtual reality technology in that its application in the realm of entertainment concerns the transmission of images, representations and ideas which essentially derive their meanings from social constructs of 'reality'. The question remains: whose concept of reality is used and will this reality simply reflect the perceptions of dominant groups in society? Before this question is addressed we must evaluate what role (if any) technology itself plays in shaping social processes.

   Previous discussion rejected the viewpoint that technology alone shapes society, embodied in 'technological determinism'. An opposite viewpoint would suggest that society shapes technology (Mackenzie and Wajcman, 1985), but in doing so this antithesis to technological determinism leaves little space for the analysis of the impact of technology itself. Winner (1983) argues that a viewpoint concentrating exclusively on the social shaping of technology ignores the fact that artefacts in themselves may embody social and political properties. Again this suggests that technology cannot be viewed as value-free. In using the term 'politics', Winner suggests that artefacts embody forms of power and authority which arise from the 'agenda' established by technological innovation. This agenda establishes a framework which then affects decision-making on social issues and technological innovation from thereon. Winner himself uses the analogy of Acts of Parliament, which he argues are similar to technological innovations because decision-making is then framed by their parameters. Similarly, it could be argued that technological innovations have an impact on the way people behave for generations; for example, the telephone has revolutionized communications and consequently social relations in business and personal life.

   Winner's (1977, 1985) proposition that 'artefacts have politics' has not, however, escaped criticism. In trying to steer a middle course between two schools of thought (technological determinism and the social construction of technology), Winner is accused by Grint (1991) of 'designer determinism'. Winner is anxious to challenge any notion that technology is neutral, and in effect is saying that technology is embedded within the social and political preferences of its designers, and this control over the design and construction process of the artefact ensures the particular effects. Conversely, Grint (1991, p. 291) argues that:

> Winner's assumptions and others of 'technicist' inclinations. . .do not lead in any immediate way to particular outcomes, for the position and power of current users mediate between design and outcome and are channelled in part by the unintended consequences of social life.

Grint (1991) thus believes we must retain a sense of the 'ambiguity of technology' in the sense that social relations are determined neither by technology nor by social agency but are a contingent result of changing networks of human and non-human actors. In other words, if there is intent on the part of particular actors in the technological process to embody politics into technological development,[3] this may be successful and available for analysis, but it is not necessarily a fixed outcome: it is contingent and negotiable.

It may be argued that the most useful way of analysing technology and society is therefore to view technology as a process. To concentrate specifically on its design and political issues related to that process, obscures the fact that technology can be applied to different social contexts. This is certainly true of virtual reality, which is seen to have a wide range of applications in both industry and entertainment. The context in which the technology is used will therefore be a contingent factor in the meanings that are eventually associated with it. The implications of employing virtual reality technology in the aerospace industry to train pilots and to entertain people in their own homes will inevitably be different.

10.8 How might technology (and virtual reality) influence society?

This discussion focuses primarily on the use of virtual reality in the realm of entertainment, and addresses a number of questions which emerge from the use of other forms of media which have established themselves as legitimate forms of cultural transmission. By examining other forms of media associated with different technological developments, it is possible to draw out relevant themes that need addressing with respect to virtual reality. These themes are largely 'content-based', but require analysis in the context of the capacity of virtual reality to enhance social and cultural value, through its potential to create a more realistic representation of 'reality', that is, a more high-fidelity simulation.

Before moving on to these content-based issues our final illustration can be made regarding the ability of technology to affect meanings in its relationship with society which is appropriate to virtual reality. Scarborough and Corbett (1992, p. 70) argue that: 'the influence of technology and its ideological underpinning is greatest when we live it'. They present the analogy of a mirror, for when a person looks into a mirror, they do not see the mirror itself; rather they see 'reality'. The development of high-quality glass in the fourteenth century later encouraged innovations such as the microscope and telescope, and affected the way humankind viewed the world and themselves. As Mumford (1934, pp. 125–6) observed:

> glass helped put the world in a frame: it made it possible to see certain elements of reality more clearly and it focused attention on a sharply determined field – namely that which was bounded by the frame.

Given that a prime objective of virtual reality technology is to create a sense of 'reality' that is also lived, we must address its potential for framing our perceptions of meaning in terms of the world and ourselves.

10.9 Images and ideology

> The meaning of modern technology cannot be found in the artefacts themselves but in relating the symbolism of the artefacts to the cultural and ideological context within which their meaning is located. (Scarborough and Corbett, 1992, p. 75)

Ideologies are representations of aspects of 'realities' reflected in, and further reinforced by, socio-historical conditions. Ideologies cannot be defined as either true or false, but represent particular views of the world and often serve to reflect or reinforce existing social relations. Virtual reality technology has the ability to influence our definitions of 'reality', in the sense that our constructions of the world can be mediated by the technology. In the realm of entertainment, the ideological content of the virtual world, the ways in which it represents 'reality', and how the user interacts in an immersive sense with that reality, could all prove powerful at a socio-cultural level.

In some respects virtual reality has elements in common with other media. Studies of television and cinema, for example, are concerned with issues such as realism, ideology, representation, and images, with regard to the construction of meanings which enter into our everyday cultural discourse. Interest in these issues has been prominent amongst feminist researchers who have concerned themselves with social constructs of femininity and how these representations have become 'naturalized' into mainstream culture. The usefulness of such analyses is that they focus on the construction of meaning, and in doing so they address issues of 'reality' as an ambiguous rather than a fixed concept.

Most visual media concerned with realism and realist forms have the appearance (at least) of reality in common. Kuhn (1982, p. 132) writing about fictional and non-fictional realism in cinema argues: 'a certain kind of credibility in relation to the "real world" is set up and the spectator undergoes a "suspension of disbelief" '. What is seen on the television or cinema screen must be perceived as in some sense constructed in the real world. But this does not necessarily mean that the media are in some way value-free and present an unproblematical picture of reality; rather they present a selective, constructed view of 'reality'. This representation of 'partial reality' may depend on a number of things, but inherently is a reflection of the dominance of certain power-wielding groups. With reference to television, for example, Baehr and Dyer (1987, p. 6) contend: 'Men own and control the media and it is their ideas, viewpoints and values which dominate the system of production and presentation in broadcasting.' Baehr and Dyer are in effect arguing that, in the case of politics and ideology, the controlling position of men in broadcasting and wider society has affected the technological process in terms of the outcomes achieved. The somewhat ambiguous notion of 'reality' being discussed here has much to do with the concept of meaning, and to this extent we need to consider how images construct meanings and what influences these images. Virtual reality, in common with other forms of media, is engaged in the construction of images. Feminist studies of cinematic images have examined the ways in which the spectator may become influenced via representations. Meanings in themselves are not embodied in images, but are, in the cinematic example, as Kuhn (1985, p. 63) has described them: 'circulated between representation, spectator and social formation'.

Kuhn goes on to describe the image as occupying various contexts: cultural in terms of spectatorship, together with institutional, social and historical contexts of production and consumption. The central problem here, in comparing cinematic representations of images and virtual reality, is that the role of spectator is replaced with that of participant: it is an immersive experience. Moreover, in conventional media the construction of reality is mediated through television or film by the dislocation of the viewer or 'looker' from the image or the action portrayed by it. Immersion thus replaces 'spectatorship' and the addition of sensory experience (sound and tactile feedback) aimed at replicating physical reality may obscure representation and in this sense appear more real. A question posed by Shapiro and McDonald (1992) is significant in relation to this, for they ask whether the user's memory of entering virtual reality will mean that the experience is not perceived as real? This may indeed be the case and raises issues concerning the interaction of people with technology.

10.10   The user's relationship with virtual reality technology

The relationship between user and technology is more direct and involved in the case of video games. Unlike television and cinema, the 'spectator' is also a 'performer'; in this sense, video games (and virtual reality) may be described as an active medium. Video games represent microworlds with their own rules and culture; therefore we must view them as 'cultural texts' (Provenzo, 1991) embodying meanings which can mediate a person's (usually a child's) understanding of culture. As a mediator of culture, Provenzo argues, video games are not neutral, for within them are specific symbolic and social constructs and as such they are powerful instruments of cultural transmission.

An identifiable difference between television, video games and some cinematic images is the extent to which realism is adhered to. Television, and in particular news reporting, although not value-free in the presentation and selection of 'news-worthy' stories, does present (however de-contextualized) snapshots of reality. Much debate has surrounded the production and construction of news (see the Glasgow Media Group's work (1976, 1980)) and the ways in which the global village (McLuhan, 1967) is portrayed through our screens. Similarly, cinematic and television drama adheres to a form of story-telling which aims to incorporate realism. Video games, however, are divergent in their content and form, for, as Skirrow (1990) discusses, part of the attraction of video games is that they break down the barriers between fantasy and science, and between play and real-life. In effect, they can make things happen in the microworld with which the player is engaged, which cannot happen in the 'real world'. Nevertheless, the increasing sophistication of computer-generated images has brought with it an increased sense of realism in the images portrayed, and alongside this is increased concern about the role of the user as an active participant in a more realistic computer-generated world. This concern has been fuelled by conflicting reports about the the possible detrimental impact of video worlds on a child's psychology.

Concern relating to the long-term effects of video game playing on children's behaviour has been investigated from a number of perspectives. Research into the links between social behaviour and video games has produced conflicting results. Some researchers have argued that game-playing produces addictive and compulsive patterns in behaviour (Griffiths and Hunt, 1993), whilst other research (Gibb *et al.*, 1983) has concluded that video-game playing is a social activity and that there is no significant relationship between negative behaviour and social interaction. This debate, though inconclusive, will continue to court controversy in the shadow of recent coverage of the James Bulger case and the subsequent trial judge's comments citing video violence as a possible influential factor. The absence of a provable link in previous studies may be due to the methodologies being employed. This has been suggested recently by researchers at Leicester University who have argued:

> The relationship between graphic representations and reality is not a simple process of proving cause and effect. Rather it involves a more complicated process of evaluating, from a variety of perspectives, the potential for different types of graphic scenes to influence the varying circumstances.[4]

In a critique of previous studies, the researchers argue that these have concentrated largely on psychological factors, examining individuals outside their social context. The outcome of the Leicester research is unpublished at the time of writing. But a central problem has been highlighted by the researchers, which is the need to identify the influence computer games have had in developing new subcultures amongst children: subcultures which are usually inaccessible to adults and parents and therefore difficult to understand and interpret.

One means of access to understanding these developing subcultures might be to learn the 'informal' language and 'rules of the game' inherent within them which derive from the most popular games. A detailed study by Provenzo (1991) sought to address such issues by 'Making sense of Nintendo'. In this analysis, Provenzo touches on a number of themes of cultural and sociological interest. One such theme which parallels earlier discussion on television and cinema is the gender implications of video games. The content of most games in Provenzo's analysis provided 'storylines' and images that actively excluded women. Essentially video games are designed by males for males, and provide a male preserve in public spaces such as arcades. Video games offer few positive stereotypes of women in their scenarios. Indeed research (Toles, cited in Provenzo, 1991) analysing a sample of 100 video games found 92% did not include any female roles at all. Of the remaining 8%, 6% portrayed women in passive roles and only 2% in active roles (which have limited impact as one was a female kangaroo and the other a feminized blob!).

The marginalization of women in the cultural content of video games is amplified further by the aggressive character of much of the action taking place, and by the roles male stereotypes adopt. In the video world, aggression is both acceptable and desirable, since most of the games require the player to defend 'good' from 'evil' by whatever means possible. Constructions of 'good' and 'evil' become confused with their relationship to violence, but what is certain is that destruction is inevitable if control is lost. Control is the key concept underlying the scenarios of such games, and in the video world, control equates with male aggression. Passivity is the female preserve; women are essentially victims reinforcing the cultural hegemony existing around the representation of women, and which eventually can be felt as 'real', in that it acts to 'saturate our consciousness' (R. Willimas, cited in Provenzo, 1991 pp. 115–16). The debate again returns to the issues of images and ideology, in that the content of representations in video games, as social and cultural 'texts', reinforce dominant stereotypes, e.g. male aggression and female passivity. As Provenzo (1991, p. 115) asserts:

> Video games and their content represent symbolic universes that are spontaneously consented to by the general population. In this context it can be argued that video games are instruments of a larger social, political and cultural hegemony.

Although it is apparent that video games and other representative media cannot be entirely value-free, they could be used to amplify positive images of cultural and social concern on issues such as gender, race or even international cooperation, rather than warfare. The fact that such negative images are tolerated has, in the case of video games, much to do with their location in the private sphere. Their content remains largely unregulated and increasingly consumption is within the privatized realm of the home. This trend of consumption moving from the public space of the arcade to the private space of the home usually escapes even the regulation of parents, as a survey carried out for the television programme 'Check Out 93' revealed.[5] The survey found that whilst 57% of adults controlled what children watched on television, only 3% regulated the content of video games. Only recently has the British government awoken to the problems associated with computer-generated images because of concern over computer pornography, and responded by setting up a select committee to investigate possible regulation.

The notion of retrospective regulation, however well intended, may have limited impact. Moreover, a central problem is the regulation of private space such as the home. Computer generated pornography is but one extreme in this dilemma, since video games represent the 'harmless' opposite but also remain unregulated. Video games have created new and significant subcultures for children which are beyond parental regulation. The only substantive bar to this subculture for children themselves is ability to pay, since access to the privatized technological market-place is expensive.

10.11 Who decides the content of virtual reality?

The preceding discussion has highlighted that technology and society are not independent of one another and that social relations are formed by changing networks of human and non-human actors. Simulations, including virtual reality, embody certain aspects of 'reality' and consequently one's perception of reality may be mediated by such technology. We can view video games, television etc., and hence also virtual reality, as an instrument through which people mediate their understanding of culture (especially children who are composing their cultural model for the first time). The representation of gender roles, social and racial stereotypes and the permissibility of violence and aggression may have an important impact on a child's expectations of the 'real' world[6] by such selection and amplification of culture. (See also the discussion in section 2.4.6.)

The designers of virtual environments can completely control what appears in such micro-worlds, and this brings with it a measure of responsibility.

10.12 Ethics and application

As with any evolving science, the application of the technology has a large influence on whether that technology is seen as 'good' or 'bad', and the applications chosen say a lot about the society which creates them:

> Any technology is an expression of the framework of meaning within which a particular human society lives. (Helsel and Roth, 1991)

The development of virtual reality technology was until quite recently driven by military funding, and consequently was viewed with specific goals in mind, such as the improvement of fighter-pilot performance. Commercial and other non-military application domains, however, have adopted virtual reality with both neutral and non-neutral goals in mind. Consider the application of virtual reality to medical domains. Virtual reality has obvious applications as a visualization tool for surgical operations, but in addition there has been speculation about the use of virtual reality to treat mental disorders such as phobias and even as a form of psychotherapy (Whalley, 1993). Although such uses may initially be seen as positive applications of technology, an editorial article in the *Lancet* (Anon., 1991) raises some interesting points:

> Continuous exposure to virtual reality will impoverish those aspects of life that determine social development, interpersonal insight and emotional judgement. Vulnerable patients should not be exposed to virtual reality until the full extent of its likely impact can be reliably anticipated... At one level virtual reality becomes a terrifying instrument of torture, at another a powerful means of education.

Whalley (1993) discusses the ethical issues involved in using virtual reality to treat patients with mental disorders, and highlights societies' adoption of technology as 'cures' which do not, in retrospect, have a wholly unblemished reputation (electroconvulsive therapy, for example). The treatment of fears, phobias and stress, however, are seen as likely areas for the positive application of virtual reality in the near future.

Moving back to virtual reality as a form of commercial media or entertainment, all the issues affecting television and video games are still present. How long will it take for virtual reality to be used as a form of advertising,[7] and how effective will it be, given that the ability to tailor the advertising to the particular user will be high (Brand, 1987)? Again the ability to control the content of such application will influence the possible social effects.

The power and social impact of virtual reality will be determined by its intelligent application.

10.13   Directions for future research

A wide range of issues has been examined in this chapter, many of which could provide a framework for future research. Any such research in this stage of the development of virtual reality should aim at further discussion which is constructive and proactive in terms of raising social concerns.

A major problem in evaluating the future implications of virtual reality is the current lack of substantiative research in established areas of technological development. This is particularly true with regard to mass-marketed entertainment technology such as computer games, which have become increasingly sophisticated and 'realistic' but present problems of research access because of their social context. The increase in individual consumption of such games, and their location within the private sphere of the home, create methodological difficulties. Most studies cited in this chapter have relied either on deconstructing the cultural content of computer-generated images or scenarios (see Provenzo, 1991) or on interviewing children outside the home about the context of games and their influence on outlook. Both approaches, though useful, do not provide a context in terms of the importance of meanings and whether these meanings are influenced by the situation in which the game images are mediated. This context of meanings may differ between public and private spaces; it may also depend on individual and collective game-playing and, in the case of children, parental influence.

It is apparent that further research is required even in areas of mass-marketed entertainment, given the conflicting evidence available on the psychological and sociological effects of images portrayed in both the conventional media and in computer-generated forms. Recent debates around television censorship, particularly with relevance to the

Private Member's Bill put forward by MP David Alton, confirm a widespread concern for the effects of media content. If virtual reality is to become the fantasy medium of the future, it will need to steer clear of 'moral panics' and technophobic reactions. This may require an acknowledgement of the social context of technological innovation and application. Similarly, when considering a future research agenda aimed at investigating the social implications of virtual reality, it is important not only to examine the technical potential of virtual reality itself but also to place it in a social context. The aims and objectives of future research should therefore include:

1. Attempts to fill gaps in existing literature which has often been psychologically based. This would incorporate two central objectives. Firstly, drawing on themes raised in this chapter, an analysis should be made of socio-cultural images and ideologies generated by current media, with particular emphasis on computer-generated images and the context given to them (e.g. storyline, game, market niche). This would attempt to evaluate *content-based* issues. Secondly, it is important to acknowledge *context-based* issues, that is, the structural aspects of technology use (e.g. the home, the arcade, school, individual or collective user involvement).
2. An investigation of the market potential of virtual reality technology. This would serve to identify areas of social life which may be affected by virtual reality in the future. Again the issue of *context* is raised. For example, the implications of using virtual reality technology in medicine might lead to an investigation of professionalism, and professional training and status amongst the medical fraternity, which would be a narrow field of enquiry. The implications of the potential market of virtual reality within the sphere of entertainment or consumption (e.g. tele-shopping) would, however, create a broader research agenda.
3. In conjunction with 2 above, an investigation of companies involved in research and development issues associated with virtual reality. This would aim not only to establish future market trends but also to uncover meanings given to the social context of virtual reality by those primarily involved with technical and developmental issues.
4. A further objective might embrace issues of the construction of 'reality'. Unlike previous research aims, this would present problems in terms of the collection of solid, quantifiable data. An investigation at this level would necessarily involve examining political, cultural and ethical questions associated with virtual reality technology. Again the *context* in which the technology is employed may influence these value-laden questions.

10.14 Concluding comments

The aim of this chapter was to stimulate further debate, and as such any firm conclusions would be inappropriate. Only the future will provide concrete answers to many of the questions raised here about virtual reality. We can learn from the past and the present, however, and in doing so accept some responsibility for the future we may help to shape. Underlying our debate has been an attempt to forge a relationship between past, present and future by examining historical, political, social and cultural aspects of technology. We have contextualized technology, and by doing so rejected standpoints which prefer to view technology as neutral or 'value-free'. Virtual reality is currently embodied by a form of technology, and like most technologies creates images of fascinating potential but simultaneously raises fears about its application. The power and social impact of virtual reality will therefore be determined by its intelligent application and the responsibility for the dream or nightmare lies with us all.

Notes

1. This chapter is based on a paper presented by John F. Meech and Michelle Baker at the Applied Vision Association Conference 1993, Bristol.

2. In this chapter we will not address the issues surrounding text-based virtual environments such as multi-user dungeons (MUDS) etc.
3. See Cockburn's (1985) study of the British printing industry where retention of heavy machinery prevented women entering employment without the support of male trade unionists.
4. K. Livingstone quoted in A. Utley, 'Crimewatch' may cause crime waves, *Times Higher*, 3 December 1993.
5. Channel 4, 23 November 1993.
6. Although we have drawn attention to the negative impact of such representations in video games, it is equally possible for representations to show a more 'positive' effect.
7. At the time of writing it is not known if VR has been the subject of product placement, for example where a particular brand of canned drink is explicitly shown in a micro-world.

References

Anon., 1991, Being and believing: ethics of virtual reality (edtorial), the *Lancet*, **338**, 283–4.
Baehr, H. and Dyer, G., 1987, *Boxed In: Women and Television*, Swansea: Pandora Press.
Baudrillard, J., 1983, *Simulations*, New York: Semiotext(e).
Brand, S., 1987, *The Media Lab*, New York: Viking.
Cockburn, C., 1985, *Machinery of Dominance: Women, Men and Technical Know-How*, London: Pluto Press.
Gibb, G. D., Bailey, J. R., Lornbirth, T. T. and Wilson, W. P., 1983, Personality differences between high and low electronic video game users, *Journal of Psychology*, **114**, 143–52.
Gibson, W., 1986, *Neuromancer*, London: Grafton Books.
Glasgow University Media Group, 1976, *Bad News*, London: Routledge & Kegan Paul.
Glasgow University Media Group, 1980, *More Bad News*, London: Routledge & Kegan Paul.
Griffiths, M. D. and Hunt, N., 1993, The acquisition, development and maintenance of computer game playing in adolescence, paper presented at the British Psychological Society Conference, Blackpool, 15 December 1993, Leicester: British Psychology Society.
Grint, K., 1991, *The Sociology of Work*, Oxford: Polity Press.
Heim, M., 1991, The metaphysics of virtual reality, in Helsel, S. K. and Roth, J. P. (Eds), *Virtual Reality: Theory, Practice and Promise*, pp. 27–34, London: Meckler.
Helsel, S. K. and Roth, J. P., (Eds), 1991, *Virtual Reality: Theory, Practice and Promise*, London: Meckler.
Kuhn, A., 1982, *Women's Pictures: Feminism and Cinema*, London: Routledge & Kegan Paul.
Kuhn, A., 1985, *The Power of the Image: Essays on Representation and Sexuality*, London: Routledge & Kegan Paul.
MacKenzie, D. and Wajcman, J., 1985, *The Social Shaping of Technology*, Milton Keynes: Open University Press.
McGuinness, B. and Meech, J. F., 1992, Human factors in virtual worlds: (1) Information structure and representation, in *IEE Colloquium on Using Virtual Worlds*, Digest No: 1992/093, pp. 3/1–3, London: IEE.
McLuhan, M., 1967, *Understanding Media: The Extensions of Man*, Harmondsworth: Penguin.
Mumford, L., 1934, *Technics and Civilization*, London: Harcourt Brace Jovanovich.
Pacey, A., 1983, *The Culture of Technology*, Oxford: Blackwell.
Pimentel, K. and Teixeira, K., 1993, *Virtual Reality: Through the New Looking Glass*, Singapore: McGraw-Hill.
Provenzo, E. F., 1991, *Video Kids: Making Sense of Nintendo*, Cambridge, MA: Harvard University Press.
Rendings, W., 1991, *Introducing Lyotard: Art and Politics*, London: Routledge.
Scarborough, H. and Corbett, J., 1992, *Technology and Organization, Power Meaning and Design*, London: Routledge.
Schroeder, R., 1993, Virtual reality: social impacts and cultural dimensions, in *Proceedings of Virtual Reality '93, The Third Annual Conference on Virtual Reality*, pp. 7–14, London: Meckler.
Shapiro, M. A. and McDonald, D. G., 1992, I'm not a real doctor, but I play one in virtual reality: implications of virtual reality for judgements about reality, *Journal of Communication*, **42** (4), 94–114.

Skirrow, G., 1990, Hellivision: an analysis of video games, in Alvarado, M. and Thompson, J. O. (Eds), *The Media Reader*, London: British Film Institute.

Whalley, L. J., 1993, Ethical issues in the application of virtual reality to the treatment of medical disorders, in Earnshaw, R. A., Gigante, M. A. and Jones, H. (Eds), *Virtual Reality Systems*, pp. 273–87, London: Academic Press.

Winner, L., 1977, *Autonomous Technology*, Cambridge, MA: MIT Press.

Winner, L., 1985, Do artefacts have politics?, in Mackenzie, D. and Wajcman, J. (Eds), *The Social Shaping of Technology*, Milton Keynes: Open University Press.

Wollaston, A., 1960, *Descartes' Discourse on Method*, London: Penguin.

Woolley, B., 1992, *Virtual Worlds: A Journey in Hype and Hyperreality*, Oxford: Blackwell.

11

Summary and conclusions

Karen Carr

A brief summary of the main points made in each chapter is presented here, followed by a discussion of future directions in research for virtual reality.

Stephen Ellis outlines the history of the various branches in the evolution of virtual environments and provides an overview of many applications being developed today. With this broad point of view, he sees the need for a systematic approach to future development and suggests the following taxonomies:

1. Definitions of virtual space, virtual images and virtual environments and the concept of 'virtualization'.
2. An analysis of the components of an environment, real or virtual.
3. A functional breakdown of the technology required for virtual environments.

In addition he emphasizes the need for human factors data to be generated in a form which can guide the creation of virtual environments. He presents a starting point for a human factors database which meets that need. Finally he discusses some sociological concerns for virtual environments.

Chris Christou and Andrew Parker provide a useful introduction to the philosophy and psychology of perception, highlighting the major themes of empiricism, nativism, the theory of 'direct' perception, and psychophysics. They then demonstrate that an understanding of the physics of natural images and the constraints of the display medium, as well as the physiology and the psychology of visual perception, is necessary to provide a realistic image in virtual environments. Realism in computer graphics is considered in perceptual terms. Christou and Parker contrast the task of artists using paint to represent pictorial images with that of the designer of a virtual environment using computational models of light. Most importantly, they use psychophysical techniques to evaluate the 'realism' of the result, thus combining physics and psychology as tools to create a visual experience.

Graham Edgar and Peter Bex select two contrasting areas of research which have significant implications for the specification of displays. They provide a detailed example of the extensive studies of temporal aliasing, illustrating the need to understand the underlying perceptual mechanisms and the needs of a particular application. It is emphasized that solutions must be sought for specific problems and contexts. They contrast this highly focused area of research with the current, much more vague understanding of ocular accommodation in virtual environments. Existing literature about accommodation warns us of many possible problems with displays for immersive virtual reality.

Richard So and Michael Griffin have extensive experience in researching the problem of display lag with head-mounted displays. This problem is fundamental to the achievement of virtual reality, and should be a research priority. Although So and Griffin's work has been directed mostly at displays providing virtual cross-wires and other information against a real world, most of their results are applicable to any other head-coupled virtual environment. They provide a brief summary of the main issues and evaluate a technique which they have proposed as one way of reducing the very fundamental problems caused by head-slaved images lagging behind head movements.

The chapter by John Findlay and Fiona Newell discusses the current understanding of how people are able to recognize objects from visual information. The fact that we are able to recognize objects from very variable and often partial information suggests that, while accurate imagery might be essential for the perception of distance, orientation or motion, for recognition purposes simple representations may be sufficient. Findlay and Newell review the scientific work on object recognition, and indicate factors which affect how quickly or accurately we can recognize visually perceived objects. Some of these factors are 'top-down' processes, while others are 'bottom-up'. They conclude their chapter with the hope that some time in the future virtual reality could itself be a tool to help investigate human perception.

Rupert England's chapter highlights the fact that if we wish to do anything very much in a virtual environment we shall be using our various senses to interact with virtual objects. England emphasizes the need firstly to define requirements according to what the virtual environment will be used for. He shows that by understanding the physiological and psychological bases of behaviour we can identify how the senses work together for specific activities and use this knowledge to enhance interaction in the virtual environment. He discusses the fundamentals of behaviour in terms of perception and response, and the feedback necessary for interaction, and provides an overview of the neurophysiology of perception. Interaction with virtual environments will frequently require visuo-motor coordination, and often manipulation using the hands. A review of tactual perception and tactual virtual environments is presented by England in the context of multi-sensory perception, showing that we cannot consider our perceptual senses as separate and independent processes.

Mark Williams describes the physiology of auditory perception and discusses factors which affect how we perceive sounds to have varying qualities and to be coming from different directions. He includes a discussion of speech synthesis, and a review of auditory display technology. He gives examples of auditory and visual integration, and of the use of auditory icons in human–computer interfaces.

Gerda Smets, Pieter Jan Stappers, Kees Overbeeke and Charles van der Mast show how simulation, telepresence and computer models can all be integrated with virtual reality to provide an optimum human–computer interface for computer-aided design. They describe the psychological concepts underlying their approach and the requirements for the interface which arise from these concepts. They show how they have made use of the tools of virtual reality technology and computational neural networks to achieve their goals.

Deborah Foster and John Meech consider the social impacts of hypothetical virtual reality technology which can, by whatever means, provide some sense that the virtual world is 'real'. The concepts of simulation and reality are explored, and virtual reality is considered as simulation, media and technology in a social context. They suggest that getting perceptual 'cues' right might not be critical for achieving realism, given that our sense of reality is determined by society, media and other factors extrinsic to ourselves. They pay particular attention to virtual reality as entertainment, and suggest major areas for sociological research.

This book has shown that although there is a large awareness of the human characteristics which can affect our perceptions of virtual reality, there is still much to be learned about these characteristics so that they can be appropriately stimulated by technology. There is a need for much fundamental research into human perception and response both in the real world and in the virtual environment. A human factors database, such as suggested by Ellis (2.4) or the one suggested by England (7.1.3), has the scope to be very large indeed. If a suitable format for the collection of such information is identified, then once we start to try to fill it in with current knowledge, it will soon be apparent where the holes in our knowledge exist. If we also devise the means for analysing what we want from a virtual reality for specific applications, this would be a suitable way of identifying and prioritizing research.

Although we can already create virtual environments of some sort which can convey three-dimensional, interactive information, it is not clear how accurately we can control the perceptual 'message' or whether we can really achieve virtual reality. It would seem necessary, therefore, that we should have some objective means for assessing what perceptual results a virtual environment can elicit. As England suggests (7.1.2), measures of performance on specific tasks can be used as functional assessments of a virtual reality. These functional measures will not, however, allow the more phenomenological aspects, such as 'a sense of realism' or a 'suspension of disbelief', to be investigated. While these aspects of the perceptual experience might not affect our abilities to perform certain tasks in a virtual environment (especially after we have learned the particular conditions of the environment), it is likely that they can have an effect on longer-term behaviour. For example, the facility to transfer learned behaviour from virtual reality to reality, whether desired or undesired, may depend upon some quality of virtual reality. Similarly, there may be holistic qualities of virtual reality, and the context in which it is used, which could cause changes in social behaviour and culture. A fascinating aim could be to link electrical activity in the brain, and other physiological measures, to such phenomenological aspects of virtual reality. There may also be certain behaviour patterns, particularly the lower-level, automatic responses (England 7.8.1), which could be indicative of a type of perceptual experience. Thus one key area for research is to develop methods of measuring what kind of virtual reality can be created by different technologies and different virtual environments.

An important assumption underlying many of the themes in this book is that we can define perceptual processes in ways which are common to all humans. This assumption is supported by many results which show commonality of physiology and of behaviour. There are, however, well-known differences between people in their physiological apparatus, for example, various degrees of 'colour blindness', 'stereo blindness', or pitch discrimination. These will hold true in the virtual environment as they do in reality. There are also many ways in which people differ in their perceptual interpretations (for example, a review of cross-cultural differences is given by Gregory (1987, pp. 601–3); individual differences in the effects of context upon perception is reported by Witkin (1949); and a review of personality and perception is given by Arndt (1974, pp. 267–304)). Often the same people will apparently perceive differently at different times (as when one notices something for the first time). These differences will make it more difficult to provide the same virtual reality for different people. Given the unique set of experiences that we each undergo, it is quite possible that such perceptual diversity is attributable to experience. It is also possible that there are some physiological bases for differences, such as genetics or ontogeny. In any case, the task of tracking all these variations in experience and physiology would be unending, and if we wish to control perceptions in virtual reality we need a more useful and practical approach to dealing with individual differences. We should assess the 'robustness' of so-called generic findings by validating them with as broad a range of people as possible. Where individual differences are apparent, large patterns of variation should be looked for, such as those associated with age, culture, or even personality traits. Only such broadly predictable individual differences have the potential to be generally catered for in virtual reality; further refinements would need to be achieved on a much more costly individual basis.

The suggestion by Findlay and Newell (6.4), that virtual reality may help in psychological research, leads to an interesting prospect, in which quite complex aspects of reality could be systematically changed in virtual reality in order to extract the subjectivity of reality as we know it. By experimenting with 'reality' we could identify perceptual and cognitive processes which control our ability to understand. Thus, while the perceiver of virtual reality may, as suggested in the Introduction (1.1), achieve a more subjective understanding of objective information, those in the real world who observe the virtual reality perceiver may be able to identify some of the subjectivity in understanding. At a more basic level, as intended by Findlay and Newell, virtual environments could become experimental laboratories. But only when the technological interface to the human senses has become transparent (and we have 'realism') will experimentation in these virtual laboratories be relevant to the real world. In the meantime, however, if the development of virtual reality is based on psychological study only within current virtual environments, not only will the technological interface be a variable affecting the results, but also we shall be developing a self-existent environment in which virtual reality is both the tool and the product. The product would then be not so much 'virtual' as 'parallel', as it will be self-contained and separate from reality. Research in the real world is thus a very important requirement for the development of virtual reality.

The general conclusion of this book is that there is a wide diversity of knowledge required in order to control the development and use of virtual reality. From basic laboratory experiments investigating perception to studies of sociological trends, this broad range of information needs to be coordinated before we can know what technology is needed to provide the virtual realities we want. We even need methods of identifying what virtual realities we want. Faced with such a daunting amount of research, it would be all too easy to rely on technological developments to lead the way ahead. Commercial developers of virtual reality have no choice but to take this 'easy' way, until they are directed by market requirements or regulations. It is the responsibility of the research institutions to coordinate and carry out the research needed to identify the human basis for virtual reality, and to specify the technological requirements which commercial developers should fulfil.

References

Arndt, W. B., 1974, *Theories of Personality*, New York: Macmillan.
Gregory, R. L., 1987, *The Oxford Companion to the Mind*, Oxford: Oxford University Press.
Witkin, H. A., 1949, The nature and importance of individual differences in perception, *Journal of Personality*, **18**, 145–60.

First Author Index

Plain: citation. Bold: full reference.

Abernethy, B. 134, 159, 160, **170**
Adams, J. A. 159, **170**
Adelson E. H. 87, 89, **98**
Adelstein, B. D. 28, 31, 35, 39, 43, **44**
Airey, J. M. 27, **44**
Albrecht, R. E. 108, **110**
Alpern, M. 96, **99**
Allen, J. H. 104, 105, 108, **110**
Allport, D. A. 90, 91, **99**
Amanatides, J. 87, **99**
Ando S. 179, **186**
Anstis, S. M. 66, **82**
Arndt, W. B. 227, **228**
Asanuma, H. 163, **170**
Attneave, F. 59, **82**

Bahill, A. T. 40, **44**
Baehr, H. 216, **222**
Baker, C. L. 92, **99**
Baker, J. R. 98, **99**
Barfield, W. 20, **44**
Barlow, H. B. 91, **99**
Barnes, J. 28, **44**
Barrette, R. 23, 37, **44**
Basalla, G. 39, **44**
Bassett, B. 23, **44**
Batteau D. W. 181, **186**
Baxter, D. W. 166, **170**
Baudrillard, J. 211, **222**
Beggs, W. D. A. 161, **170**
Bejczy, A. K. 13, 18, 31, 36, 37, **44**
Belendiuk, K. 182, **186**
Bendat, J. S. 106, **110**
Benson, A. J. 104, **110**
Benson, C. 74, **82**
Berkeley, G. 55, 58, **82**, 97, **99**, 164, **170**
Bergault, D. R. 85, **186**
Berger, K. W. 181, **186**
Bergman, A. S. 181, **186**
Bernstein, N. A. 158, 161, **170**

Bertin, J. 24, **44**
Bex, P. J. 90, 92, **99**
Biederman, I. 117, 119, 121, 122, 126, 127, **128**
Bischof, W. F. 92, **99**
Biocca, F. 205, **206**
Bishop, P. O. 18, **44**
Bizzi, E. 161, **170**
Blakemore, C. 117, **128**
Blattner, M. M. 186 **187**
Blauert, J. 36, **44**
Blinn, J. F. 26, **44**
de Boer, I. 181, **187**
Boff, K. R. 36, **44**, 145, **170**
Bootsma, R. J. 199, **206**
Borah, J. 36, **44**
Borden, G. 179, **187**
Boring, E. W. 56, **82**
Bouma, H. 123, **128**
Braddick, O. J. 117, **128**
Brand, S. 36, **44**, 220, **222**
Bravo, M. J. 119, **128**
Brehde, D. 11, **44**
Breitmeyer, B. G. 91, **99**
Broadbent, D. 156, **170**
Brooks, F. Jr. 11, **44**, 203, **206**
Brooks, M. J. 74, **82**
Brooks, T. L. 31, 36, **45**
Brooks, V. B. 141, **170**
Bruce, V. 127, **128**, 141, 142, **170**
Bruner, J. S. 114, 116, **128**
Bryson, S. 104, 105, **110**
Bui-Tuong, P. 68, **82**
Bullimore, M. A. 97, **99**
Burger, J. F. 181, **187**
Burgess, P. R. 148, **170**
Burr, D. C. 86, 89, 90, 93, **99**
Bussolari, S. R. 13, 24, **45**

Calhoun, G. 185, **186/187**
Campbell, F. W. 64, **82**, 94, **99**

Cardullo, F. 13, 20, 23, **45**
Carlton, L. G. 161, **170**
Carpenter, R. H. S. 136, 139, 146, 148, 150, 151, 153, 160, 162, **170**
Carr, K. 3, **9**
Carroll, L. 22, **45**
Casagrande, V. A. 143, 157, **170**
Cellier, F. 25, **45**
Chang, J. J. 92, **99**
Chapanis, A. 59, **82**
Chapman, R. M. 4, **9**
Chernikoff, R. 161, **170**
Christou, C. G. 76, **82**
Churchland, P. S. 119, **128**
Clark, F. J. 150, 165, **170**
Clark, J. H. 23, **45**
Clark, W. C. 59, **82**
Cleary, R. 92, **99**
Cohen, M. F. 72, **82**
Collewijn, H. 18, **45**
Comeau, C. P. 28, 30, **45**, 136, **171**
Connolly, K. 156, **171**
Cooper, G. E. 19, **45**
Corballis, M. C. 120–2, **128**
Cordo, P. J. 161, **171**
Cornsweet, T. N. 66, **88**, 98, **99**
Cotman, C. W. 135–6, 139, 141, **171**
Cotter, C. H. 24, **45**
Cowdry, D. A. 24, 40, **45**
Cowley, K. C. 184, **187**
Crampton, G. H. 19, 24, **45**
Crowe, S. J. 180, **187**
Cruz-Neira, C. 23, **45**
Curry, R. E. 24, **45**
Cutting, J. E. 114, 118, **128**

Daley, B. 23, **45**
D'Arcy, J. 11, **45**
Deecke, L. 163, **171**
Deering, M. 18, **45**
DeFanti, T. 136, **171**
Descartes, R. 97, **99**
Di Lollo, V. 91, **99**
Dippe, M. A. Z. 87, **99**
Dickson, R. A. 163, **171**
DiZio, P. 165, **171**
Donders, F. C. 97, **99**, 161, **171**
Durlach, N. I. 36, **45**, 182, 186, **187**
Dyson, F. 1, **9**, **189**

Eckert, R. 137, **171**
Edelman, S. 120, 122, **128**
Edgar, G. K. 95, 98, **99**
Elkind, J. I. 36, **45**
Elliott, J. M. 163, **171**
Ellis, S. R. 11–14, 18, 22, 24, 36, 38, 41, 45–6, **207**
Ellis, A. W. 114, **128**
England, R. D. 3, 9, 163, **171**

Enns, J. T. 124, **128**
Enroth-Cugell, 142, **171**
Erkelens, C. J. 18, **46**
Evarts, E. V. 141, 163, **171**

Fagan, B. M. 22, **46**
Fahle, M. 86, **100**
Fant, G. 184, **187**
Farrell, J. E. 90, 91, **100**
Feldon, S. E. 18, **46**
Fels, S. S. 202, **207**
Fincham, E. F. 94, **100**
Findlay, J. M. 115, 124, **128**
Fisher, P. 31, **46**
Fisher, S. S. 23, 38, **46**, 136, **171**
Fitts, P. M. 160, **171**
Flach J. M. 160, **171**, 194, **207**
Flanagan, J. L. 183, 184, **187**
Foley, J. D. 23, **46**, 69, **82**, 164, 168, **171**
Foley, J. M. 18, **46**
Forsyth, D. 71, **82**
Foster, S. H. 185, **187**
Franke, E. K. 180, **187**
Fransson-Hall, C. 163, **171**
Freeman, R. A. 59, **82**
Friedman, A. 127, **128**
Friedmann, M. 105, **110**
Fritsch, G. 161, **171**
Fulton, J. F. 162, **171**
Furness, T. A. 23, **46**, 104, **110**, 136, **171**

Gale, A. 3, **9**
Gaver, W. W. 181, 186, **187**, **207**
Gibb, G. D. 217, **222**
Gibbs, C. B. 161, **171**
Gibson, C. P. 103, **110**
Gibson, J. J. 7, **9**, 15, 45, **46**, 57, 59, 81, **82**, 117–8, **128**, 135, 149, 155, 161, 169, **171**–2, 194–5, 200, **207**
Gibson, W. 23, **46**, 131, **172**, 212, **222**
Gilmartin, G. 93, **100**
Glassner, A. 69, **82**
Glencross, D. J. 161, **172**
Goertz, R. C. 27, 28, **46**, 136, **172**
Gogel, W. C. 59, **82**
Gombrich, E. H. 6, **9**, 80, **82**
Goodale, M. A. 36, **46**, 114, **128**, 143, **172**
Goodwin, G. M. 165, **172**
Goral C. M. 72, **82**
Gore, A. 42, **46**
Gould, J. D. 124, **128**
Graham, C. H. 91, **100**
Grant, C. W. 87, **100**
Greaves, C. 180, **187**
Green, J. H. 141, **172**
Green, P. 27, **46**, 124, **128**
Greenberg, A. 22, **46**
Greenberg, D. P. 27, **46**

Greening, S. 156, **172**
Gregory, R. L. 4, **9**, 15, 17, **46**, 76, **82**, 126, **128**, 227, **228**
Griffiths, M. D. 217, **222**
Griffiths, H. E. 163, **172**
Grillner, S. 159, 160, **172**
Grint, K. 215, **222**
Grudin, J. 12, **46**
Grunwald, A. J. 14, 26, **46**-7, 106, **110**
Guski, R. 182, **187**

Haber, R. N. 124, **128**-9, 195, **207**
Hannaford, B. 31, **47**
Hart, S. G. 19, **47**
Hashimoto, T. 13, 37, **47**
Hatada, T. 37, **47**
Hay, J. C. 155, **172**
Haykin, S. 183, **187**
Head, H. 153, **172**
Heeley, D. W. 66, **82**
Heeger, D. J. 15, **47**
Heilig, M. L. 36, **47**, 169, **172**
Heim, M. 213, **222**
Heinemann, E. G. 97, **100**
Held, J. R. 143, **172**
Held, R. 15, 19, **47**
Heller, M. A. 155, **172**
Helmholtz, H. von, 63, **82**, 161, **172**
Helsel, S. K. 210, 219, **222**
Henderson, S. E. 135, **172**
Henderson, J. M. 123-4, **128**-9
Hennessy, R. T. 95-7, **100**
Henry, F. M. 159, **172**
Hering, E. 56, **82**
Hershenson, M. 3, **9**
Hess, R. A. 15, **47**
Hillebrand, F. 97, **100**
Hirose, M. 38, **47**
Hitchner, L. E. 24, **47**
Hochberg, J. 15, 113, **128**-9
Hoffman, D. D. 118, **128**-9
Hogben, J. H. 91, **100**
Holst, E. von, 97, **100**
Hooper, S. L. 157, **172**
Horn, B. K. P, 74, **82**
Hottel, H. C. 72, **82**
Howard, I. P. 36, **47**, 144, 155, **172**
Howell, J. R. 72, **82**
Howell, P. 184, **187**
Howells, J. 169, **172**
Hubel, D. H. 141, 142, **172**
Hummels C. 191, **207**
Humphrey, G. K. 122, 125, **128**-9
Humphreys, G. W. 123, **128**-9
Hung, G. 18, **47**
Huxley, A. 22, **47**

Iggo, A. 146-51, 153, **172**
Ingle, D. 143, **172**, 200, **207**

Ishizaka, K. 184, **187**
Ittleson, W. H. 97, **100**
Iwasaki, T. 95, 98, **100**

Jacobson, S. C. 31, 32, **47**
Jacobus, H. N. 31, **47**
Janin, A. L. 18, 38, **47**
Jaschinski-Kruza, W. 95, **100**
Jenkins, C. L. 23, **48**
Jex, H. R. 15, 36, **48**
Johansson, R. S. 146-7, 164, 165, **172**-3
Johansson, G. 87, **100**, 114, **129**
Johnson, C. A. 96, **100**
Jolicoeur, P. 120-3, **129**
Jones, G. J. 73, **83**
Jones, G. M. 15, 19, **48**
Jones, R. K. 17, **48**

Kaiser, M. K. 26, **48**
Kalawksy, R. 11, 38, **48**, 203, **207**
Kalman, R. E. 15, **48**
Kahneman, D. 156, **173**
Kanisza, G. 125, **129**
Kant, I. 56, **83**
Katz, D. 145, 155, 164, **173**
Keele, S. W. 159-61, **173**
Kellman, P. J. 125, **129**
Kelly, D. 65, **83**
Kelso, J. A. S. 159-60, **173**
Kenshalo, D. R. 147, **173**
Kepler, J. 93, **100**
Kiang, N. Y. S. 180, **187**
Kim, W. S. 13, **48**
Kleffner, D. A. 74, **83**
Kleinman, D. L. 15, **48**
Knoll, H. A. 94, **100**
Koenderink J. J. 15, **48**, 89, **100**, 207
Koffka, K. 56, **83**
Kolers, P. A. 87, **100**
Korein, J. 87, **100**
Koriat, A. 121, **129**
Krueger, M. W. 12, 23, **48**, 136, **173**
Krueger, L. E. 155, **173**
Kruger, P. B. 94, **100**
Kuffler, S. W. 142, **173**
Kugler, P. N. 159, 160, **173**
Kuhl, P. K. 179, **187**
Kuhn, A. 216, **222**

Lackner, J. R. 16, **48**, 166, 171, **173**
Landsmeer, J. M. F. 163, **173**
de Lange, H. 65, **83**
Larimer, J. 36, **48**
Lashley, K. S. 159, **173**
Laszlo, J. I. 141, 148, 158, 161, **173**
Laural, B. 22, **48**
Lederman, S. J. 162, **173**, 181, **187**
Lee, D. N. 133, 135, 153, 155, **173**
Legge, D. 136, 161, **173**

Lennie, P. 142, 143, **173**
Levene, J. R. 96, **100**
Levine, M. 12, **48**
Levit, C. 29, **48**
Lewis, C. H. 104, 106, **110**
Licklider, J. C. R. 11, **48**
Linggard, R. 183, **187**
Lippman, A. 36, **48**
Lipton, L. 22, **48**
Lisberger, S. G. 145, **173**
Lishman, J. R. 153, 155, **173**
List, U. H. 108, **110**
Lofqvist, A. 182, **187**
Long, C. L. 163, **173**
Lorenz, K. Z. 136, **173**
Luria, A. R. 141, **173**
Lypaczewski, P. A. 23, 37, **48**

Mach, E. 3, **9**
MacKenzie, D. 214, **222**
MacNeilage, P. F. 160, **174**
Maddox, E. E. 95, **100**
Malmstrom, F. V. 95, **100**
Mandelbaum, J. 94, **100**
Mandelbrot, B. 12, **48**
Mark, L. S. 100, **207**
Marr, D. 4, 9, 57, **83**, 115, 118, 126, **129**
Martin, L. 19, **48**
Masterton, R. B. 144, **173**
Matthews, P. B. C. 148, **174**
Max, N. L. 87, **100**
Mayhew, J. E. W. 58, **83**
McAdams, S. 181, **187**
McBride, E. D. 163, **174**
McCloskey, D. I. 148, **174**
McDonnell, J. R. 145, **174**
McDowall, I. E. 29, **48**
McGreevy, M. W. 14, 20, 24, **49**
McGuinness, B. 211, **222**
McGurk, H. 179, **187**
McKinnon, G. M. 23, 30, **49**
McKinley, R. L. 185, **187**
McLuhan, M. 210, 217, **222**
McMahon, A. P. 74, **83**
McRuer, D. T. 15, **49**
Meagher, D. 23, 26, **49**
Meijer, O. G. 134, **174**
Meredith, M. A. 157, **174**
Meyer, D. E. 160, **174**
Milliken, B. 122, **129**
Minsky, M. 119, **129**
Mitchell, D. P. 87, **100**
Mohler, C. W. 141, **174**
Monheit, G. 36, **49**
Monmonier, M. 24, **49**
Mon-Williams, M. 97, **101**
Morgan, M. J. 65, **83**, 89, 101, 155, **174**
Muller, H. J. 156, **174**

Mumford, L. 215, **222**
Myers, T. H. 23, **49**

Naish, P. L. N. 181, **187**
Nakayama, K. 125, **129**
Napier, J. R. 163, **174**
Navon, D. 154, 156, **174**
Neary, C. 97, **101**
Neisser, U. 116, 120, 126, **129**, 156, **174**
Nemire, K. 19, **49**
van Nes, F. L. 64, **83**
Netrovali, A. N. 24, **49**
Neuckermans, H. 192, **207**
Newell, F. 121, 122, 126, **129**
Newell, K. M. 156, **174**
Newell Jones, F. 58, **83**
Nielsen, S. H. 182, **187**
Nisly, S. J. 90, **101**
Nomura, J. 27, 39, **49**
Norman D. A. 201, **207**
Norton, A. 87, **101**

Oman, C. M. 24, 37, **49**, 105, **110**
O'Neill Y. V. 179, **187**
Owens, D. A. 94, **101**
Ouh-young, M. 31, **49**

Pacejka H. B. 189, **207**
Pacey, A. 214, **222**
Paillard, J. 158, **174**
Palmer, J. 124, **129**
Palmer, S. 122, 125, **129**
Papert, S. 119, **129**
Pausch, R. 145, **174**
Pedotti, A. 26, **49**
Pellieux, L. 185, **187**
Penfield, W. 162, **174**
Pentland, A. P. 74, **83**
Perret, D. I, 143, **174**
Petzold, P. 3, 4, **9**
Pew, R. W. 160, **174**
Phillips, C. 36, **49**
Pimentel, K. 211, **222**
Pinker, S. 116, **129**
Poggio, T. 120, 122, **129**
Pollack, A. 11, **49**
Pollack, I. 182, **187**
Pollatsek, A. 123, **129**
Posner, M. I. 155-6, **174**
Post, P. B. 97, **101**
Postman, L. 146, 149, **174**
Potmesil, M. 87, **101**
Poulton, E. C. 20, **49**
Power, R. P. 155, **174**
Praske, V. 148, **174**
Prazdny, K. 58, **83**
Pritsker, A. A. B. 25, **49**
Provenzo, E. F. 217-8, 220, **222**

Raab, F. H. 28, **49**
Rabiner, L. R. 183, **187-8**
Ramachandran, V. S. 74, **83**, 93, **101**, 118, **129**
Randle, R. J. 98, **101**
Ratoosh, P. 59, **83**
Rayner, K. 123, **129**
Reed, E. S. 134, **174**
Regan, D. 15, **49**
Regan, E. C. 105, **110**
Reichardt, W. 87, **101**
Rendings, W. 212, **222**
Revesz, G. 145, **174**
Rice, D. L. 184, **187-8**
Richards, O. W. 96, **101**
Richter, G. M. A. 59, **83**
Robinett, W. 25, **49**
Robinson, A. H. 24, **49**
Robson, J. G. 89, **101**
Rock, I. 155, **175**
Rogers, D. F. 68-9, **83**
Roland, P. E. 162, **175**
Rolfe, J. M. 23, **50**
Romanes, G. J. 138, 142, **175**
Rolland, J. 18, 38, **50**
Rosenbaum, D. A. 135, 145, 160-3, **175**
Rosenblatt, F. 119, **129**
Rosenbloom, L. 163, **175**
Rosenfield, M. 95, **101**
Ross, H. E. 133, **175**
van Rossum, J. H. A. 160, **175**
Rothwell, J. C. 159, **175**
Rumelhart, D. E. 119, **130**, 202, **207**

Sakitt, B. 90, **101**
Sams, M. 154, **175**
van Santen, J. P. H. 87, 89, **101**
Satava, R. M. 27, **50**
Scarborough, H. 215-6, **222**
Schiffman, H. R. 144, 147, 149, **175**
Schmandt, C. 155, **175**
Schmidt, R. A. 159, 160, **175**
Schmidt, R. C. 159, **175**
Schneider, W. 156, **175**
Schneider, G. E. 143, 157, **175**
Schober, H. 96, 98, **101**
Schroeder, R. 209, **222**
Schuffel, H. 23, **50**
Sedgwick H. A. 195, **207**
Sekuler, R. 133, 142, 144, 147, 166, **175**, 181, **187-8**
Senden, M. V. 17, **50**
Seymour, P. K. 120, **130**
Shadle, C. H. 183, **187-8**
Shaffer, L. H. 156, **175**
Shapiro, M. A. 213, 217, **222**
Shaw, E. A. G. 181, **188**
Shaw, R. E. 200, 202, **207**
Shepard, R. N. 116, **130**

Shepherd, M. 124, **130**
Shepp, B. E. 120, **130**
Sheridan T. B. 13, 27, **50**, 189, **207**
Sherrington, C. S. 135, 161, **175**
Shik, M. L. 154, 159, **175**
Siegel, R. 72, **83**
Skirrow, G. 217, **223**
Slater, M. 4, **8**
Slocum, D. B. 163, **175**
Smets G. J. F. 190-1, 196, **207-8**
Smith, W. M. 160, **175**
Smith, D. C. 12, **50**
Smith, B. R. Jr, 108, **110**
So, R. H. Y. 104-9, **110-11**
Sperber, R. D. 127, **130**
Sperling, G. 86, **101**
Sprague, J. M. 143, **175**
Stein, B. E. 141-2, 154, 156-8, **175**
Stevens, J. C. 150, **176**
Stewart, D. 11, **50**
Stone, R. J. 38, **50**
Stritzke, J. 23, **50**
Sullivan, A. H. 149, 150, **176**
Summers, J. J. 159, **176**
Sutherland, I. E. 12, 23, 30, **50**, 136, 166, **176**

Tachi, S. 13, 38, **50**
Takeda, T. 40, **50**
Tanaka, K. 117, **130**
Tastevin, J. 155, **176**
Taub, E. 159, **176**
Taylor, R. H. 27, **50**
Thelen, E. 160, **176**
Thompson, R. F. 153, 161-2, **176**
Thornbury, J. M. 147, **176**
Tobler, W. R. 24, **50**
Tomovic, R. 32, **50**
Trachtman, J. N. 98, **101**
Treisman, A. 117, 123, **130**, 156-7, **176**
Trevarthen, C. B. 143, 157, 159, **176**
Tufte, E. R. 24, **50**

Ullman, S. 87, **101**, 116-7, **130**
Ungerleider, L. G. 114, **130**, 143, 157, **176**

Vallbo, Å. B. 146, **176**
Van den Berg, J. W. 183, **188**
Veldhuyzen, W. 23, **50**
Vertut, J. 27, **50**, 132, **176**
Vickers, D. L. 136, **176**
Vince, M. A. 161, **176**
Von Bekesy, G. 180, **188**

Wade, N. J. 77, **83**, 133, **176**
Wald, G. 98, **101**
Walker J. 198, 199, **208**
Wallach, H. 97, **101**

Walraven, J. 4, **9**
Walsch, K. W. 140, **176**
Wang, J.-F. 28, 39, **50**
Warren W. H. 200, 201, **208**
Warrington, E. K. 123, **130**
Watson, A. B. 63, 65–6, **83**, 86–7, 89, **101**
Watt, A. 67, 69, **84**
Weale, R. 97, **101**
Weintraub, D. J. 40, **51**
Weiskrantz, L. 143, **176**
Welch, R. B. 15, **51**, 182, **188**
Wells, M. J. 37, **51**, 104–5, 108–9, **111**
Wenzel, E. M. 18, 36, **51**, 181–2, 184–5, **188**
Wesner, M. F. 96, **101**
Westheimer, G. 65, **84**, 94, **102**
Westling, G. 165, **176**
Whalley, L. J. 220, **223**
Wheatstone, C. 58, **84**
White, K. D. 18, **51**
Whiting, H. T. A. 161, **176**
Whitted, T. 69, **84**

Williams, L. G. 124, **130**
Wickens, C. D. 20, **51**
Wiesendanger, M. 162, **176**
Wightman, F. L. 15, 18, **51**
Winn, B. 95, **102**
Winner, L. 214, **223**
Winstein, C. J. 135, 147, 165, **176**
Witkin, H. A. 3, **9**, 227, **228**
Witkin, A. 26, **51**
Woodworth, R. S. 135, 161, **176**
Woolsey, C. N. 162, **176**
Wollaston, A. 213, **223**
Woolley, B. 212, **223**
Wysecki, G. 63, **84**

Yonas, A. 74, **84**

Zangemeister, W. H. 15, **51**
Zeki, S. 117, **130**, 143, **176**
Zeltzer, D. 26, **51**
Zimmerman, T. 32, **51**
Zipf, G. K. 12, **51**

Subject Index

Abstract
  data 1, 2
  properties 11, 12, 13, 14, 42, 123, 211
accommodation 18, 40, 58, 64, 85, 93-8,
    154
  empty field myopia 94
  instrument myopia 96
  mandelbaum effect 94
acuity (see resolution)
adaptation
  human 12, 19, 115, 144-5, 146, 147-8,
    165-8
affordances 7, 118, 198-201
after-effects (see side-effects)
aliasing
  anti-aliasing (spatial) 63, 64, 65, 67
  anti-aliasing (temporal) 86-7, 92
  spatial 64
  spatio-temporal 86
  temporal 5, 65, 85-93
apparent motion 86, 90, 91
application demands (see requirements;
    task)
artificial intelligence 113, 201-2
arousal 139, 144, 152, 158
attention 115, 123, 124, 155-6, 159, 167
  attentional capture 155
  conjunctions 117, 123, 156, 168
audition 143-4, 179-85
autonomic nervous system 93, 141

behaviour (see also violence) 133-7, 141,
    154, 156, 157-61, 168-9, 213, 214,
    217, 227
belief 6, 209
blur 5, 86, 87, 92, 93, 94, 95, 96, 97,
    103, 104, 155
brightness (see luminance)

central nervous system 137-41, 150-3, 159
cognitive effort/resources 95, 98, 156
colour 4, 63, 65, 117, 141, 143, 157
  chromatic aberration 94, 95-6
communicate/communication 3, 6, 11, 12,
    36, 41, 42, 43, 114, 179, 198, 214

constancy/invariance 117, 118, 119, 120,
    122
  lightness 79
  object 115, 116
  size 116, 117
context 3, 4, 115, 124, 126-7, 182, 206,
    209, 215, 217, 218, 220-1, 225, 227
contrast
  luminance 64, 89, 90, 97
  colour 65
  sensory 153
controls/input devices see coupling;
    technology
convergence of the eyes see vergence
cortex 114, 117, 123, 142-3, 144, 151-2,
    153, 157, 161
  motor coordination 161-3
  sensory areas 140-1
coupling of actions to environment
  technology 22, 30-8, 132, 136
  psychology 194, 196, 205
cues/stimuli (see also degraded cues,
    information, perspective, stereopsis,
    texture, features) 56, 58-9, 80, 94,
    114, 116, 118, 133, 145, 150, 154,
    157, 165, 181-2, 191, 198, 226
  artificial 32
  horizon 19, 194
  motion parallax 18, 54, 66, 79, 108,
    194
  multi-modal 157, 165, 168, 186,
  physiological 18, 58, 94-5, 97
  pictorial 17, 57, 58, 59
  relative motion 54
  reverse shading 74
  shading 57, 59, 74, 81, 114
  shadow 69, 79
cyberspace 5, 131, 169, 212
database (see also requirements)
  ergonomic factors
    human 36, 132, 227
    technology 38, 40-1, 227
degraded cues/impoverished information 76,
    96, 115, 127, 132-3, 154, 158-9,
    164-5, 167, 191

degraded/impaired human performance (*see
    also* misperception) 19, 158, 163, 165,
    167, 168, 169, 220
direct perception 7, 56, 57, 118
disparity (*see* stereopsis)
display 5, 12, 14, 20, 23, 28, 35–6, 62–3,
    66, 85–6, 92–3, 95–8, 124, 136, 225
    head-mounted 5, 11, 18, 22–3, 30, 37–8,
        40, 103, 108
    distortion 6, 18, 22, 24–5, 76, 86, 108,
        167, 213

ecological perception (*see also* evolution) 7,
    54, 74, 118, 161, 169, 182, 194, 205
egocentric thinking/subjectivity (*see also*
    viewpoint) 1, 2, 42, 212, 228
empiricism 1, 8, 54, 55
eye movements (*see also* vergence) 19, 40,
    115, 123–4, 141, 144
evaluation/assessment 19, 132, 185, 191,
    197, 205–6, 227
evolution 3, 7, 80, 132, 134, 202, 213, 225
expectations 76, 213, 219
experience (*see also* phenomenology) 4, 8,
    11, 12, 14, 22, 36, 54–6, 97–8, 138,
    169, 206, 210, 213, 227

features/primitives 118–9, 122, 124, 200,
    206
feedback 134–7
    error 15, 166
    force 31
    sensory 26, 155, 159–61, 166, 168–9,
        190, 203, 217
    terminology 134–6
feedforward 135
fidelity (*see also* realism)
    of simulation 12, 19, 31, 43
    of stimulus 15, 20, 31–2, 37, 39, 210
field-of-view 18, 21, 24, 28, 36, 37, 40,
    97, 108, 110, 113, 124, 190
frame of reference (*see also* viewpoint) 3,
    19–20, 26–7, 40, 215–6
frequency (temporal and spatial; *see*
    resolution)

Gestalt psychology 43, 56, 57

haptic perception (*see* taction)
human capacities and performance (*see*
    requirements; user)

illumination 68–72, 115
    direct/indirect 70
    radiosity 55, 63, 72
    global/local 68–81
illusion 6, 12, 18, 20, 24, 36, 41, 53,
    167–8
    haptic 15, 148, 150, 166–7

kinaesthetic 165–6, 169,
    self-motion 145
    visual 4, 66, 76–7, 81, 87–8, 154, 205
individual differences/variation 19, 227–8
inference (*see* interpretation)
information (perceptual *see also* cues) 6,
    12, 15, 17, 53–4, 57, 59, 103, 113,
    155, 198
inhibition
    perceptual 91–2, 136, 141, 158
integration
    feature 124, 156–7
    perceptual 36, 92, 131, 138, 141, 154–8,
        167, 226
    receptor 149, 153
    sub-cortical 157
    synaesthesia 157
interaction/interface (*see also* metaphor)
    between cues 154, 182
    between display variables 62, 64, 66, 98
    between light and surface 55, 60, 68
    between human and
        man-made system 2, 11–2, 14, 23,
            26–7, 31, 36–7, 41–2, 96, 113, 186,
            192–4, 198, 210, 226
        virtual reality 8, 11, 19
        world 4, 14, 20, 22, 56, 98, 133,
            158–9, 164–9, 192, 196–8, 202, 205
    sensory interaction 57, 151–2, 162, 164–6
    social 212–3, 217
    technology and society 209, 214
interpretation 4, 15–7, 56–7, 76, 124–7,
    133, 227
    cinematographic 210
    perceptual language 3

kinaesthetic perception (*see* taction)
kinematics 13–4, 26, 30, 32, 37, 160
knowledge 1, 6, 8, 15, 43, 53, 55, 95, 97,
    113, 124, 126, 162, 202–3, 206,
    213–14, 228
    metaphysics 2

lag (display/time) 5, 19, 37, 98, 104–10,
    145, 165, 205, 210, 226
    compensation 107–10
learning/development of perception (*see also*
    training) 8, 56, 120, 122, 124, 162,
    165–6, 202, 210, 213
light (physics of *see also* interaction) 54,
    59–61
luminance 62–7, 96

manipulation
    direct 131, 163, 192, 194, 197
    of objects 26–7, 31–2, 35–6, 41, 131,
        136, 139, 147, 154, 156, 162–6, 226
manipulators (*see* coupling technology)
media 5, 11, 209–21

metaphor (interaction) 12, 22, 42, 186
memory 116, 118–19, 121, 127, 217
mental effort (*see* cognitive effort)
misperception 5, 76, 97, 179
models 4, 6
  cultural/sub-cultural 219
  human performance 36, 202
  illusion 4, 24
  interaction 198
  mental 15, 136–7
  motor behaviour 159–61
  perception 4, 7, 57, 74, 113, 115–20, 181–2, 202
  prototype design 189
  real world 4, 55, 67–74
  speech 182–4
motion/movement perception 65, 86–92, 114, 143
  displacement limit 92
motion sickness 19, 22, 24, 39, 105, 110, 145
movement perception (*see* motion perception)

nativism 8, 54–6
neural networks 7, 119, 202, 206

optic array 17, 113, 196, 199–202

perceptual processes (*see also* integration) 3, 54–6, 66, 81, 114–5, 117
  'blindsight' 142
  bottom-up/top-down 4, 15, 124–7, 153, 155, 167–8, 226
  early processes 115, 117–8
  enhancement 153–4, 157–8
  focal/ambient 143, 154, 159
  interference 123, 125–6, 156, 165, 179
  multiple tactual systems 153
  multiple visual systems 7, 114, 118, 123, 142–3, 157
  parallel/serial 115, 156
  perceptual capture/dominance 19, 155, 166, 179
  transformations 116, 119–121, 156
perceptuo-motor/sensory-motor behaviour 15, 19, 42, 134, 136, 140, 154, 158–63, 168–9, 192, 198
peripheral nervous system 138
peripheral vision 92, 115, 123–4, 142
perspective
  artificial/linear 59
  foreshortening 59–60
  natural 59–60
  projection 6, 59
phenomenology 54, 56–7
physiological/neurophysiological studies 4, 7, 57, 113–14, 116, 119, 131, 137–53, 161–3

auditory structures 143–4, 180, 226
contralateral processing 139, 142, 144, 150–2, 161
multi-modal structures 157, 168
retinotopy 142
speech structures 184
tactual structures 145–53
vestibular structures 144–5
visual structures 141–3
practice (*see* training)
preferences (perceptual) 57, 75, 77, 200
presence 4, 19–20, 210
psychophysics 3, 57–8, 74–6

realism 3, 6, 36, 53–4, 60, 62, 67, 73, 80–1, 85, 110, 169, 210–11, 217, 225
  unrealism 6
reality (-ies) 3, 4, 6, 14–6, 17, 22, 53, 56, 189, 205, 209, 210–3, 216, 226–7, 228
real world 6, 53, 85, 97–8, 104, 132, 219, 228
receptive fields 57, 87–9, 120, 142–3, 146–9, 150–1, 157, 166
recognition/form perception 3, 113–27, 143, 154, 226
redundancy/multiple cues 53, 76, 81, 114, 135, 154, 168, 179
reflex 17–18, 24, 134, 141–2, 154, 159, 160, 165, 170
  vestibulo-ocular 15, 104, 135, 144–5
representation 6, 20, 54, 57, 60, 64, 67, 113, 115, 119–21, 169, 209, 211, 216, 218
resolution/frequency sensitivity (*see also* fidelity)
  auditory 144, 180
  hyperacuity 65–6
  spatio-temporal 65, 89
  tactual 32, 141, 146–50, 164, 166, 168
  vernier acuity 65
  visual spatial 23–4, 28, 32, 36, 40, 64–7, 86–7, 89–90, 92, 95–6, 117, 124, 141, 190, 205
  visual temporal 65, 67, 89–90, 92–3
response times 91, 120, 161
retina 7, 55–6, 64–5, 115, 124, 141, 142, 145
requirements (*see also* database)
  user 2, 12, 36, 38, 40, 62–3, 66–7, 92, 132, 164, 185, 194, 226, 228
  task 12, 39–40, 42, 53–4, 66–7, 81, 106, 113, 132, 163, 169, 193, 201, 228

sensors (*see* coupling; technology)
side-effects 42–3, 134, 165
simulation (*see also* visualization; models) 3, 5–6, 20, 23–6, 53–4, 110, 113, 169, 189, 209–12
  environment/real world 12, 80, 86

flight/ aircraft 1, 12, 19, 20, 39, 185, 189
molecules 31
perceptual stimuli 4
speech 183–4
telerobotic/manipulation 12, 27, 166
skill acquisition (*see* training)
social effects 43, 210
spatial frequency (*see* resolution)
specifications (*see* requirements)
standards/regulation 39, 219,
stereopsis/disparity 18, 40, 58, 64, 66, 79, 97, 113–14, 194
stimulus (*see* cues, information, features)

tactile perception (*see* taction)
taction 15, 133, 145–58, 164–9
definition 145
task demands (*see* requirements; task)
taxonomy/classification 225
computer interaction 198
environment 13
manipulation 163
technology 20
virtualization 16
technology (*see also* displays, coupling; technology) 2, 4–7, 11, 20, 28, 36, 39, 85, 131–2, 137, 190, 209, 210, 214–20
auditory feedback technology 184–5
tactual feedback technology 164, 166, 168–9
teleoperation/telerobotics 11–12, 20, 27, 30, 38, 105, 107, 132, 136, 191
telepresence 18, 30, 35, 103, 131, 189, 196, 203
temporal frequency (*see* resolution)
texture
auditory 181, 186
tactile 145, 154, 162
visual 19, 57, 59, 97, 194–6, 199
thresholds
(in perception) 57–8, 64–5, 89, 149, 150, 153

just noticeable difference 58
tracking (of targets) 20, 106, 108, 110, 144
training/practice/skill (*see also* learning) 23, 36, 41, 98, 107, 119–21, 136–7, 156, 158–9, 168–9, 189, 227
de-skilling 137

up-date rate (display/image) 5, 36, 85–6, 89–90, 92, 98

vergence 18, 40, 58, 94–5, 97
vestibular perception 19, 24, 144–5, 155
vestibulo-ocular reflex (*see* reflex)
vibration 103–4
image deflection 108–10
viewpoint (view/vantage point/viewing angle) 4, 13, 17–8, 20, 41, 55, 57, 59–61, 69, 72–3, 76–7, 115, 118, 120–3, 145, 210, 214, 216
egocentric 12, 20, 36, 40
exocentric 12, 20
violence/aggression 42, 218–9
virtual image 11, 15–16
virtualization 5, 15–16, 40, 211, 225
vision 53–127, 133, 141–3, 154–8, 161, 179, 182, 194–203, 225–8
visual persistence 90–1
visual search 3, 103, 106, 110, 117, 119, 124
visual systems (*see* perceptual processes)
visualization 26, 189, 203
aerodynamic 29
educational/scientific 26
manoeuvre planning 26, 28
medical 26, 29
planetary 24,
product design 191
visuo-motor coordination (*see* perceptuo-motor behaviour)

'window of visibility' 5, 66, 89